P9-DWG-594

D0037766

MICROBIAL TECHNOLOGY:
CURRENT STATE,
FUTURE PROSPECTS

Other Publications of the
Society for General Microbiology*
THE JOURNAL OF GENERAL MICROBIOLOGY
THE JOURNAL OF GENERAL VIROLOGY

SYMPOSIA

* Published by the Cambridge University Press, except for the first Symposium, which was published by Blackwell's Scientific Publications Limited.

MICROBIAL TECHNOLOGY: CURRENT STATE, FUTURE PROSPECTS

EDITED BY

A. T. BULL, D. C. ELLWOOD AND C. RATLEDGE

TWENTY-NINTH SYMPOSIUM OF THE
SOCIETY FOR GENERAL MICROBIOLOGY
HELD AT
THE UNIVERSITY OF CAMBRIDGE
APRIL 1979

Published for the Society for General Microbiology
CAMBRIDGE UNIVERSITY PRESS
CAMBRIDGE
LONDON · NEW YORK · MELBOURNE

Published by the Syndics of the Cambridge University Press
The Pitt Building, Trumpington Street, Cambridge CB2 1RP
Bentley House, 200 Euston Road, London NW1 2DB
32 East 57th Street, New York, NY 10022, USA
296 Beaconsfield Parade, Middle Park, Melbourne 3206, Australia

First published 1979

Printed in Great Britain by
Western Printing Services Ltd, Bristol

Library of Congress Cataloguing in Publication Data
Society for General Microbiology
Microbial technology
(Symposium of the Society for General Microbiology; 29)
Includes index
1. Industrial microbiology–Technique–Congresses. I. Bull, Alan T.
II. Ellwood, Derek C. III. Ratledge, C. IV. Title. V. Series: Society
for General Microbiology. Symposium; 29.
QR1.S6233 no. 29 [QR53]   576′.08s   [660′.62′028]   78–12206
ISBN 0 521 22500 0

CONTRIBUTORS

AHARONOWITZ, Y., Department of Microbiology, Tel-Aviv University, Tel-Aviv, Israel

ATHERTON, K. T., Corporate Research Laboratory, ICI Ltd, Runcorn Heath WA7 4QE, Cheshire, UK

BEALE, A. J., Wellcome Research Laboratories, Beckenham BR3 3BS, Kent, UK

BRIERLEY, C. L., New Mexico Bureau of Mines and Mineral Resources, Socorro, New Mexico 87801, USA

BULL, A. T., Department of Applied Biology, University of Wales Institute of Science and Technology, Cardiff CF1 3NU, UK

BU'LOCK, J. D., Weizmann Microbial Chemistry Laboratory, Department of Chemistry, University of Manchester M13 9PL, UK

BYROM, D., Agricultural Division, ICI Ltd, Billingham TS23 1LB, Cleveland, UK

DART, E. C., Corporate Research Laboratory, ICI Ltd, Runcorn Heath WA7 4QE, Cheshire, UK

DEMAIN, A. L., Department of Nutrition and Food Science, MIT, Cambridge, Massachusetts 02139, USA

ELLWOOD, D. C., Microbiological Research Establishment, Porton Down, Salisbury SP4 0JG, UK

FIECHTER, A., Mikrobiologisches Institut, Eidgen Technische Hochschule, Weinbergstrasse 38, CH-8006, Zurich, Switzerland

FINN, R. K., School of Chemical Engineering, Cornell University, Ithaca, New York 14853, USA

HARRIS, R. J. C., Microbiological Research Establishment, Porton Down, Salisbury SP4 0JG, UK

HIGGINS, I. J., Biological Laboratory, University of Kent, Canterbury CT2 7NJ, UK

HILL, H. A. O., Inorganic Chemistry Laboratory, University of Oxford, Oxford OX1 3QR, UK

KELLY, D. P., Department of Environmental Sciences, University of Warwick, Coventry CV4 7AL, UK

KENNEL, Y. M., Rhone-Poulenc Industries, 94400 Vitry-sur-Seine, France

KOSSEN, N. W. F., Biotechnology Group, Department of Chemical Engineering, Delft University of Technology, Jaffalaan 9, The Netherlands

LAWRENCE, R. C., New Zealand Dairy Research Institute, Palmerston North, New Zealand

MATELES, R. I., Laboratory of Applied Microbiology, Institute of Microbiology, Hebrew University – Hadassah Medical School, P.O. Box 1172, Jerusalem, Israel

NEIJSSEL, O. M., Laboratorium voor Microbiologie, Universiteit van Amsterdam, Plantage Muidergracht 14, Amsterdam, The Netherlands

NORRIS, P. R., Department of Environmental Sciences, University of Warwick, Coventry CV4 7AL, UK

RATLEDGE, C., Department of Biochemistry, University of Hull, Hull HU6 7RX, UK

SLATER, J. H., Department of Environmental Sciences, University of Warwick, Coventry CV4 7AL, UK

SOMERVILLE, H. J., Shell Biosciences Laboratory, Sittingbourne Research Centre, Sittingbourne ME9 8AG, UK

SUTHERLAND, I. W., Department of Microbiology, University of Edinburgh EH9 3JG, UK

TEMPEST, D. W., Laboratorium voor Microbiologie, Universiteit van Amsterdam, Plantage Muidergracht 14, Amsterdam, The Netherlands

THOMAS, T. D., New Zealand Dairy Research Institute, Palmerston North, New Zealand

CONTENTS

EDITORS' PREFACE

The increasing industrial utilization of biological processes supports our belief that microbial technology will be a major growth industry in the near future. If this assertion proves correct then microbial technology will affect the lives and welfare of people all over the world.

Our belief is based on the trends that are currently taking place in harnessing biological activities to achieve a variety of aims. These encompass the production of industrially important materials, including bulk chemicals as well as antibiotics and compounds of pharmaceutical value; the manufacture of single-cell protein from diverse substrates; coping with what is loosely called the energy crisis; and the recycling or processing of waste materials. Future developments have received a recent powerful stimulus for expansion with the advent of novel genetic-manipulation techniques, and it now becomes possible to contemplate the production of many biological materials which hitherto have only been available at an extremely high cost. Indeed, it is now theoretically possible to design a micro-organism to carry out almost any specific biological task.

Microbial technology, as one may gather, is a subject which covers many fields. It involves the interaction and close coordination of many disciplines: the knowledge of basic microbiology and microbial chemistry must be married to the skills and techniques of the geneticist, the engineer, and even the mathematician and computer expert. Never far away from this amalgam of activities are the economists and industrialists in whose hands resides the ultimate power of deciding on a particular venture. So far absent from the scene, at least in the United Kingdom, is the direct influence of government. Perhaps the time is now fast approaching when this situation should change and active steps be taken for the planning of an overall strategy for microbial technology. This would be of vital significance and give impetus to the established industries, confidence to the many companies who are thinking about entering the field, and encouragement to universities and other institutions to develop this discipline in a more committed fashion.

This symposium of the Society for General Microbiology naturally focusses attention primarily on the microbiological aspect of this subject. The programme covers many fields of microbial technology and its main purpose is to illustrate the importance of understanding microbial behaviour, in the broadest sense, and to emphasize that this is the key to the development of this new technology. Hopefully this symposium will bring together experts in the many disparate fields and will allow a lively and stimulating interaction between them. We asked each of the contributors to give an objective appraisal of the current state of their respective subjects and, wherever appropriate, to speculate about the future, as the future of this subject has much to offer not only to society but to the science of microbiology itself. We hope that this, too, will become evident during the course of this symposium.

It is for these reasons that we think it is most opportune that this symposium is being held now. We hope the intellectual challenges and excitement of this aspect of microbiology will be conveyed to those who attend the symposium itself and to the readers of this book. We hope that by illustrating some of the prospects for future development this will act, in some small way, as a simulus for achieving these aims.

Although editors of these volumes customarily thank their contributors for getting their manuscripts prepared in good time, and also the staff of Cambridge University Press for their help with the editing of the book, this time it is really meant!

Alan T. Bull
Derek C. Ellwood
Colin Ratledge

THE CHANGING SCENE IN MICROBIAL TECHNOLOGY

ALAN T. BULL,* DEREK C. ELLWOOD† AND COLIN RATLEDGE‡

Department of Applied Biology, University of Wales Institute of Science and Technology, King Edward VII Avenue, Cardiff CF1 3NU, UK

† *Microbiological Research Establishment, Porton Down, Salisbury SP4 0JG, UK*

‡ *Department of Biochemistry, University of Hull, Hull HU6 7RX, UK*

INTRODUCTION

Fermentation industries are a development of the twentieth century, but for at least 5000 to 6000 years prior to this development man had produced fermented foods and beverages for his own consumption while knowing nothing or very little about the causative agents. Of course, in common with most generalities, one can identify notable exceptions, none more dramatic than that of sake brewing. Written records from the mid sixteenth century, for example, reveal a remarkably sophisticated technology including heat-processing practices that anticipated pasteurization by 300 years. Nevertheless, there were few conscious developments of processes utilizing micro-organisms before Pasteur proved that alcohol fermentation was caused by yeast. Since that time a somewhat different type of empiricism has been followed. The activities of micro-organisms have mainly been harnessed on an *ad hoc* basis and improvements in process productivity again have come about mainly as a result of empirical discoveries.

Over the last thirty years or so, our knowledge of micro-organisms has increased enormously and with it a growing appreciation of their potentials and how best they may be realized. We are, therefore, at a point in the history of microbiology where it is possible to indicate, with some degree of accuracy, whether or not a given product could be produced microbiologically and, moreover, what type of organism would have to be used; where it might be found if it is not yet known; and under what conditions it must be grown to express its desirable properties. We can also say how the productivity of a micro-organism might be improved: by genetic manipulation; by judicious choice of culture condition; by addition, or subtraction, of a particular nutrient.

This rational approach to microbial technology is now with us and is ripe for exploitation.

The purpose of this introductory chapter is to illustrate how various forces – scientific, economic and even political – have shaped and continue to shape the development of microbial technology, and why we believe that, now especially, it will have a major impact on our everyday lives and an ever-increasing influence on national economies. This symposium, however, is not just concerned with microbial technology and the UK: the countries of origin of the various contributors immediately give lie to that. Microbial technology is a global pursuit. Each national picture, though, is perforce different to any other and the constraints and opportunities which exist in one country will not apply, or will apply with different emphases, in others. What is striking about microbial technology, however, is that it has something to offer for us all; industrially developed and under-developed countries alike, irrespective, to an extent, of the abundance of natural resources.

MICROBIAL TECHNOLOGY: AN INTERACTION OF DISCIPLINES

Microbial technology is concerned with the industrial processing of materials by micro-organisms to provide desirable products or serve other useful purposes. Such a definition, albeit a broad one, provides a convenient working reference for the subject of this symposium. In the sense in which we use the term, microbial technology encompasses a wide variety of fermentations for the synthesis of complex biological materials; the conversion of renewable resources into chemicals, food and energy; waste and water treatment; and recycle and recovery processes – all of which are mediated by micro-organisms *per se*: however, it excludes such related activities as enzyme technology, food technology and biodeterioration. Without question the essence of microbial technology is its interdisciplinary nature – it is a technology that brings microbiology together with genetics, biochemistry, chemistry and engineering and demands intimate collaboration between scientists in these disciplines. In addition, the scientist working in this area needs to be conscious of diverse economic, sociological and legal constraints that impinge on the success or failure of the operation under consideration. While the major emphasis in this symposium is rightly placed on microbiological principles, the interfaces of microbiology and disciplines referred to above are explored in many of the contributions.

The interdisciplinary character of microbial technology is clearly brought out by reference to a hypothetical but, we hope, typical research and development programme for a new product or process. Such a programme is initiated by a phase of exploratory research during which the microbiologist makes the primary discovery, a microbial activity or product, and proceeds with preliminary feasibility studies. Subsequently, process and product researches constitute a second phase of the programme: microbiologists become concerned with defining the conditions for optimum production, biochemists and chemists characterize the chemical nature of the product or process. Finally, if an investment decision is made, a phase of process and product development ensues. The contribution of the engineer becomes dominant at this point, with problems of mass and energy transfer, process scale-up, instrumentation and automation and plant design needing to be resolved. At the same time such matters as process safety and product–process properties of interest to potential customers require attention. Naturally, these various activities do not proceed in isolation; for example, the choice of micro-organism may be dictated in part by the type of reactor system to be used and vice versa, or by the requirement for genetic manipulation during the research and development programme.

The desirability of genetic manipulation of industrial micro-organisms will be a recurring theme in this symposium and the input of the microbial geneticist during process development is often a critical and continuous one. Whereas genetic manipulation in the past has largely had the objective of product yield improvement, the advent of genetic engineering techniques opens the way to novel microbial products and processes. Indeed, the major consideration for the future of microbial technology is ultimately a genetic one, and the continued success and diversification of the fermentation industry will be dependent upon achievements in this field. Reference to the improvements in penicillin and streptomycin fermentations since the time of their discovery (see Demain, 1973) is evidence of this claim but it is noticeable that the mutation–selection strategy on which these improvements have been made is crude and empirical compared with the array of hybridization, recombination and cloning technologies which are now becoming available.

One further point – which is rarely absent from the thinking of the industrial microbiologist but is even more rarely the concern of his academic colleague – requires comment and this is the question of economics. Economic considerations arise at all stages of a research and development programme: for example, in fermentation-medium design what substrates (especially carbon) are most advantageous? In the face of interrupted supplies of raw materials, can the fermentation be operated with alternative substrates? What are the merits of batch, continuous-flow or other modes of processing? What are the economics of various 'downstream' operations such as culture harvesting, product recovery and purification? Effective answers to these and related questions only come from the continual interaction of the microbiologist with other members of the microbial technology team.

In the fermentation industry, as in many other industries, analyses suggest that economies of scale can be expected. For most chemical industries the scale factor* is in the range 0.60 to 0.75 and for certain fermentation industries such as brewing it may be considerably better (less than 0.40) (Pratten, 1970). It has been argued that in order to produce animal-feed grade single-cell protein economically, plants having an output of the order of 10^5 tonnes per year ($t\ y^{-1}$) are essential (Maclennan, 1976). We believe, however, that it is equally important to stress the dynamic nature of economics and to realize that the viability of a fermentation or fermentation-related process will also be dependent on the economic climate of the day and the economic and political status of the country or bloc in which the process is to be operated. Thus, the thinking on a common problem might be quite distinct in, say, Western Europe–North America–Japan, the communist bloc and the Third World countries. In our view microbial technology-based activities seem to be particularly well suited to developing countries, where the philosophy of economies of scale has a less fettering influence; we will return to this point later.

STIMULI FOR CHANGE AND INNOVATION

Historians declaim that an examination of the past enables us to put present events into perspective and to gauge what courses future developments may take. Consequently, we may ask what can be learnt from the recent past of the fermentation industries which may help us to judge what activities are likely to be influential in shaping the future? What are the major stimuli which have operated in this field? Because the basic over-riding principle is that any process put into operation should ultimately be profitable, economic considerations are paramount in deciding what may or may not be done.

Perlman (1978) considers that there have been three major stimuli

---

* In estimating total costs of a process, if the capacity is multiplied by a factor X, the capital cost increases as X^n, where the exponent n, the scale factor, varies with the process and range of size. Thus a scale factor of 1 indicates zero economies of scale.

for innovation: (1) the occurrence of crises and their resolution; (2) the availability of 'new' technology (directly or indirectly designed for the fermentation industries); and (3) interference of the 'normal pattern' by outside influences.

We would take a somewhat more lateral view of events and suggest the following as influences of major importance: (1) raw material availability; (2) product scarcity; (3) exploitation of a novel product; and (4) environmental awareness. (Perlman's views and ours are not in conflict, they merely provide alternative views of the same scene.)

(1) *Substrate availability*

If we go back far enough, the development of ale making in the UK can be ascribed to the failure of the Norman barons to grow the grape in England following their uninvited arrival in 1066 (Hastings – the man not the place – 1971). A new 'technology' was started, therefore, using the natural abundant substrate of the time and place, rye. In the same way, 900 or so years later, the appreciation that n-alkanes, unwanted in petroleum refining, could be converted into food opened up new horizons which still have not been fully evaluated. As with alkanes, so with methanol and methane: the product (single-cell protein, SCP) is of less importance than the consumption of the substrate. Such an assertion may appear outrageous but companies involved with the exploitation of these substrates urgently have to find large-volume products for new markets which will enable them to run their own refineries or production units on larger scales and thus bring unit costs down within the whole industry. SCP production must, therefore, be related to the total activities of the producing company. The apparent economics cannot be assessed on the immediate data; 'hidden' cost factors are very much in evidence (Ratledge, 1975).

These SCP processes have contributed considerable spin-off benefits to the whole of microbial technology. Thus, technological innovations on the handling of water-insoluble substrates, commissioning the first large-scale chemostat and designing the 'pressure-cycle' fermenter all have followed in the wake of the initial concept and have stimulated interest and activity in most universities and in many countries. In keeping with these developments, projects like the current attempts to convert cellulose to SCP or the Brazilian National Alcohol Programme (Jackman, 1976; Hammond, 1977) can be put into context. With the latter, the need to convert starch and sugar to alcohol is again leading to technological innovation: the vacuferm fermenter (Ramalingham & Finn, 1977) is probably the forerunner of several designs which will have to be tried before the most effective system for converting sugar into alcohol is found (see Bu'Lock, this volume pp. 309–25). Similarly, once an effective and efficient microbial system which can degrade cellulose on an industrial scale has been found there could be rapid technological innovation to overcome the problems concerned with handling this difficult substrate.

(2) *Product scarcity*

Many microbial products have come into existence by being able to compete with established commodities derived from different sources. Examples which come readily to mind include citric and other organic acids, ascorbic acid and vitamin B_{12}, most of the amino acids, cortisone and the other sterols produced by microbial transformations. All of these microbial products were originally obtainable from animal or plant sources or by chemical synthesis. Consequently appropriate markets were already in existence before the microbial process was devised. The industrial microbiologist was able to recognize a potential market and realize that the product could be obtained microbiologically with greatly increased efficiency.

Market capture, though, is not all one-sided. Competition is always fiercest in this section as alternative means of production do exist. The rise and subsequent fall of the acetone–butanol fermentation industry is a case in point. It has been argued by Bu'Lock (1974; see also this volume, pp. 309–25) that this now-redundant industry could make a comeback and improvements in microbial technology over the past thirty years could be a crucial factor in aiding the re-emergence of this process. Each rise and fall has been, and will be, dictated by the economics of the competing chemical process. The economics of ethanol production is probably similarly finely balanced between chemical and microbiological systems (Ratledge, 1977).

(3) *Novel products*

Before the advent of antibiotics there were few effective anti-microbiological agents. Only the sulphonamides had any standing as therapeutic agents, while arsenicals and mercurials were recognized as crude agents with only partial effectiveness. With the discovery of antibiotics, completely new markets were created and the fermentation industry took a new lease of life.

Micro-organisms can thus open up new and perhaps unthought of horizons. Less dramatic than antibiotic production, but of considerable economic importance, is the current interest in enzymes, which are finding increasing applications in the food and drink industry and in medicine. Enzymes are being used in many processes which hitherto have been accomplished only slowly, or are being used in their turn to produce new products. An obvious example is fructose production by the action of glucose isomerase on glucose syrups. The demand for fructose is enormous as it has a high sweetening value with none of the problems associated with the use of cyclamates or other artificial sweetening agents.

Similarly one can anticipate that plant and animal cell culture, which are essentially identical in technique to the culture of micro-organisms, will have much to offer in the near future. Pharmaco-logically-active compounds and other secondary metabolites (Puhan & Martin, 1971; Jones, 1974) have obvious outlets. The production of specific antibodies by the manipulation of animal cell lines (Galfre et al., 1977) has an important future as there are many clinical problems which will require the production of reliable and standardized quantities of anti-histocompatability antigens. Applications in the treatment of carcinomas and in aiding tissue acceptance after trans-plantation are already considered feasible. The fermentation industry will undoubtedly be called on to aid in the provision of techniques and experience in the exploitation of these cell lines.

(4) *Environmental awareness*

Without doubt, control of environmental pollution is now a central theme in many governmental programmes. Microbiological solutions to enviromental problems, unfortunately, are not the swiftest to get acceptance and invariably find themselves in competition with more conventional solutions, such as incineration or chemical treatment. The activated-sludge process for sewage treatment is now widely practised and is accepted as an efficient process. But this is hardly true as no useful product is generated. A more cost-effective way would be to offset the costs of processing by producing methane from the anaero-bic digestion process.

Pollution, though, is not just sewage. All effluents and wastes are potential pollutants and one may anticipate government and supra-government legislation to become increasingly more exacting, with the penalties for evasion to be severe. Micro-organisms offer a genuine means for recycling many waste products and some schemes, notably the Bioplex Concept for dealing with urban and rural wastes, have reached advanced stages of planning (Forster & Jones, 1976). On a slightly wider front, the chapter in this book by Slater and Somerville provides much food for thought, as it deals with the possibilities of handling industrial as well as domestic wastes. Extending the concept of recycling even further, the possibilities of recovering metals from low-grade ores, from the oceans or from industrial spoil and effluents by the use of micro-organisms is now under active consideration in many laboratories throughout the world and these concepts are discussed in detail by Kelly, Norris & Brierley (this volume, pp. 263–308).

CONCEPTUAL INPUTS TO MICROBIAL TECHNOLOGY

It will be evident from what we have said already that microbial technology has inputs from several distinct disciplines. In this section we will highlight certain microbiological, genetic and engineering concepts which, we believe, have a critical bearing on this evolving technology. Space does not allow exhaustive expositions of these concepts but the reader will discover many of them illustrated and developed in the chapters which follow.

(1) *Microbiology*

(*a*) *Organism isolation and selection*

Industrial microbiology is still very much occupied by the selection of organisms having desirable attributes and this constitutes a very critical aspect of exploratory research. In sharp contrast to the usual requirements of academic research, organism selection for an industrial process is dependent on a range of criteria that are relevant to the optimization of the process. Therefore, selection may be made on the basis of best compromise between (i) *nutritional characteristics*, e.g. vitamin requirements might argue in favour of bacteria rather than yeasts; cost and the energetics of assimilation may recommend the use of ammonia and not nitrate; (ii) *temperature characteristics*, cooling of large-scale fermenters constitutes a large proportion of the total operating costs and, because of the ambient temperatures of sea and river water, it becomes desirable to ferment above 40 °C. The selection of thermophilic species has obvious attraction in that their use may also substantially reduce risks of contamination; (iii) *type of process*, the merits of organisms will vary depending on whether they are to be used in batch, continuous, aseptic, septic or 'protected' culture, and whether growth is to be made in other than a conventional stirred tank fermenter (see (3) Engineering, below); (iv) *reaction of the organism with equipment*, the susceptibility of an organism to shearing stress or a propensity to attach to surfaces or, as a consequence of its growth, to produce foam, are examples of undesirable reactions; (v) *organism stability*, in terms of morphology, genotype, resistance to bacteriophage (or other viruses) and contamination: considerations of this sort are particularly important in the design of continuous processes; (vi) *amenability to genetic manipulation*; (vii) *high product yield* (on the basis of substrate conversion); (viii) *high productivity* (on the basis of product yield per unit time) and (ix) *product recovery*, the ability of an organism to grow in a mycelial form or to flocculate may be apposite to the cost of biomass recovery, while intracellular or extracellular location will be relevant to the recovery of metabolites and enzymes.

Traditionally, screening procedures based on agar plate reactions have been important, for example in the detection of enzyme-producing and antibiotic-producing organisms. However, such approaches are labour intensive and very largely empirical. Various types of enrichment culture have been adopted for organism isolation which offer the promise of rationality. Until quite recently such enrichments were made in batch (closed) systems which, due to the imposition of a substrate-excess condition, almost invariably selected organisms on the basis of maximum specific growth rate. Unfortunately, this characteristic may not be the most significant selection criterion for the process being researched. In addition, because conditions change as a function of time in a batch enrichment, it is all too easy to miss sampling at a stage when the most desirable organisms are dominant or present in sufficiently large numbers to guarantee their isolation.

An attractive alternative, therefore, is the continuous-flow (open) enrichment strategy. Following Veldkamp's (1970) advocacy of this approach, several reports have demonstrated its usefulness in selecting organisms for single-cell protein (Wilkinson, Topiwala & Hamer, 1974) and enzyme (Rowley & Bull, 1977) production and the biodegradation of environmentally harmful chemicals (Munnecke & Hsieh, 1975; Senior, Bull & Slater, 1976). Continuous-flow enrichments enable organisms to be selected on the basis of their substrate affinity (by deploying a chemostat), maximum specific growth rate (by deploying a turbidostat), resistance to toxic materials or bacteriophage, and so on. Consequently, selection programmes can proceed in a logical and systematic way at the dictate of the experimenter. The value of selecting organisms in this way for subsequent continuous processes is also of significance. Chemostat theory predicts that in the absence of species interactions the outcome of competition is decisive and determined by Monod kinetics (Powell, 1958). However, chemostat enrichments tend to select communities of interacting micro-organisms which themselves may excite the interest of industrial microbiologists.

(b) Mixed cultures

Although mixed-culture technology is not a recent innovation – precedents include numerous fermented food products, waste treatment, the Amylo process for ethanol production from starch and the Symba process for food yeast production from starch – until recently industrial microbiologists have neglected its potential, presumably thinking that such systems would be unstable and difficult to control and being sceptical of any advantages that might ensue. To the contrary, it has been argued (Bull & Brown, 1978) that activity may be enhanced within a multispecies system due to synergistic or complementary reactions occurring by virtue of a greatly increased gene pool. It may be impracticable to attempt to achieve a similar enhanced reactivity with an individual species as this would lead to genetic instability. Consequently, a community of organisms would be more effective in metabolizing chemically complex or otherwise 'difficult' substrates. To date few tests of this assertion have been made, but work on the production of single-cell protein from methane and methanol provides elegant support for it. Thus, the yield from and maximum specific growth rate on methanol increased by 1.7-fold and 3.2-fold respectively when a five-membered community, isolated by continuous-flow enrichment, was compared to the single methanol-utilizer present in the community (Harrison, Wilkinson, Wren & Harwood, 1976). Additional benefits of using mixed cultures in this context included culture stability, resistance to contamination and decreased foam production.

(c) Batch or continuous fermentation

The merits of batch and continuous fermentation systems have been debated on several occasions but the arguments can be usefully restated here because they remain unfamiliar to many microbiologists and because they are taken as read by the contributors to this symposium. In general terms, continuous fermentations possess all the inherent advantages that continuous industrial operations have over batch counterparts, namely, ease of control, greater uniformity, reduced reactor volumes and reduced down-time (harvesting, sterilizing, recharging). Moreover, chemostat theory predicts that significantly higher productivity is possible compared with batch culture, a feature that was stressed by Maxon nearly twenty years ago (Maxon, 1960). Thus, the biomass productivity of a chemostat operating at a dilution rate of 0.35 h^{-1} is approximately five times higher than comparable batch systems, assuming a down-time of six hours.

A second advantage to chemostat users is the potential for varying the enormous phenotypic expression of micro-organisms. It is well known that growth conditions exert profound effects on structure, gross chemical composition, chemical modification and replacement, metabolite production and enzyme synthesis (Bull & Brown, 1978). The infinite and unique range of defined, reproducible environments that can be established in chemostats enables this phenotypic variability to be exploited to the full. Many illustrations of this point exist in the literature and it is sufficient here to note its significance for vaccine production in the context of achieving optimum antigenicity, for example.

Chemostat-type continuous fermentations impose strong selective pressures on a population via competition for the limiting substrate, a property which some would claim is a serious drawback to their use (faster-growing, lower-yielding strains, for example, may be selected and displace the original strain). However, by the same token, the chemostat can be used to select for (i) over-producing strains in which duplication of desired genes is managed (see Hartley, 1974) and (ii) to cause the evolution of enzymes in certain prescribed directions (the studies of Francis & Hansche (1972) on yeast acid phosphatase are an encouraging illustration of this approach). Similarly, the adaptation of producer organisms to higher concentrations of potentially toxic metabolites, such as organic solvents, may be achieved in continuous culture. Thus, the tolerance of *Clostridium acetobutylicum* to n-butanol was increased three-fold by this means (to 2.5%) whereas attempts to make similar improvements via batch cultivation were abortive (Ierusalimski, 1958).

It would be sanguine to state only the advantages of continuous fermentations: problems of strain stability, low product (other than biomass) concentrations and the consequences of contamination are among the most serious disadvantages. However, it may be feasible to circumvent problems of these types by, for example, appropriate use of multistage or feedback systems to improve productivity. And perhaps the chief reason why continuous fermentations have not been adopted more widely by industry remains the excess capacity for batch fermentations.

(d) Concerning choice of organisms

To date, the majority of commercial fermentations are made with monocultures of aerobic heterotrophs (fungi, eubacteria, actinomycetes). Certain anaerobic fermentations, though, are (alcoholic beverages) or have been (organic solvents) operated on very large scales and the current global research into methane production might stimulate interest in anaerobic systems. However, it is perhaps the greater experimental difficulty of growing anaerobes, especially obligate anaerobes, and the concomitant preoccupation of biochemists with aerobic metabolism, that is in part responsible for the comparative neglect of anaerobes. There is also an inherent belief among some biologists that the metabolic versatility of anaerobes is rather limited. Evans' (1977) critique of anaerobic catabolism is a timely caution that such a view might be quite misleading.

We have already mentioned the potential of mixed cultures in fermentation and suggested that concurrent growth of several species may have advantages and properties not attainable in monocultures of the constituent organisms. The utilization of mixed and recalcitrant substrates is also likely to be a promising area for mixed-culture research, indeed many traditional fermented foods and waste treatments are illustrations of these very situations. However, the scope of mixed-culture technology is considerably wider and ranges from metabolite production (β-carotene is manufactured from fermentations of mixed mating-types) to metal leaching and recovery. For example, a three-membered bacterial community, selected by chemostat enrichment, has been shown to have greatly superior tolerance to, and a capacity for bioaccumulation of, silver than any of the component species (R. C. Charley & A. T. Bull, unpublished experiments, 1978). For further discussions of mixed microbial systems see Bailey & Ollis (1977) and Bull & Brown (1978).

Finally in this section we wish to make brief reference to photosynthetic systems and to plant cell culture. Considerable work has been done on the use of algae and phototrophic bacteria in waste management and many domestic, agricultural and industrial wastes can be treated by these organisms with solar conversion efficiencies in excess of 5% (i.e. close to the practical maximum) and annual yields greater than 100 t ha^{-1}.

The problem of harvesting algae can be resolved by using rotating screen filters and recycling a proportion of the biomass to maintain the dominance of easily recovered species, e.g. those with spiny and filamentous morphologies. Currently, the outlets for photosynthetically derived microbial biomass are in pisciculture, as feed supplements for poultry and cattle, as a feedstock for methane production and, directly, as a soil fertilizer. The recovery of algal products has received comparatively little attention. Biophotolytic generation of hydrogen using algae and photosynthetic bacteria is being quite widely studied but very considerable research and development work is required to achieve reliable, continuous production. The exploitation of purple membranes from *Halobacterium* to generate electrical current, ATP or hydrogen is at a similar stage of basic research (see Schlegel & Barnea, 1977).

A major factor in the development of plant cell fermentations has been the dwindling supply of many medicinal plants due to over-exploitation and the cost of field collection (appropriate wild plants often have proved difficult to cultivate). Among the medicinals that can be produced in good yield from cell cultures are diosgenum (used for subsequent conversion to corticosteroids and steroid contraceptives), L-dopa (used in the treatment of Parkinson's disease), quinones (for use as laxatives and burn and skin treatments), ginsenosides (for multifarious uses as ginseng) and certain alkaloids. Plant cell cultures also have great potential in bioconversions, e.g. phenols to corresponding glucosides and digitoxin to more potent cardioactive compounds.

Apart from medicinals, plant cell and tissue cultures offer opportunities for the production of many diverse materials including food additives, essential oils, plant virus inhibitors, antibiotics, pyrethrum and proteinase inhibitors. From a biotechnological viewpoint, unstable productivity and high costs have delayed the industrial development of plant cell cultures, added to which plant cell suspensions tend to be sensitive to shearing stress and to produce high-viscosity broths that create aeration difficulties. A particularly valuable assessment of fermenter design in relation to plant cell growth and metabolite synthesis was made by Wagner & Vogelmann (1977). These authors assessed the performance of shake flasks, fermenters with paddle and perforated-disc impellers, and draft-tube and air-lift fermenters in the production of anthroquinones by cell cultures of *Morinda citrifolia*. The air-lift reactor proved much superior to any other in terms of yields and productivities and offers considerable possibilities for scale-up.

(2) *Genetics*

The *ad hoc* approach to the application of genetics in industrial microbiology was pointed out by Pontecorvo in his Presidential Address to the Second International Symposium on the Genetics of Industrial Micro-organisms in 1974 (Pontecorvo, 1976):

The main technique used is still a prehistoric one: mutation and selection. In 1944, independently, Demerec in the USA and myself in Scotland proposed this technique only as a war-time emergency measure for improving penicillin yields. Penicillin was desperately needed then and even an approach intellectually crude and, *a priori*, not very likely to be successful, was worth a trial. The success was so unexpectedly good that, unfortunately, since then most industrial laboratories have been contented with its *exclusive* use.

Nature in its remarkable ways of coping with the improvement of living organisms – what we call evolution – has given up the exclusive use of mutation and selection at least one billion years ago. It has supplemented mutation and selection with a wonderful variety of mechanisms for the transfer of genetic information. Sexual reproduction, combined with diploidy, is the most highly elaborate of these mechanisms.

About the time that Demerec and Pontecorvo developed the use of mutation and selection to improve penicillin yields, geneticists were becoming aware of the potential of fungi, bacteria and viruses as organisms for study. Once they had been shown to be amenable to genetic analysis, their remarkably short life-cycles made them much more attractive subjects for study than higher organisms; experiments could be completed in days or even hours rather than in weeks, months or years.

The discovery of novel methods of genetic exchange in fungi, bacteria and viruses is too well documented to need repeating. The explosion which has taken place in our genetic knowledge during the past three decades or so has been due almost completely to work with micro-organisms. Why, then, have studies on genetic recombination in industrial micro-organisms lagged behind? Apart from the reason given by Pontecorvo, i.e. the success of the mutation and selection method, there are, perhaps, two other main reasons:

(1) Most micro-organisms used in industrial processes are asexual. By the time these had been shown to possess parasexual processes, the yields of commercially important metabolites had been raised sufficiently by mutation and selection for there to be no over-riding necessity for genetic recombination processes to be investigated in depth, although some recombination work has been done.

(2) None of the micro-organisms being studied by academic scientists was of immediate industrial use, so the large amount of genetic knowledge accumulated could not be applied directly. The spade work had to be done with industrial microbes and a comparatively long-term approach was needed to carry out the necessary genetic work before parasexual processes, for example, could be exploited industrially.

So, despite the remarks of Pontecorvo – which are true in essence, in that mutation and selection is the normal method of strain improvement in industry – fermentation industries have not been unaware of the possible benefits of genetic recombination. While their investment in mutation and selection has paid off handsomely, it becomes progressively more difficult to isolate improved mutants. Apart from the obvious reason that fewer genes at which beneficial mutations could take place will be available from the total gene pool, another possibility is that cryptic, deleterious mutations, which reduce an organism's genetic potential for further beneficial change, could also be picked up unintentionally. Such mutations could ultimately lower the yield of product below its potential level, or otherwise affect the industrial performance of the micro-organisms. An inherent drawback of the method of mutation and selection is that is does not allow the removal of deleterious genetic material from an otherwise desirable genome.

The availability of genetic recombination, so that a cross could be made between a terminal strain (in a sequential mutation and selection programme) and its pregenitor strain, might produce recombinants which retained the beneficial mutations of the terminal strain whilst losing the harmful cryptic mutations. Crosses between divergent lines of mutation and selection could lead to the isolation of recombinants bearing beneficial mutations from both parents.

Published results on the practical utilization of parasexual methods are few but, both in the USA and the UK, strains derived by parasexual manipulation have been used for the industrial production of penicillin (Elander, Epenshade, Pathak & Pan, 1973; Calam, Dalglish & McCann, 1976). This more rational approach of using the parasexuality as the method of genetic exchange has been extended to exchange of genetic information between protoplasts. Protoplast-fusion techniques are likely to have special significance for fungi in which heterokaryon formation is difficult to achieve, and the ability to induce protoplast fusion in such industrially important species as *Penicillium chrysogenum* and *Cephalosporium acremonium* is well established (Peberdy, 1976). The strategy of protoplast fusion has also been advocated as a means for enabling high-frequency recombination in actinomycetes, and the first reports of such work suggest great promise for the technique (Hopwood, Wright, Gibb & Cohen, 1977). It should also be noted that protoplast fusion and mutation offer considerable scope for de-repressing secondary metabolite synthesis in plant cell and tissue cultures; so far most attempts to improve productivities have relied on manipulating culture conditions. Some progress has been made in the selection of biochemical plant cell culture mutants in recent years (see Widholm, 1977, for a review) with a number of hormone-autotrophic and visual mutants having been obtained, as well as some auxotrophic and resistant mutants which have proved invaluable for recombination-via-fusion programmes.

Finally, the immense potential of the genetic manipulation approach that specifically identifies, isolates, inserts and amplifies genes is one of the subjects of this symposium (see Atherton *et al.*, this volume pp. 379–405). Without question, the emerging field of molecular cloning can be regarded as one of the most significant discoveries in modern science; its development is particularly propitious for microbial technology.

(3) *Engineering*

The ultimate objective of fermentation development is to increase the efficiency with which micro-organisms convert their substrates to economically valuable products. The definition of efficiency must in the last analysis be based upon operational costs which include those of all materials, power and labour, as well as return of borrowed capital, payment of interest and the depreciation of plant. This strictly utilitarian objective has been redefined as the need to subvert the natural tendency of an organism which, as a living control system, has as its objective self-preservation and survival (Hockenhull, 1971). An understanding of the commercial objectives and how they may be attained once more highlights the interdisciplinary nature of microbial technology. The efforts of the microbiologist, which must come first, have to rely on the talents of the engineer to bring them to fruition. The industrial or applied microbiologist cannot be expected to translate his research at the laboratory level into a finished process or even to see how the latter may be accomplished. He should, nevertheless, be cognisant of the ultimate feasibility of what he is proposing; this means possession of some rudimentary knowledge of engineering, leavened with a little economics and market awareness so that he may appreciate the broad implications and possibilities of his work.

Much of microbial technology is directly derived from chemical engineering principles and, not surprisingly, chemical engineers are very much to the fore in the development of large-scale fermentation process. The engineer is instrumental in converting the basic microbiology of a process into the final production unit. This, though it is a key role, is strictly a development function.

A marked change in attitude of engineers towards the microbiologist has taken place over the past 25 years. The prevailing view in the 1950s, typified by the 1952 Presidential Address to the Institution of Chemical Engineers (see Hartley, 1967) suggested that the development of industrial microbiology should be based on chemical engineering, with the microbiologist being introduced solely as a help to the engineer. The fallacy of this is now appreciated, or should be. When the microbiologist has discovered the conditions permitting efficient operation in the laboratory, it is then the (bio)chemical engineer's task to develop the complex design and control system needed for industrial production. Clearly, if the microbiology or biochemistry is not right, or is badly understood at the laboratory level, then technological innovations in the design of equipment will be to little avail.

The contributions of the engineer, of course, extend beyond the design and construction of the pilot and large-scale production units. While appreciating the importance of the engineer, experience shows that his role must be carefully coordinated with that of the biologist: each without the other is useless. Integration is therefore essential and this, or its lack, has been commented on by many eminent workers in this field. Perlman (1969) went so far as to consider that the technology of the fermentation industry had usually lagged behind the microbiological aspects though Malek (1974) argued that this was '... not due to an intrinsic failure of engineering and technology but reflected the *lack of integration* [our italics] of microbiological and technological solutions'.

Where one can point to the establishment of successful projects, such as the processes designed by British Petroleum and ICI for the production of single-cell proteins from, respectively, alkanes and methanol, the close cooperation between engineer and microbiologist is very much in evidence. Although Finn & Fiechter in their chapter later in this book (pp. 83–105) fully discuss the influence which microbial physiology has on reactor design, the point is worth repeating here: in the development of the ICI 'Pruteen' project, a thorough understanding of factors influencing the rate of growth of a bacterium on methanol indicated that oxygen transfer was probably the critical factor in achieving high biomass densities in the fermenter. Therefore, if the oxygen concentration could be raised by the simple expedient of increasing the operating pressure this would be beneficial to the organism. Instead of designing a pressurized fermenter, the innovation was to design a very tall fermenter wherein the pressure at the bottom, where the air was to be introduced, was increased by several atmospheres by the head of the fermentation broth. Although this system would also serve to keep the substrate in solution, this was a lesser consideration, as the system operates under carbon limitation and consequently the concentration of methanol is always very low. This fermenter is without doubt the most efficient system yet developed for the growth of a micro-organism but has as its basis a knowledge of fundamental microbiological principles.

In a similar manner, the development of dialysis fermenters illustrates how the natural tendency of a micro-organism to slow its own growth rate in batch cultures can be overcome by a simple technological device. In the course of growth many micro-organisms produce extracellular metabolites which may accumulate to toxic proportions. If the organism could be grown within a dialysis chamber these metabolites, providing they are of low molecular weight, could be continuously removed. Two types of fermenter have been designed to effect this principle: the dialysis fermenter of Gallup & Gerhardt (1963) (see also Pirt, 1975; Rivière, 1977) which has the dialysis unit remote to the fermenter and operates batch-wise; and the Rotorfermenter of Margaritis & Wilke (1972) in which the dialysis unit and fermenter are integrated. Both types of fermenter have proved useful in the production of vaccines, enzymes and even viruses. They enable a dense population of cells to be produced and, therefore, decrease the centrifugation capacity needed for harvesting. Production of toxic metabolites, such as lactic acid from glucose by *Lactobacillus delbrueckii* and salicylic acid by naphthalene cleavage by *Pseudomonas aeruginosa*, has been achieved by these methods (see Rivière, 1977).

Other examples of how engineers have adapted their methods to overcome the perversity of micro-organisms can be found throughout this book and in more specialized works dealing solely with reactor design (see, for example, Atkinson, 1974). However, it should be appreciated that engineering contributes more than just reactor design to microbial technology. There is increasing emphasis on the accurate control and monitoring of all the various activities which go to make up a fermentation process. An awareness that factors such as the redox potential or dissolved oxygen tension can be monitored and regulated *in situ* in large as well as small fermenters has helped to increase productivity of many processes (Sikyta, Prokop & Novak, 1974). In time, as instrumentation becomes more sophisticated, additional parameters, such as substrate concentrations within a fermenter, will be able to be measured as a routine. To carry out such measurements the creation of sterilizable electrodes is needed, but this will probably be no more difficult than the production of reliable sterilizable pH and oxygen electrodes 20 years ago.

Movement, then, is clearly towards better control of fermentations taking into account not only physicochemical parameters but also biochemical ones. It is a logical development of this trend that computers are now becoming part of the fermentation panoply, for it is only by their application that it will be possible to integrate the multitudinous events which occur in a fermentation process. For example, the control of growth rate by the maintenance of dissolved oxygen tension through control of agitation and aeration, and thus regulation of the output from a continuously operating fermenter (see Nyiri *et al.*, 1974), is not difficult. But the output from a fermentation process is also dependent on the pre- and post-fermentation facilities, which are part of the whole process. The rate at which centrifuges and other downstream systems can handle the output from a fermenter have to be balanced not only with the output of the fermenter but also with the rate at which medium can be prepared and sterilized. Fermentation productivity is literally in the middle of all operations. One can, therefore, anticipate that computers will become a recognized part of microbial technology and will play a key role in the regulation not only of the fermentation itself but of all the other unit processes which go to make up the entire operation. This will apply to both continuous and batch processes, particularly where these require manipulations to be made during the course of the fermentation. Such is the case with antibiotic production in batch culture where additions of appropriate precursors to the fermentation broth require accurate timing.

Concomitant with the introduction of computers is the need to translate microbial growth into mathematical terms. The complex interplay of variables which leads to predictions of product yield, for example, is discussed by Kossen later in this symposium (pp. 327–57). Mathematical modelling is a subject with which all those involved in microbial technology need to be familiar, as its ramifications spread far and wide.

It should not be forgotten that biotechnology is now moving beyond the whole-cell phase, and the use of isolated enzymes on an industrial scale is now very important. The use of enzymes, immobilized on inert supports to aid stability and ease of recovery, has developed to an extent where it is no longer strictly an aspect of microbiology; it is a discipline in its own right having been born out of an interaction between enzymology, microbiology and engineering. Microbiology is now merely the servant of the discipline, acting as the provider of the cells from which the enzymes are derived. It is for this reason alone that an account of immobilized-enzyme technology does not appear in this symposium: not for reasons of believing that it has yet to prove itself or is of only minor importance – nothing could be further from the truth.

Codicil

Although we have sought to illustrate some of the conceptual inputs which are necessary to make a success of microbial technology, it should be remembered that this is not a one-way process. Microbial technology has acted as priming agent for many far-reaching biological investigations. To explain alcohol production by yeasts, glycolysis was studied and pathways of glucose catabolism were unravelled; in seeking an explanation of how and why citric acid was accumulated by *Aspergillus niger*, the tricarboxylic acid cycle was discovered; the production of antibiotics led to fundamental studies being made on genetics of fungi and of the regulation and production of secondary metabolites. More recently, the exploitation of alkanes, methanol and methane as substrates for single-cell protein production has prompted the discovery of new metabolic pathways. Thus, the exploitation of the potential benefits of micro-organisms has acted as one of the main spurs to unravelling their behaviour.

ENVIRONMENTAL MANAGEMENT

In a later contribution to this symposium, Slater and Somerville (pp. 221–61) analyse certain features of waste and effluent processing. Without trespassing on their ground we wish to make a few brief observations on the philosophy of this subject. Frequently a basic understanding of biodegradation, toxicology (acute or chronic) and the microbiology of treatment systems is lacking, although such knowledge is essential for rational design of these processes. There is a real distinction between the *management* and the *treatment* of waste and effluent, and too often the producer of such materials is concerned solely with their rapid disposal without assessing the possibilities for recovery and recycling. The technology for these two activities is unlikely to be the limiting factor: economic considerations will largely determine the viability of recovery–recycle schemes. It is worth considering the nature of biodegradable waste materials: 'waste' has been defined as a resource arising in say agriculture, or from urbanization, where the cost *of* using it is greater than the return *from* using it. This definition embodies a vital concept because it implies that the nature of what is termed 'waste' is not constant, but will be dependent on the economic climate and state of technology at any given time. Consider sewage sludge, the excess of which in the UK (and many other countries) is an embarrassment (1 to 2 Mt y^{-1}) and the disposal of which is expensive (about £30 t^{-1}). One strategy would be to optimize, rather than minimize, sludge production and adapt the treatment process for animal feed production. A major drawback when applying this approach to industrial effluents might be the undesirable concentration of non-biodegradable materials such as heavy metals and chlorinated hydrocarbons. However, with few exceptions, current thinking and technology is of the 'throw-away' type and the end-product of numerous ingenious treatment processes is discharged at sea or into rivers, is incinerated or used for land-fill. Many feasibility studies have been made on sludge as a basis for the production of animal feed supplement, fine chemicals, gas, oil, paper, plastics and compost. Of these, feed supplement and gas production are the most attractive, while current economics have largely led to the dismissal of the remaining alternatives.

In addition to sewage sludge, a whole variety of agricultural wastes can be considered for fermentation feedstocks, or as suitable materials for composting or ensilaging. However, economic constraints again will determine the viability of waste-processing schemes. Thus, the waste must (i) be biodegradable, (ii) have a constant and predictable quality, (iii) be cheap or have zero cost, (iv) be transportable to the plant at low cost, (v) be available in quantities sufficient to operate an economically sized plant and (vi) be available on a reliable, preferably year-round, basis. Numerous consequences follow on from these arguments such as seasonal availability necessitating the design of processes that can operate on more than one waste, or transportation problems dictating the development of small, local processes. But one additional point is, perhaps, the most critical of all in determining the economics of waste management: this is the desirability of developing simultaneous processes, in other words, attempting to integrate waste treatment *per se* (pollution control), resource recycling (for example, water) and recovery (for example, biomass). A typical scenario might show a primary waste treatment step by algae leading to water purification and biomass production (algal growth could be enhanced by the input of waste heat and CO_2 from a power plant) and the biomass being used as single-cell protein (especially carp, shellfish and chicken feed supplementation) or as a digester feedstock (CH_4, soil fertilizer).

COMPETITIVENESS OF MICROBIAL TECHNOLOGY

It would be quite wrong to conclude these introductory comments without tempering the euphoric view of microbial technology with a few hard commercial facts. Foremost is the necessity that microbiological processes and products are competitive with their chemical counterparts otherwise they will exist only as interesting but fossilized options. Thus, of the hundred most important organic chemicals, for only six have alternative large-scale microbiological processes been developed (ethanol, acetic acid, isopropanol, acetone, n-butanol and glycerol) (Pape, 1977); of these only ethanol production is mainly by fermentation. Most bulk chemicals produced by and for the chemical industries are relatively simple and stable and can be manufactured at low cost via petrochemical technology. Pape (1977) illustrates this point with reference to acetic acid production and compares the carbonylation of methanol at 250 °C and 700 atm (45% yield) with the oxidation of ethanol at 30 °C by acetobacter (15% yield). Production costs based on feedstocks and other materials, utilities, labour and depreciation are nearly three times greater for the fermentation than the chemical process. Capital investment costs are even more disadvantageous to the fermentation process.

The current situation in industrially developed countries might be expressed as follows: (i) with the exception of certain metal-recovery operations, microbiological processes are uncompetitive in the production of simple bulk chemicals. Such microbiological processes might become viable if they are integrated with waste treatment, or can rely on agricultural surpluses for their feedstocks, or, are enforced by political or economic policies (e.g. monopoly and fiscal measures within the EEC dictate the proportions of ethanol produced chemically and by fermentation in different countries); (ii) microbiological processes are competitive for the production of an enormous range of chemically complex and labile materials. Production of such materials may be difficult or impossible via chemical means, and the cost totally prohibitive.

We emphasized earlier the dynamic nature of economics and its significance for the exploitation of fermentation processes. Thus, the conclusions above have to be set in the context of an oil-based economy (at least for industrially developed countries) and until supplies of oil and other fossil energy reserves are considerably more depleted the incentive to develop alternative fermentation routes to bulk organic chemicals will remain low. However, the development of new, and the renaissance of old, fermentation processes will be hastened in those countries that lack indigenous petroleum resources or the means of purchase. The most dramatic instance of this is Brazil's National Alcohol Programme which seeks to reduce petroleum imports by blending ethanol in motor fuel and by processing it to ethylene (Jackman, 1976).

Another factor influencing the competitiveness of fermentation processes is the availability of feedstocks, notably photosynthetically renewable raw materials, wastes and chemical feedstocks. Renewable materials essentially are those of agricultural and forestry origin – grains and other starch crops, sugar crops, cellulose and lignocellulose. Revolutionary developments in enzyme-based production of sugar syrups from maize can have a major impact on traditional cane sugar producing countries, but in turn the prospect of developing indigenous fermentation industries based on molasses or cane juice is an attractive one for many countries in Central and South America, Africa and Asia.

Much has been written about the opportunities for using cellulose as a basic fermentation substrate; we will not continue the debate here except to reiterate that, at present, the cost of converting cellulose to fermentable materials prevents its commercial exploitation. In contrast, relatively scant attention has been given to lignin-derived materials as industrial feedstocks. Microbial processes for the controlled degradation of lignin should be explored for the possibility of generating valuable phenolic materials. The utilization of waste materials for fermentation has already been alluded to – the viability of any microbiological process has to be gauged against alternative processes such as incineration (heat recovery), pyrolysis or gasification (fuel production), direct recovery (enzymes, vitamins, pectin, fibre, etc.), direct recycle through livestock (e.g. poultry faeces in cattle feed), or simply sanitary land-fill schemes.

Finally chemical and petrochemical feedstocks are now counted among important fermentation substrates, a situation encouraged by the success of methanol as a substrate for single-cell protein production. Methanol is produced from methane, crude oil or coal and also synthesized from carbon dioxide and hydrogen. World capacity for synthetic methanol production is high and recent feasibility studies of its production from air and water using controlled thermonuclear fusion power (Dang & Steinberg, 1977; Steinberg & Dang, 1977) suggest that it could become a major economic feedstock for many fermentation processes in addition to single-cell protein production.

EPILOGUE

In concluding this chapter we wish to highlight the national importance of microbial technology and to suggest that the participation of the scientific community in formulating a national policy towards it should be very much greater. All too frequently advice from the scientific community only comes after the need for such advice is recognized and requested by government: we share the view expressed by Melnick, Melnick & Fudenberg (1976), applied by them to biomedical research but equally applicable to microbial technology, that '... scientists must direct more of their efforts towards policy innovations that will benefit the public (and) take responsibility for bringing to the attention of the public ... the tangible benefits that emanate from progress'.

The case for microbial technology, of course, has been presented on several occasions; Sir Harold Hartley, considered by many to be the father of biochemical engineering, voiced the importance and urgency of the matter twelve years ago.

The national importance of the development of industrial microbiology is so great that it deserves the attention of the Minister of Technology who now carries responsibility for the welfare of the science-based industries of this country and for supporting new developments of the growth industries of the future, among which biochemical engineering must take a high place.... Successful development...must take place at universities in order to attract and train the young men and women who will be needed in these industries in the future. We have the ability and the enthusiasts if they are given support on an adequate scale for the complex operations involved. What we need now are two or three strong centres of teaching and research directed specifically to industrial microbiology in order to safeguard the future of these industries in this country. Now is the time to retrieve the position that has been allowed to slip (Hartley, 1967).

If this analysis was correct in 1967 it is pertinent to ask what has happened during the intervening years. The answer, unfortunately, is very little and one which accords with the basic conclusion of the survey commissioned by the Institution of Chemical Engineers. 'our future technical abilities in these areas (education and research) look very thin' (Emery, 1975).

The situation, as briefly outlined here, is bad enough, but when considered against the background of commitment being made by countries in a position similar to Britain, its seriousness is compounded. The progress of microbial technology in Japan has been of the most impressive kind and contributes in a major way to that country's economic success (the size and activities of Japanese fermentation industries together with detailed sales, investment and manpower statistics have been recently assembled (International Technical Information Institute, Tokyo, 1977; Yamada, 1977). Yamada cites eager government promotion of microbial technology and the wide-scale establishment of applied microbiology laboratories throughout the university sector as chief factors for Japan's pre-eminence in this field. In the Federal Republic of Germany, on the basis of a commissioned survey made by the Deutsche Gesellschaft für Chemisches Apparatewesen, the government has recognized the long-term importance of biotechnology and has authorized a major programme of research in fields such as fermentation technology, waste disposal and water purification, cell culture, enzyme technology and unconventional food technology. On the other hand, although the European Community has been developing common policies for science and technology (Schuster, 1978), the impact of biotechnology and fermentation on Community programmes so far has been very limited. Within Britain the collective activities of a few industrial, government and university groups provide a good base from which to launch an effective long-term programme of research aimed at a major expansion of microbiologically-based industries. However, we cannot help but recall the remark of The Red Queen: 'It takes all the running you can do to keep in the same place. If you want to get somewhere else, you must at least run twice as fast as that' (Carroll, 1872).

One of our hopes for this symposium and this book is that it will provoke a wider consideration of microbial technology and further encourage appropriate industrial development.

REFERENCES

ATKINSON, B. (1974). *Biochemical Reactors*. London: Pion.

BAILEY, J. E. & OLLIS, D. F. (1977). *Biochemical Engineering Principles*. New York: McGraw-Hill.

BULL, A. T. & BROWN, C. M. (1979). Continuous culture applications to microbial biochemistry. In *Microbial Biochemistry*, ed. J. R. Quayle. (*International Review of Biochemistry*, Series II). Lancaster: MTP. (In press.)

BU'LOCK, J. D. (1974). Acetone–butanol, butandiol, and other fermentations. In *Large-scale Fermentations for Organic Solvents*, eds. A. J. Powell & J. D. Bu'Lock, pp. 5–19. *Octagon Papers No. 2*. Manchester: Department of Extra-mural Studies, Manchester University.

CALAM, C. T., DAGLISH, L. B. & McCANN, E. P. (1976). Penicillin: tactics in strain improvement. In *Proceedings of the Second International Symposium of the Genetics of Industrial Micro-organisms*, ed. K. D. Macdonald, pp. 273–87. London; New York: Academic Press.

CARROLL, L. (1872). *Through the Looking-Glass, and what Alice found there*. (Chapter 2.) London: Macmillan.

DANG, V. D. & STEINBERG, M. (1977). Production of synthetic methanol from air and water using controlled thermonuclear reactor power – II. Capital investment and production costs. *Energy Conservation*, 17, 133–40.

DEMAIN, A. L. (1973). Mutation and the production of secondary metabolites. *Advances in Applied Microbiology*, 16, 177–202.

ELANDER, R. P., ESPENSHADE, M. A., PATHAK, S. G. & PAN, C. H. (1973). The use of parasexual genetics in an industrial improvement program with *Pencillium chrysogenum*. In *Genetics of Industrial Micro-organisms: Actinomycetes and Fungi*, eds. Z. Vanek, Z. Hostalek & J. Cudlin, pp. 239–53. Prague: Academia.

EMERY, A. N. (1975). Biochemical engineering. *The Chemical Engineer* (November), 682–3 and 697.

EVANS, W. C. (1977). Biochemistry of the bacterial catabolism of aromatic compounds in anaerobic environments, *Nature, London*, **270**, 17–22.

FORSTER, C. F. & JONES, J. C. (1976). The bioplex concept. In *Food from Waste*, eds. G. G. Birch, K. J. Parker & J. T. Worgan, pp. 278–91. London: Applied Science Publishers.

FRANCIS, J. C. & HANSCHE, P. E. (1972). Directed evolution of metabolic pathways in microbial populations. I. Modification of the acid phosphatase pH optimum in *Saccharomyces cerevisiae*. *Genetics*, **70**, 59–73.

GALFRE, G., HOWE, S. C., MILSTEIN, C., BUTCHER, G. W. & HOWARD, J. C. (1977). Antibodies to major histocompatibility antigens produced by hybrid cell lines. *Nature, London*, **266**, 550–2.

GALLUP, D. M. & GERHARDT, P. C. (1963). Dialysis fermentor systems for concentrated culture of micro-organisms. *Applied Microbiology*, **11**, 506–12.

HAMMOND, A. L. (1977). Alcohol: A Brazilian answer to the energy crisis. *Science*, **195**, 564–5.

HARRISON, D. E. F., WILKINSON, T. C., WREN, S. J. & HARWOOD, J. H. (1976). Mixed bacterial cultures as a basis for continuous production of SCP from C_1 compounds. In *Continuous Culture 6: Applications and New Fields*, eds. A. C. R. Dean, D. C. Ellwood, C. G. T. Evans & J. Melling, pp. 122–34. Chichester: Ellis Horwood for the Society for Chemical Industry.

HARTLEY, B. S. (1974). Enzyme families. In *Evolution in the Microbial World*, eds. M. J. Carlile & J. J. Skehel, pp. 151–82. *24th Symposium of the Society for General Microbiology*. Cambridge University Press.

HARTLEY, H. (1967). *Process Biochemistry*, **2** (4), 3.

HASTINGS, J. J. H. (1971). Development of the fermentation industries in Great Britain. *Developments in Applied Microbiology*, **16**, 1–45.

HOCKENHULL, D. J. D. (1971). Observations on fermentation development. *Progress in Industrial Microbiology*, **9**, 113–53.

HOPWOOD, D. A., WRIGHT, H. M., GIBB, M. J. & COHEN, S. N. (1977). Genetic recombination through protoplast fusion in streptomyces. *Nature, London*, **268**, 171–4.

IERUSALIMSKI, N. D. (1958). A study of the process of development of micro-organisms by the continuous flow and exchange of media technique. In *Continuous Culture of Micro-organisms*, ed. I. Malek, pp. 62–74. Prague: Czechoslovak Academy of Sciences.

International Technical Information Institute, Tokyo, Japan (1977). *Japan's Most Advanced Industrial Fermentation Technology and Industry*.

JACKMAN, E. A. (1976). Brazil's National Alcohol Programme. *Process Biochemistry*, **11** (5), 29–30.

JONES, L. H. (1974). Plant cell culture and biochemistry: studies for improved vegetable oil production. In *Industrial Aspects of Biochemistry*, ed. B. Spencer, pp. 813–33. *FEBS Special Meeting, Dublin, 1973*. Amsterdam: North-Holland.

MACLENNAN, D. G. (1976). Single cell protein from starch. In *Continuous Culture 6: Applications and New Fields*, eds. A. C. R. Dean, D. C. Ellwood, C. G. T. Evans & J. Melling, pp. 69–84. Chichester: Ellis Horwood for the Society for Chemical Industry.

MALEK, I. (1974). Present state and perspectives of biochemical engineering. In *Advances in Biochemical Engineering*, eds. T. K. Ghose, A. Fiechter & N. Blakebrough, pp. 279–90. Berlin: Springer.

MARGARITIS, A. & WILKE, C. R. (1972). Engineering analysis of the Rotorfermenter. *Developments in Industrial Microbiology*, **13**, 159–76.

MAXON, W. D. (1960). Continuous fermentation. *Advances in Applied Microbiology*, **2**, 335–49.

MELNICK, V. L., MELNICK, D. & FUDENBERG, H. H. (1976). Participation of biologists in the formulation of national science policy. *Federation Proceedings. Federation of American Societies for Experimental Biology*, **35**, 1957–62.

MUNNECKE, D. M. & HSIEH, D. P. H. (1975). Microbial metabolism of a parathion-xylene pesticide formulation. *Applied Microbiology*, **30**, 575–80.

NYIRI, L. K., JEFFERIS, R. P. & HUMPHREY, A. E. (1974). Application of computers to the analysis and control of microbiological processes. In *Advances in Microbial Engineering*, eds. B. Sikyta, A. Prokop & M. Novak, part 2, pp. 613–18. *Biotechnology and Bioengineering Symposium No. 4*. New York: Wiley.

PAPE, M. (1977). The competition between microbial and chemical processes for the manufacture of basic chemicals and intermediates. In *Microbial Energy Conversion*, eds. H. G. Schlegel & J. Barnea, pp. 515–30. Oxford: Pergamon.

PEBERDY, J. F. (1976). Isolation and properties of protoplasts from filamentous fungi. In *Microbial and Plant Protoplasts*, eds. J. F. Peberdy, A. H. Rose, H. J. Rogers & E. C. Cocking, pp. 39–50. London, New York: Academic Press.

PERLMAN, D. (1969). Fermentation Industry – Evolution. *Process Biochemistry*, **4**, (6), 29, 30 and 34.

PERLMAN, D. (1978). Stimulation of innovation in the fermentation industries, 1910–80. *Process Biochemistry*, **13**, (May), 3–5.

PIRT, S. J. (1975). *Principles of Microbe and Cell Cultivation*. Oxford: Blackwell Scientific.

PONTECORVO, G. (1976). Presidential address. In *Proceeding of the Second International Symposium on the Genetics of Industrial Micro-organism*, ed. K. D. Macdonald, pp. 1–4. London, New York: Academic Press.

POWELL, E. O. (1958). Criteria for the growth of contaminants and mutants in continuous culture. *Journal of General Microbiology*, **18**, 213–32.

PRATTEN, C. (1970). *Economics of Scale in Manufacturing Industry*. Department of Applied Economics, Occasional Paper No. 28. Cambridge University Press.

PUHAN, Z. & MARTIN, S. M. (1971). Industrial potential of plant cell culture. *Progress in Industrial Microbiology*, **9**, 13–33.

RAMALINGHAM, A. & FINN, R. K. (1977). The vacuferm process: a new approach to fermentation alcohol. *Biotechnology and Bioengineering* **19**, 583–9.

RATLEDGE, C. (1975). The economics of single-cell protein production; Substrates and processes. *Chemistry and Industry*, pp. 918–20.

RATLEDGE, C. (1977). Fermentation substrate. *Annual Reports on Fermentation Processes*, **1**, 49–71.

RIVIÈRE, J. (1977). *Industrial Applications of Microbiology* (English edition). Surrey University Press in association with International Textbook Company.

ROWLEY, B. I. & BULL, A. T. (1977). Isolation of a yeast-lysing *Arthrobacter* species and the production of the lytic enzyme complex in batch and continuous flow fermenters. *Biotechnology and Bioengineering*, **19**, 879–99.

SCHLEGEL, H. G. & BARNEA, J. (1977). *Microbial Energy Conversion*. Oxford: Pergamon.

SCHUSTER, G. (1978). European community policy in the field of science and technology. *Endeavour*, **2**, 22–6.

SENIOR, E., BULL, A. T. & SLATER, J. H. (1976). Enzyme evolution in a microbial community growing on the herbicide Dalapon. *Nature, London*, **263**, 476–9.

SIKYTA, B.. PROKOP, A. & NOVAK, M. (eds.) (1974). *Advances in Microbial Engineering* (*Biotechnology and Bioengineering Symposium No. 4*, parts 1 and 2). New York: Wiley.

STEINBERG, M. & DANG, V. D. (1977). Production of synthetic methanol from air and water using controlled thermonuclear reactor power – I Technology and energy requirement. *Energy Conservation*, 17, 97–112.

VELDKAMP, H. (1970). Enrichment cultures of prokaryotic organisms. In *Methods in Microbiology*, eds. J. R. Norris & D. W. Ribbons, vol. 3A, pp. 305–61. London, New York: Academic Press.

WAGNER, F. & VOGELMANN, H. (1977). Cultivation of plant tissue cultures in bioreactors and formation of secondary metabolites. In *Plant Tissue Culture and its Biotechnological Applications*, eds. W. Barz, E. Reinhard & M. H. Zenk, pp. 245–52. Berlin: Springer.

WIDHOLM, J. M. (1977). Selection and characterization of biochemical mutants. In *Plant Tissue Culture and its Biotechnological Applications*, eds. W. Barz, E. Reinhard & M. H. Zenk, pp. 112–22. Berlin: Springer.

WILKINSON, T. G., TOPIWALA, H. H. & HAMER, G. (1974). Interactions via mixed bacterial population growing on methane in continuous culture. *Biotechnology and Bioengineering*, 16, 41–59.

YAMADA, K. (1977). Bioengineering Report. Recent advances in industrial fermentation in Japan. *Biotechnology and Bioengineering*, 19, 1563–1621.

THE PHYSIOLOGY OF SINGLE-CELL PROTEIN (SCP) PRODUCTION

RICHARD I. MATELES

Laboratory of Applied Microbiology, Institute of Microbiology, Hebrew University – Hadassah Medical School, P.O. Box 1172, Jerusalem, Israel

INTRODUCTION

There have been a number of books which have dealt extensively with various aspects of SCP production (Mateles & Tannenbaum, 1968; Gounelle de Pontanel, 1972; Gutcho, 1973; Davis, 1974; Tannenbaum & Wang, 1975; Wagner, 1975) relieving me of the need to be comprehensive in this review. It is especially fitting that the Society for General Microbiology is holding this Symposium, as a considerable part of the development of processes for SCP production has occurred in British industrial laboratories (British Petroleum, Shell, ICI, Rank Hovis McDougall) and much of what we know about the microbial metabolism of hydrocarbons and C_1 compounds is due to the efforts of British scientists.

SCP production is yet another example of an actual or potential large-scale application of science which has had the very beneficial effect of stimulating a great deal of scientific research, much of it of a seemingly pure nature, which most probably would not have occurred without the economic motivations. Thus, since the early 1960's, when the idea of SCP as a non-agricultural means of producing foods or feeds first became prominent, there has been a flowering of scientific investigation on the physiology, genetics, and biochemistry of organisms which might have a role in the production of SCP. Not only have there been extensive investigations into the growth of micro-organisms on hydrocarbons, but the entire area concerned with the metabolism of methane and methanol has received a most significant impetus from the application. Although in the early years the several companies that were first involved apparently devoted very large programmes primarily to the engineering side of the problem of SCP production, and in some cases reached pilot programmes rather quickly, it soon became apparent that the biological problems had been neglected, and in many cases the SCP programmes were halted, reduced in scope, or re-examined. As one consequence of this re-examination much good work has been done by industrial groups, and at least a small part of this research is now reaching the scientific literature. In retrospect, it is now apparent to many observers, what some people claimed even then: that as SCP is inherently a biological solution to a biological problem, it is too important to be left to engineers. From personal acquaintance with some of the programmes now being conducted at different companies, I believe that the original over-emphasis on engineering has been corrected and many of the programmes clearly reflect the contributions of microbiologists and biochemists, as well as engineers, to the creative solution of an important problem.

CONTINUOUS CULTURE FOR SCP PRODUCTION

Although in the years after the first introduction of continuous culture (Herbert, Elsworth & Telling, 1956) it appeared that the proponents of continuous culture expected its major applications to be industrial (Society for Chemical Industry, 1961), this has been a disappointment. Virtually the only serious industrial applications of continuous culture have been in SCP production, where there are well-defined advantages (Mateles, 1968) in terms of productivity per unit volume per unit time. On the other hand, the application of continuous culture as a tool in microbial physiology has been enormously fruitful (see, for example Dean, Ellwood, Evans & Melling, 1976; or Dean, Pirt & Tempest, 1972). In this review it is assumed that the reader is familiar with the principles and terminology of continuous culture and virtually all of the examples and discussion are oriented to the use of continuous culture for SCP production. It is suggested that the reader who is not already acquainted with the principles of continuous culture should consult Herbert, Elsworth & Telling (1956), or one of the other expositions (Malek & Fencl, 1966; Kubitschek, 1970) of this subject.

YIELDS AND STOICHIOMETRY

Substrate yields

Since the subject of SCP production has distinct economic overtones, it is not surprising that the question of yield, i.e. how much SCP can be obtained from a unit of substrate, is so central. Generally speaking, experience suggests a maximum value for a particular type of substrate, but the minimum value obtained depends only on how badly one grows the micro-organism. Table 1 gives examples of the yields which may be obtained using good growth conditions (continuous culture with carbon-substrate-limited growth) aimed at producing cells with a high protein content. Clearly the production of cells containing large amounts of storage materials, which are often slightly modified or polymerized forms of the substrate, may lead to apparent yields higher than recorded in Table 1, but such cells are consequently rather low in protein. This problem will be dealt with again later in this review.

Table 1. *Representative yields of SCP on various substrates[a] and calculated[b] oxygen requirement*

Substrate	Organism	Y (g SCP per g substrate)	Oxygen requirement (g O_2 per g dry wt)	Reference
Glucose	*Escherichia coli*	0.53	0.4	Schulze & Lipe (1964)
	Candida utilis	0.54	0.6	Brown & Rose (1969)
Methanol	*Pseudomonas* C	0.54	1.2	Goldberg *et al.* (1976)
	Hansenula polymorpha	0.36	2.6	Levine & Cooney (1973)
Ethanol	*Saccharomyces cerevisiae*	0.63	2.0	Nagai, Nishizawa & Aiba (1969)
Octadecane	*Pseudomonas* sp.	1.07	1.6	Wodzinski & Johnson (1968)
Octane	*Pseudomonas* sp.	1.06	1.7	Wodzinski & Johnson (1968)
Methane	Mixed bacterial culture	0.90	3.0	Brewersdorff & Dostálek (1971)
	Mixed bacterial culture	0.62	4.8	Sheehan & Johnson (1971)
	Mixed bacterial culture	0.80	3.7[c]	Wilkinson, Topiwala & Hamer (1974)
	Mixed bacterial culture	0.99[d]	2.6	Wilkinson, Topiwala & Hamer (1974)

[a] The data are from experiments conducted in continuous culture with the carbon-substrate as the limiting nutrient.

[b] The oxygen requirement was calculated according to Mateles (1971). For cases in which a cell composition was not reported, the average data reported by Humphrey (1968) were used.

[c] Data on oxygen requirements were reported and were 25% higher than this calculated value. However, the reported carbon balance was also about 20% in error.

[d] In this case the culture was oxygen-limited. The carbon balance was at least 10% in error.

The high yield obtained by replacing carbohydrate substrates with hydrocarbon substrates attracted great attention in the early days of SCP. However, it was soon realized that the much higher yield of cells growing on a relatively reduced hydrocarbon substrate was countered by the greater oxygen requirements of these cells per unit of SCP produced. That is, the high carbon yield was offset by a low oxygen yield, with the mechanical complications of having to provide for enhanced oxygen transfer. The relationship between yield based on the carbon source and the oxygen requirement, for substrates of any com- position and cells of any composition, can be calculated quite easily assuming only that the products of metabolism are cells, water and carbon dioxide (Mateles, 1971). These assumptions are usually valid for SCP production, and the equation resulting for cells of 'normal' composition (C, 53%; N, 12%; O, 19%; H, 7%) is given below; in this equation C, H and O represent the number of atoms of carbon, hydrogen and oxygen in each molecule of carbon source; Y represents the yield of cells based on the carbon source (g cells (dry wt) per g carbon source consumed) and M is the molecular weight of the carbon source:

$$\frac{\text{g oxygen}}{\text{g cell}} = \frac{32C + 8H - 16O}{Y \times M} - 1.58$$

As SCP production, like all aerobic fermentations, is highly exother- mic, the question of oxygen requirement has implications not only in terms of the mechanical arrangements required for ensuring the supply of oxygen but also in terms of the complications arising from the need to remove large amounts of heat from the fermenter to maintain a constant temperature. Thus, another unwanted result of the high yield on hydrocarbons as compared to carbohydrates was the high heat production on hydrocarbons.

While the heat of fermentation can be estimated by various thermo- dynamic calculations for cases in which the substrate and cell com- position are well defined, a general theoretical solution is not available. However, the heat production can be estimated very well by using the quotient 110 kcal per mole O_2 taken up. This figure was obtained experimentally using different substrates and micro-organisms (Cooney, Wang & Mateles, 1969) and over the years has been found to be a good estimate by many groups. It has the advantage of being useful also for organisms growing on complex media and producing various metabolites and requires only a knowledge of oxygen consumption during fermen- tation, a variable readily computed from the gas flow-rate through the fermenter and the difference between inlet and outlet oxygen con- centrations in the gas stream. This result has sufficient accuracy for the design of cooling systems but should not be expected to compete with microcalorimetry in terms of throwing light on metabolic pathways.

The yield can also be expressed in terms of cell carbon per gram of substrate carbon (Herbert, 1976) and this form is often useful in comparing yields on substrates of different degrees of oxidation. Attempts (Payne, 1970) have been made to correlate the yields of micro-organism obtained with the energy available through oxidation of the substrate (heat of combustion). In a recent paper, Linton & Stephenson (1978) found a good correlation between energy content of substrate and yield of micro-organism for substrates with heats of combustion up to about 11 kcal per g substrate carbon, as is shown in Fig. 1. This

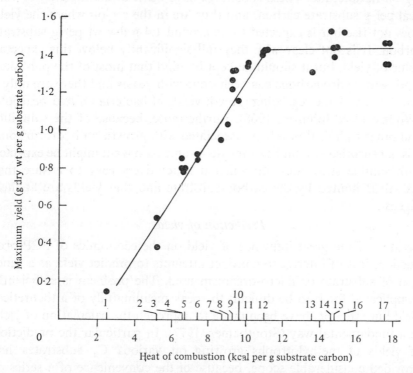

Fig. 1. The relation between growth yield and heats of combustion for various heterotrophic growth substrates. 1, oxalate; 2, formate; 3, citrate; 4, malate; 5, fumarate; 6, succinate; 7, acetate; 8, benzoate; 9, glucose; 10, phenylacetic acid; 11, mannitol; 12, glycerol; 13, ethanol; 14, propane; 15, methanol; 16, ethane; 17, methane. (Linton & Stephenson, 1978.)

value is equivalent approximately to glycerol, while ethanol, methanol, and methane are all richer in energy. Above this apparent critical value the yield did not increase with heat of combustion. The authors suggest that investigations of problems of coupling between ATP production and ATP use in cells growing on energy-rich substrates will be confused by the potential excess energy available to fix the substrate carbon, and that investigations of microbial energetics be undertaken using poorer substrates in which it is likely that the growth process is energy-limited.

Linton & Stephenson's data do not include examples of hydrocarbons other than the gases methane, ethane and propane. However, yield values of 1.29 g dry wt per g substrate carbon were found by Sukatsch & Johnson (1972) for a mixed bacterial culture growing on hexadecane, while Johnson (1967) reported a value equivalent to 1.29 for a *Coryne-bacterium* growing on octadecane and Ertola, Mazza, Balatti & Sanahuja (1969) found a value of 1.06 for *Micrococcus cerificans* growing on hexadecane. These substrates have heats of combustion of 13.3 kcal per g substrate carbon, and thus are in the region where the yield does not rise but is expected to be around 1.4 g dry wt per g substrate carbon. It is not clear why they fall significantly below this expected value of yield, but it should be kept in mind that most of the published work with hydrocarbons has been done with yeasts and there are only a few data available regarding growth yields of bacteria on hydrocarbons (Wodzinski & Johnson, 1968). Furthermore, because of the solubility and other technical problems associated with growth on hydrocarbons, it is not surprising to find a lower yield value than what might be expected with soluble substrates on which it is relatively easy to ensure that growth is limited by the carbon-substrate and that yields are at their highest.

Prediction of yields

Because of the great influence of yield on the economics of SCP production, it is of interest to consider attempts to predict yield as a function of substrate or micro-organism used. The problem is sufficiently complicated for it to be difficult to speak meaningfully of a theoretical yield but attempts have been made to approach the calculation of yield in a fundamental way (Stouthamer, 1977). In particular the prediction of yields of methylotrophs growing on various C_1 substrates has provided considerable scope, because of the convenience of a series of substrates which differ in oxidation level (methane, methanol, formaldehyde, and formic acid) and a complex variety of known or suspected assimilatory and dissimilatory pathways (van Dijken & Harder, 1975; Anthony, 1978).

Conceptually the prediction of yield requires knowledge of:

(1) the pathway of dissimilation of the carbon source in sufficient detail for an estimate to be made of ATP produced per unit of substrate oxidized;

(2) the pathways by which the carbon source is assimilated, and the resulting cell composition, so that an estimate can be made of ATP required per unit of cell synthesized;

(3) the maintenance requirement, so that an estimate can be made of the ATP required for maintaining the integrity of the cell and for all the other purposes considered as 'maintenance'; and

(4) the degree of uncoupling, so that one can estimate the ATP synthesized but broken down in non-productive ways.

Complicating the relatively simple conceptual picture presented by the above points are various practical problems represented by the following examples:

(1) Almost invariably in batch culture, and frequently even when grown in continuous culture under conditions of carbon-substrate limitation, micro-organisms excrete partly oxidized intermediates into the medium. For instance, excretion of acetate by bacteria growing with glucose as the limiting nutrient has been observed (Chian & Mateles, 1968; Mateles & Chian, 1969; Cooney, Wang & Mateles, 1976), as has excretion of methanol from a methane-utilizer growing on methane (Harrison, Wilkinson, Wren & Harwood, 1976). The phenomenon is often seen in continuous culture, particularly when the dilution rate is relatively close to the maximum specific growth rate of the micro-organism. Thus the estimate of ATP yield per unit of substrate oxidized is affected by the incomplete oxidation of part of the substrate.

(2) In some organisms there are several pathways for the oxidation of substrates which may operate simultaneously, but at different relative rates, which are affected by growth conditions. Thus, in *Pseudomonas* C growing on methanol in methanol-limited continuous culture, some of the methanol is oxidized to carbon dioxide via formate, while a larger fraction is oxidized via the cyclic ribulose monophosphate (RMP) pathway (Ben-Bassat & Goldberg, 1977). Again an error is introduced into the calculation of ATP yield unless the partitioning between the pathways is known exactly.

(3) The composition of the cells whose yields are being compared may be quite different and the composition may change significantly with growth conditions. Unless accurate data are available on the cell composition, the estimates of ATP required for biosynthesis may be significantly in error.

(4) According to Anthony (1978), under some conditions involving the growth of methylotrophs assimilating C_1 units via the serine pathway (Ribbons, Harrison & Wadzinski, 1970), the supply of NAD(P)H, rather than ATP generation may be the factor limiting cell synthesis, and therefore uncoupling will be extensive.

Calculations such as those of van Dijken & Harder (1975) and Anthony (1978) seem to be of limited use in deducing what pathways are operative in a particular case from knowledge of the yield. Because of experimental limitations and errors it does not seem possible to use yield estimates to predict P:O ratios even when pathways are known and, in general, the accuracy of the yield determination is such that several possibilities of pathway and P:O ratio fit the data. Thus, it is necessary to make rather too many assumptions for comfort. Nevertheless, one prediction of van Dijken & Harder (1975), that bacteria utilizing methanol via the ribulose monophosphate pathway can be expected to give yields significantly higher than those operating the serine pathway, has been fully confirmed (Goldberg, *et al.*, 1976) and, in the light of good agreement between theory and experiment, it is improbable that a serine-pathway organism would be chosen for SCP production.

More generally, when considering substrates other than C_1 compounds, the correlation presented by Linton & Stephenson (1978) offers an approximation of yields to be expected for growth of heterotrophs on various substrates. The spread of the data is relatively large and once again the question of cell composition must be considered in comparing yields on different substrates.

The maintenance requirement, as a factor which influences the yield,

Table 2. *Maintenance coefficients[a] for micro-organisms*

Organism	Substrate	Maintenance coefficient (mmole C per g cell dry wt per h)	Reference
Pseudomonas C	methanol	2.5	Rokem, Goldberg & Mateles
Pseudomonas methylotropha	methanol	3.9	(1978)
Pseudomonas 1	methanol	1.5	
	formaldehyde	1.5	
	formate	1.0	
Aerobacter aerogenes	glycerol	2.5	Herbert (1958)
Aerobacter cloacae	glucose	3.1	Pirt (1965)
Escherichia coli	glucose	0.5	Schulze & Lipe
Saccharomyces cerevisiae	glucose	0.5	von Meyenburg (1969)
Hansenula polymorpha	methanol	0.5	van Dijken, Otto & Harder (1976)

[a] Defined and calculated according to Pirt (1975).

seems to be of secondary importance in choosing among micro-organisms as sources for SCP. While there is a significant difference in the maintenance coefficient (see Table 2) of various organisms growing on different substrates, generally it appears that it can be said that the maintenance coefficient of yeasts is less than that of bacteria and that the maintenance coefficient does not appear to vary systematically with substrate. Recently, we have shown that the maintenance requirement of bacteria growing on methanol is in the range expected for growth of heterotrophs on carbohydrate substrates and that the advantage in yield expected for methanol-utilizers operating the RMP pathway does not depend on differences in maintenance but is seen even when the maintenance requirement is removed as a variable (Rokem, Goldberg & Mateles, 1978).

PHYSIOLOGICAL PROBLEMS ASSOCIATED WITH SCALE-UP

The scale-up of fermentations and fermenters is among the more critical problems associated with fermentation technology mainly because an important nutrient, oxygen, is required in large quantity by the cells and is not capable of being dissolved to a high concentration in the nutrient medium. Thus, an actively growing suspension of cells may have an oxygen requirement of 6000 mg l^{-1} h^{-1}, while the nutrient medium saturated with air may contain only 6 mg l^{-1} or just enough to supply the cells for several seconds. In practice, the oxygen must be transferred from the gas phase to the liquid phase and this mass transfer process is affected greatly by the degree of agitation of the air–medium–cells system and the properties of the medium. This article is not the proper place to discuss how fermentations and fermenters are scaled-up so that the large production fermenters are capable of doing their work (see Aiba, Humphrey & Millis, 1973) and the discussion will be limited, therefore, to one or two problems that may arise in SCP production from yeasts or bacteria where the liquid phase retains the viscosity and properties of water, as opposed to the very non-Newtonian liquid behaviour caused by the growth of mycelial organisms.

While in the case of a small laboratory fermenter it is relatively easy to secure nearly ideal mixing (no gradients within the liquid phase) of the contents (neglecting the problem of growth on the fermenter wall) in a large fermenter of several hundred cubic metres capacity there are likely to be gradients caused by non-ideal mixing. These gradients may relate to the concentration of dissolved oxygen or the limiting nutrient if the fermenter is operated as a chemostat, or may relate to temperature gradients caused by the exothermic nature of fermentations and the consequent need for cooling. Non-ideal mixing results in individual cells being exposed to varying concentrations of nutrients or temperatures as they move through the fermenter and it is of some importance to know that these time-varying parameters do not significantly reduce the yield of the SCP, or otherwise adversely affect the process. A good discussion of some of these factors as they relate to the scale-up of the ICI fermenter producing SCP from methanol has been published (Gow, Littlehailes, Smith & Walter, 1975).

The effect of carbon dioxide

As a consequence of the highly aerobic nature of SCP production, together with the economic factors motivating growth at relatively high cell densities and good utilitization of the oxygen supplied in the air stream (Abbott & Clamen, 1973), the carbon dioxide content of the effluent gas is relatively high. For instance, take the case of *Pseudomonas* C growing on methanol at a yield of 0.54 g cell dry wt per g methanol, with a conversion of 65% of methanol carbon to cell carbon (Goldberg *et al.*, 1976); for each mole of oxygen taken up, 0.5 mole of carbon dioxide is produced. If we assume that the aeration rate is such that 80% of the oxygen is utilized, then the composition of the gas stream leaving the fermenter is

$$pO_2 = 0.21(1 - 0.8)/(0.79 + 0.21(1 - 0.8) + 0.21(0.4)) = 0.046 \text{ atm}$$
$$pCO_2 = 0.21(0.4)/(0.79 + 0.21(1 - 0.8) + 0.21(0.4)) = 0.092 \text{ atm}$$
$$pN_2 = 0.79/(0.79 - 0.21(1 - 0.8) + 0.21(0.4)) = 0.86 \text{ atm}$$

Thus, the gas stream is more or less stripped of its oxygen and is fairly rich in carbon dioxide. Because a high rate of oxygen transfer depends partly on a high partial pressure of oxygen, it is tempting to suggest operating the fermenter under a total pressure of more than 1 atm, for instance 2 atm. This would increase the rate of oxygen transfer from the gas to the broth without requiring additional power for agitation, but would require some additional power for compression of the gas. While this optimum depends on engineering factors (Giacobbe, Puglisi & Longobardi, 1975), it is clear that any increase in the total pressure will result in partial pressures of carbon dioxide in the gas (which is in equilibrium with the broth (Yagi & Yoshida, 1977) in the range of 0.1–0.3 atm and raises the question as to the effect of this pCO_2 on the fermentation itself.

Although the literature is replete with data concerning the effect of oxygen on metabolic activities of micro-organisms, the effect of carbon dioxide has been studied rather little (Wimpenny, 1969). Production of fumaric acid by *Rhizopus nigricans* was not affected by 5% CO_2 (Foster & Davis, 1949), while 4% CO_2 inhibited respiration of *Penicillium chrysogenum* (Nyiri & Lengyel, 1965). In the case of bacterial metabolites, inosine production by *Bacillus subtilis* was reduced when the pCO_2 was above 0.05 atm (Shibai, Ishizaki, Mizuno & Hirose, 1973), while production of α-amylase by *Bacillus subtilis* in continuous culture was enhanced three-fold compared the control by 8% CO_2 (Gandhi & Kjaergaard, 1975). In connection with biomass production, the yield of *Saccharomyces cerevisiae* from molasses was not affected by up to 20% CO_2 (Chen & Gutmanis, 1976).

At the other end of the spectrum, it has long been known that many micro-organisms require small amounts of carbon dioxide for rapid growth (see Pirt, 1975). However, under the growth conditions normally contemplated for SCP production the concentration of carbon dioxide is certainly adequate. The minimum partial pressure of carbon dioxide necessary to avoid reductions in yield for the case of *Pseudomonas* C growing on methanol was estimated (Battat, Goldberg & Mateles, 1974) to be 0.003 atm, or about ten-fold higher than its normal concentration in the atmosphere.

In a large fermenter (and in the case of ICI's pressure-cycle fermenter, a fermenter height of approximately 45 m is involved), there will be a substantial hydraulic head which can result in a change of 5 atm as the cell moves from the top to the bottom of the fermenter and then up again. Thus, although the partial pressure of dissolved oxygen may not vary greatly, owing to the fermenter and process having been designed so that the dissolved oxygen is very low and nearly limiting, the partial pressure of carbon dioxide will vary by a factor of at least 3–4 from top to bottom, as will the partial pressure of the inert gas, nitrogen. Therefore, one of the requirements in choosing a micro-organism for use in large-scale SCP production is verifying that it can withstand such fluctuations without reductions in yield. This is most readily done in a small fermenter by making use of the fact that non-ideal mixing in a large fermenter is perceived by the cell as a variation with time of a particular parameter.

In a small fermenter, with close to ideal mixing, this time-variation is readily simulated by simply changing the total pressure with a cycle of several tens of seconds or even a minute or two, or by cyclically sparging carbon dioxide to change its partial pressure without affecting the total pressure. In a relatively short time it is possible to determine the range of variations within which the particular process does not suffer a significant drop in yield or other unwanted result, and knowing this one can estimate whether the organism will be able to withstand the variations of the particular parameter in the production fermenter.

Variations in the concentration of limiting nutrient

As it may be expected that large-scale SCP processes will usually be operated as continuous cultures, with the carbon source as the limiting nutrient and with a small excess of oxygen, there is a distinct possibility that there will be variations in the concentration of the limiting nutrient in different parts of the fermenter, and that these variations will be physiologically significant to the cells. Thus, it has been reported (Hansford & Humphrey, 1966) that in a small (10 l) continuous fermenter operating at a low dilution rate, increasing the number of points at which the molasses-containing nutrient medium was fed to the fermenter from one to three increased the yield of *Saccharomyces cerevisiae* by 12%. Based upon these preliminary results, Ryu (1967) undertook careful studies in which he simulated the time-dependent effects of poor mixing by using a continuous culture which, instead of being fed continuously, was fed in 'shots' at varying frequencies; the average feeding rate, and hence the dilution rate, was held constant. Using *S. cerevisiae* with glucose as the limiting nutrient, at dilution rates of 0.01 to 0.034 h^{-1}, no reduction in yield was observed when the time interval between shots was as long as 18 min. Thus, the results of Hansford & Humphrey (1966) could not be confirmed and it appeared that mixing must be very bad indeed to reduce yield, at least with *S. cerevisiae* and glucose. Similarly, using glucose-limited cultures of *Arthrobacter globiformis*, Luscombe (1974) found no effect when using shot intervals in the range 15–120 s at a dilution rate of 0.01 h^{-1}, 3–24 s at 0.05 h^{-1}, or 1.5–12 s at 0.1 h^{-1}. On the other hand, when *Pseudomonas methylotropha* was grown under methanol-limited conditions, extending the shot interval beyond 20 s led to a very significant reduction in yield (Brooks & Meers, 1973).

A reduction in yield when exposed to a short-term excess of limiting nutrient suggests that the particular micro-organism either does not couple the burst of energy production to biosynthetic processes, or is able to store only a very limited amount of energy-rich intermediates. To the extent that the results reported above for *Pseudomonas methylotropha* characterize the response of methanol-utilizing micro-organisms to variations in methanol concentration with time, it is necessary to design large-scale fermenters accordingly. Thus, for instance, the methanol could be added at several points in the fermenter, taking care to disperse the flow (Roesler, 1977). Further, it would be desirable to seek organisms in which the phenomenon was minimized.

SCP COMPOSITION

SCP is normally considered as a source of protein and it is in this context that the effects of environment and genetics on SCP composition will be discussed. However, not only does SCP contain protein, but also nucleic acids, carbohydrate and cell wall material, lipids, minerals and vitamins. Generally, the producer is satisfied if the effects of these non-protein materials are negligible, although for some of them a nutritional credit for calories, vitamins or UGF (unidentified growth factor*) effect may be taken. It is recommended that the reader interested in acquiring an understanding of the nutritional and toxicological considerations involved in evaluating SCP should consult the appropriate literature (Munro, 1968; Engel, 1972; Shacklady & Gatumel, 1972; Scrimshaw, 1975; Stringer, 1975). This section will be limited to a survey of how composition and nutritional value of SCP can be affected and what possibilities exist for obtaining SCP of improved nutritional value.

Protein and amino acids
Protein

The protein content of SCP can be measured by several techniques which give different answers, are of differing convenience and are relevant to different aims. The nutritionist is normally interested in Kjeldahl nitrogen × 6.25, which, in the case of SCP, is in error because of the large amounts of non-protein nitrogen (in nucleic acids and cell wall polymers), much higher than found in other usual animal or plant proteins (Miller, 1968). Biuret protein is a good measure of protein, simple to use, but requiring that the SCP be dissolved in alkali, which sometimes presents serious problems. Protein based on the sum of amino acids after hydrolysis is accurate if the amino acid analysis is done well and carefully, but is not a convenient assay. What is reported as 'crude protein' is usually Kjeldahl nitrogen × 6.25, while 'true protein' may be either biuret protein or the sum of amino acids, and is always

---

* The UGF effect is pertinent to animal nutrition and represents the observation that the addition of complex natural materials to an apparently adequate diet improves the growth response (rate of growth; yield of meat per unit of feed) to the diet. The effect is difficult to demonstrate in the laboratory, although many agriculturists believe that it exists and will pay a price for it. Its existence is a source of great controversy when biochemists, physiologists, animal nutritionists, and economists discuss its value, not to say its existence.

significantly lower than the crude protein figure. Thus Gow *et al.* (1975) reported for *Pseudomonas methylotropha* growing on methanol that the crude protein was 83% (on a dry basis) and the true protein was 59%, and Dostálek & Molin (1975) reported 82% crude protein and 63% total amino acids for *Methylomonas methanolica*. These results are fairly typical of bacteria grown under conditions selected to give a product with a high protein content. Yeasts grown under the same conditions can be expected to have true protein (biuret) contents around 45–50% and crude protein contents of around 60%.

Generally, the highest protein content results from the use of the carbon source as the limiting nutrient but if other factors, such as nitrogen, oxygen, phosphate or metals, are the limiting nutrient there is a tendency for the cell to form a high concentration of carbohydrate, lipid, or poly-β-hydroxybutyrate (Holme, 1957; Wilkinson & Munro, 1967; Senior, Beech, Ritchie & Dawes, 1972; Dierstein & Drews, 1974; Cooney *et al.*, 1976).

Amino Acid Content

The nutritional value of a protein is dependent on its amino acid pattern and is judged to be better the more closely it resembles the amino acid content of whole egg or the slight modification recommended as a reference by the Food and Agriculture Organization of the United Nations (FAO). In the case of SCP, the amino acids which tend to be limiting in terms of nutritional value are the sulphur-containing ones (see Table 3), while lysine, in which wheat is deficient, is present in good concentration. Bacteria generally have a higher content of methionine and cysteine than do yeasts, although these amino acids are still limiting. Thus, not only do bacteria tend to be richer in protein but the protein is usually of higher nutritional quality. In comparing published values of amino acid content a degree of caution is required, because the variation in results is quite high unless extreme care is taken in sampling, hydrolysis and treatment (Kwolek & Cavins, 1971), and many reports of amino acid content do not permit an estimate of the degree of error of the measurements.

Because of the potential economic significance of obtaining SCP with an increased level of some amino acids, particularly methionine or cysteine, consideration has been given to various strategies for altering the amino acid pattern of SCP. DeZeeuw (1968) has considered the possibility of increasing the level of methionine by obtaining mutants in which base transitions and transversions cause the substitution of methionine for other amino acids in a specific protein. Aside from the

Table 3. *Content of essential amino acids of micro-organisms considered for use as SCP Data are g amino acid per 100 g protein or 16 g N*

	FAO reference	Candida boidinii	Hansenula polymorpha	Candida lipolytica	Pseudomonas methylotropha	Pseudomonas C	Methylococcus capsulatus
Lysine	4.2	6.0	8.1	7.0	7.5	8.3	4.8
Threonine	2.8	4.4	5.2	4.9	5.9	5.9	4.2
Cysteine	2.0	–[a]	0.7	1.1	0.8	1.1	0.7
Methionine	2.2	0.9	1.5	1.8	3.1	2.6	1.9
Valine	4.2	4.6	6.2	5.4	6.7	6.7	5.5
Isoleucine	4.2	4.0	5.1	4.5	5.5	5.7	4.0
Leucine	4.8	5.4	8.3	7.0	8.6	8.1	6.7
Phenylalanine	2.8	3.4	5.0	4.4	4.4	4.0	4.0
Substrate		Methanol	Methanol	n-Paraffin	Methanol	Methanol	Methane
Reference		Sahm & Wagner (1972)	Levine & Cooney (1973)	Evans (1968)	Gow et al. (1975)	Chalfan & Mateles (1972)	D'Mello (1972)

[a] Not quoted.

technical difficulties of selecting such mutants, he drew the conclusion that the increase in methionine which might be expected from a single mutation is relatively small. Even though a single protein (enzyme) might constitute as much as 10–15% of the total protein of a cell, it would have to have a very unusual content of methionine, or any other amino acid, to change the average amino acid composition of the cell appreciably. The possible effects of changes in the levels of individual proteins on the amino acid composition of the total cell protein has been explored theoretically and experimentally by Alroy & Tannenbaum (1977a, b). In the first paper they developed an analysis of how the specific amino acid content of a protein must be related to the fraction of that protein in the total cell protein to obtain a defined change in the amino acid content of the total cell protein. On the basis of the analysis they concluded that a significant change would only be possible if there existed a set of proteins which were related by composition and presumably function. They postulated that ribosomal proteins, which are relatively rich in basic amino acids and interact with nucleic acids, might constitute such a group. In their second paper they examined the changes in amino acid content resulting from the growth of *Candida utilis* and *Enterobacter aerogenes* under different conditions, and concluded that small to moderate phenotypic modifications in the amino acid composition of total cell protein may occur. Some of these changes correlated with growth rate and may be explained by changes in the proportional amount of ribosomal protein at different growth rates leading to an increase in lysine and arginine concentration with increasing growth rate. They suggest that the maximum ratio of amino acid content that might be expected in *C. utilis* growing rapidly ($\mu = 1.0\ h^{-1}$) as compared to that at very slow growth rates would not exceed 1.54:1 for arginine and 1.27:1 for lysine.

The possibility of causing extreme over-production of one or more amino acids certainly exists, and in fact is the basis for an industry (Demain, 1971). However, these amino acids are found in the free amino acid pool and are excreted, sometimes in concentrations approaching $100\ g\ l^{-1}$, into the growth medium. The concentration of free glutamic acid in the cell pool of bacteria used for commercial production of glutamic acid can reach 20–30 mg per g cell dry wt (Fukui & Ishida, 1972) and, if one assumes that the cells are 80% crude protein, the protein containing 10% glutamic acid, the contribution from the free amino acid pool would raise the average glutamic acid content from 8% to 10%. Although this is a substantial percentage increase, and would definitely be worthwhile obtaining for amino acids of nutritional significance to SCP, such as methionine or cysteine, it does not appear that such large increases have been reported. Furthermore, and most crucially for SCP production, the content of free amino acids is of much less importance than the amino acids in protein, because free amino acids are lost during harvesting operations involving washing, are relatively labile during drying and storage and are subject to loss if the feed is prepared in a wet form. Thus, for purposes of nutrition, it is considered desirable to obtain the amino acid in a protein, although the practice of supplementing animal feed mixtures with lysine or methionine indicates that the free amino acid has definite nutritional value if it actually reaches the diet. If the free amino acid pool could be enriched in a nutritionally significant amino acid, and if this amino acid could be rendered insoluble or polymerized during or after the production stage, the resulting SCP would be relatively more desirable.

A novel approach to the problem has been proposed recently by Momose & Gregory (1978) who have obtained temperature-sensitive mutants of *Saccharomyces cerevisiae* which at the non-permissive temperature synthesized proteins containing abnormal quantities of methionine. These cells contained methionine concentrations of about 4.3% of their protein, as compared to the wild type which contained 1.8% – an impressive improvement. Although the non-permissive conditions would have to be obtained subsequent to growth, for instance by running a two-stage continuous fermentation with the second stage held at the non-permissive temperature, the concept is of considerable interest and bears further examination.

Nucleic acids

SCP is relatively rich in nucleic acids as compared to usual animal or plant protein sources and this has two implications for the nutritional value of the SCP product. For use of SCP in animal feeds the major implication is simply that nucleic acid is not protein and essentially dilutes the protein, although there are at least some possibilities of physiological effects. As far as the potential use of SCP in human feeding is concerned, the nucleic acids are undesirable because their digestion leads to unacceptably high levels of uric acid (Edozien, Udo, Young & Scrimshaw, 1970). So considerable attention has been devoted to the reduction of nucleic acid levels in SCP (Sinskey & Tannenbaum, 1975), and it is of interest to consider what the relation might be between the conditions of production of SCP and the nucleic acid content.

The bulk of the nucleic acid in micro-organisms is RNA, which has

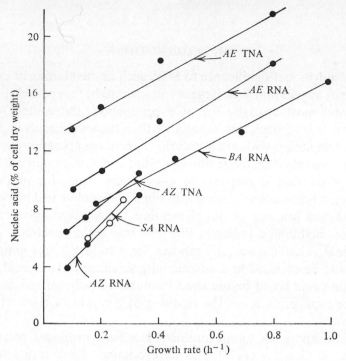

Fig. 2. Nucleic acid contents of cells as a function of growth rates. RNA, ribonucleic acid; TNA, total nucleic acids; *AZ, Azotobacter vinelandii; SA, Saccharomyces cerevisiae; AE, Aerobacter aerogenes; BA, Bacillus megaterium.* (Mateles, 1968.)

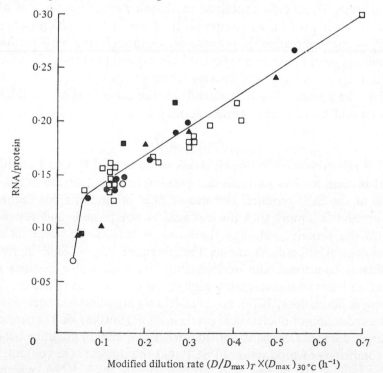

Fig. 3. RNA : protein ratio as a function of modified dilution rate in a carbon-limited chemostat; ●, *Aerobacter aerogenes;* □, *Candida utilis;* ▲, *Bacillus megaterium;* ○, *Polytomella caeca;* ■, *Azotobacter chrococcum.* (Alroy & Tannenbaum, 1973.)

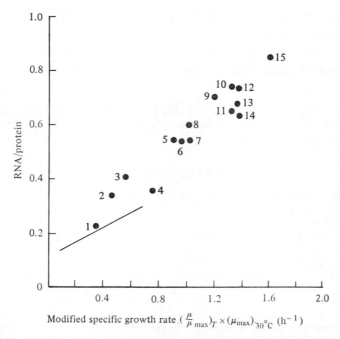

Fig. 4. RNA : protein ratio as a function of modified specific growth rate. 1, *Tetrahymena pyriformis*; 2, *Saccharomyces cerevisiae* (haploid); 3, *S. cerevisiae* (diploid); 4, *Lactobacillus bulgaricus*; 5, *Clostridium perfringens*; 6, *Micrococcus anhaemolyticus*; 7, *Pseudomonas aeruginosa*; 8, *Erwinia carotovora*; 9, *Bacillus subtilis*; 10, *Serratia marcescens*; 11, *E. coli B*; 12, *Citrobacter freundii*; 13, *E. coli*; 14, *S. typhimurium*; 15, *A. aerogenes*. The solid line represents the data of Fig. 3. (Alroy & Tannenbaum, 1973.)

a critical role in protein synthesis. Thus, it may be anticipated that the faster the rate of protein synthesis in a particular cell, the higher will be the nucleic acid content and, as Fig. 2 indicates, this expectation is fulfilled. The picture is made clearer and more general (Alroy & Tannenbaum, 1973) by considering a plot of the RNA:protein ratio of various micro-organisms against a modified growth rate (or dilution rate of a continuous culture). The modification consists of taking the ratio of the actual dilution rate (D) at a particular temperature (T) to the maximum specific dilution rate (D_{max}) at that temperature and multiplying it by the maximum specific dilution rate at 30 °C. This modification results in uniting the data as may be seen in Fig. 3, which consists of data obtained from carbon-limited chemostats. Even when growth-rate data from a variety of batch cultures are examined, the modification results in what appears to be a linear relationship (Fig. 4.) The data used for preparing Fig. 4 included low molecular weight RNA derivatives which were not included in the analyses summarized in Fig. 3, and thus there is a small discrepancy between the Figures.

Based upon these results, it is possible to predict fairly accurately the content of nucleic acids to be expected for a particular organism being considered for SCP production from a knowledge of its maximum specific growth rate at 30 °C, the temperature to be actually used for growth and the actual growth rate contemplated for SCP production. Owing to considerations of yield, productivity and economics of harvesting, this will typically be in the range of 50–80% of the maximum specific growth rate.

REFERENCES

ABBOTT, B. J. & CLAMEN, A. (1973). The relationship of substrate, growth rate and maintenance coefficient to single-cell protein production. *Biotechnology and Bioengineering*, **15**, 117–27.

AIBA, S., HUMPHREY, A. E. & MILLIS, N. F. (1973). *Biochemical Engineering*, 2nd Edition, p. 163ff. Tokyo: University of Tokyo Press.

ALROY, Y. & TANNENBAUM, S. R. (1973). The influence of environmental conditions on the macromolecular composition of *Candida utilis*. *Biotechnology and Bioengineering*, **15**, 239–56.

ALROY, Y. & TANNENBAUM, S. R. (1977a). Considerations in predicting phenotypic modifications in amino acid profiles of total cell protein of micro-organisms. *Biotechnology and Bioengineering*, **19**, 1145–53.

ALROY, Y. & TANNENBAUM, S. R. (1977b). Phenotypic modifications in amino acid profiles of cell residues of *Candida utilis* and *Enterobacter aerogenes*. *Biotechnology and Bioengineering*, **19**, 1155–69.

ANTHONY, C. (1978). The prediction of growth yields in methylotrophs. *Journal of General Microbiology*, **104**, 91–104.

BATTAT, E., GOLDBERG, I., & MATELES, R. I. (1977). Growth of *Pseudomonas* C on C_1 compounds. *Applied Microbiology*, **28**, 906–11.

BEN-BASSAT, A. & GOLDBERG, I. (1977). Oxidation of C_1 compounds in *Pseudomonas* C. *Biochimica et Biophysica Acta*, **497**, 586–97.

BREWERSDORFF, M. & DOSTÁLEK, M. (1971). The use of methane for production of bacterial protein. *Biotechnology and Bioengineering*, **13**, 49–62.

BROOKS, J. D. & MEERS, J. L. (1973). The effect of discontinuous methanol addition on the growth of a carbon-limited culture of *Pseudomonas*. *Journal of General Microbiology*, **77**, 513–19.

BROWN, C. M. & ROSE, A. H. (1969). Effects of temperature on composition and cell volume of *Candida utilis*. *Journal of Bacteriology*, **97**, 261–72.

CHALFAN, Y. & MATELES, R. I. (1972). New pseudomonad utilizing methanol for growth. *Applied Microbiology*, **23**, 135–40.

CHEN, S. L. & GUTMANIS, F. (1976). Carbon dioxide inhibition of yeast growth in biomass production. *Biotechnology and Bioengineering*, **18**, 1455–62.

CHIAN, S. K. & MATELES, R. I. (1968). Growth of mixed cultures on mixed substrates. I. Continuous culture. *Applied Microbiology*, **16**, 1337–42.

COONEY, C. L., WANG, D. I. C. & MATELES, R. I. (1969). Measurement of heat evolution and correlation with oxygen consumption during microbial growth. *Biotechnology and Bioengineering*, **11**, 769–81.

COONEY, C. L. WANG, D. I. C. & MATELES, R. I. (1976). Growth of *Enterobacter aerogenes* in a chemostat with double nutrient limitations. *Applied and Environmental Microbiology*, **31**, 91–8.

DAVIS, P. (1974). *Single-Cell Protein*. London, New York: Academic Press.

DEAN, A. C. R., ELLWOOD, D. C., EVANS, C. G. T. & MELLING, J. (1976). *Continuous Culture 6: Applications and New Fields*. Chichester: Ellis Horwood for the Society for Chemical Industry.

DEAN, A. C. R., PIRT, S. J. & TEMPEST, D. W. (1972). *Environmental Control of Cell Synthesis and Function*. London, New York: Academic Press.

DEMAIN, A. L. (1971). Microbial production of food additives. In *Microbes and Biological Productivity*, eds. D. E. Hughes & A. H. Rose, pp. 77–101. *21st Symposium of the Society for General Microbiology*. Cambridge University Press.

DEZEEUW, J. R. (1968). Genetic and environmental control of protein composition. In *Single-Cell Protein*, eds. R. I. Mateles & S. R. Tannenbaum, pp. 181–91. Cambridge, Massachusetts: MIT Press.

DIERSTEIN, R. & DREWS, G. (1974). Nitrogen-limited continuous culture of *Rhodopseudomonas capsulata* growing photosynthetically or heterotrophically under low oxygen tensions. *Archives of Microbiology*, **99**, 117–28.

D'MELLO, J. P. F. (1972). A study of the amino acid composition of methane utilizing bacteria. *Journal of Applied Bacteriology*, **35**, 145–8.

DOSTÁLEK, M. & MOLIN, N. (1975). Studies of biomass production of methanol oxidizing bacteria. In *Single-Cell Protein II*, eds. S. R. Tannenbaum & D. I. C. Wang, pp. 385–401. Cambridge, Massachusetts: MIT Press.

EDOZIEN, J. C., UDO, U. V., YOUNG, V. R. & SCRIMSHAW, N. S. (1970). Effects of high levels of yeast feeding on uric acid metabolism of young men. *Nature, London*, **228**, 180.

ENGEL, C. (1972). Safety evaluation of yeast grown on hydrocarbons. In *Proteins from Hydrocarbons*, ed. H. Gounelle de Pontanel, pp. 53–81. London, New York: Academic Press.

ERTOLA, R. J., MAZZA, L. A., BALATTI, A. P. & SANAHUJA, J. (1969). Composition of cell material and biological value of the cellular protein of a *Micrococcus* strain grown on hydrocarbons. *Biotechnology and Bioengineering*, **11**, 409–16.

EVANS, G. H. (1968). Industrial production of single-cell protein from hydrocarbons. In *Single-Cell Protein*, eds. R I. Mateles & S. R. Tannenbaum, pp. 243–54. Cambridge, Massachusetts: MIT Press.

FOSTER, J. W. & DAVIS, J. B. (1949). Carbon dioxide inhibition of anaerobic fumarate formation in the mold *Rhizopus nigricans*. *Archives of Biochemistry and Biophysics*, **21**, 135–42.

FUKUI, S. & ISHIDA, M. (1972). Biomembrane permeability and the accumulation of amino acids. In *The Microbial Production of Amino Acids*, eds. K. Yamada, S. Kinoshita, T. Tsunoda & K. Aida, pp. 123–37. New York: Wiley.

GANDHI, A. P. & KJAERGAARD, L. (1975). Effect of carbon dioxide on the formation of α-amylase by *Bacillus subtilis* growing in continuous and batch cultures. *Biotechnology and Bioengineering*, **17**, 1109–18.

GIACOBBE, F., PUGLISI, P. & LONGOBARDI, G. (1975). Economic evaluation of new trends in SCP manufacture. Presented at the meeting of the American Chemical Society, Philadelphia, April 9, 1975,

GOLDBERG, I., ROCK, J. S., BEN-BASSAT, A. & MATELES, R. I. (1976). Bacterial yields on methanol, methylamine, formaldehyde, and formate. *Biotechnology and Bioengineering*, **18**, 1657–68.

GOUNELLE DE PONTANEL, H. (1972). *Proteins from Hydrocarbons*. London, New York: Academic Press.

GOW, J. S., LITTLEHAILES, J. D., SMITH, S. R. L. & WALTER, R. B. (1975). SCP production from methanol: bacteria. In *Single-Cell Protein II*, eds. S. R. Tannenbaum & D. I. C. Wang, pp. 370–84. Cambridge, Massachusetts: MIT Press.

GUTCHO, S. (1973). *Proteins from Hydrocarbons*. Park Ridge, New Jersey: Noyes Data Corporation.

HANSFORD, G. S. & HUMPHREY, A. E. (1966). The effect of equipment scale and degree of mixing on continuous fermentation yield at low dilution rates. *Biotechnology and Bioengineering*, **8**, 85–96.

HARRISON, D. E. F., WILKINSON, T. G., WREN, S. J., & HARWOOD, J. H. (1976). Mixed bacterial cultures as a basis for continuous production of SCP from C_1 compounds. In *Continuous Culture 6: Applications and New Fields*, eds. A. C. R. Dean, D. C. Ellwood, C. G. T. Evans, & J. Melling, pp. 122–34. Chichester: Ellis Horwood for the Society for Chemical Industry.

HERBERT, D. (1958). Some principles of continuous culture. In *Recent Progress in Microbiology*, ed. G. Tunevall, vol. 7, pp. 381–96. Oxford: Blackwell Scientific.

HERBERT, D. (1976). Stoicheiometric aspects of microbial growth. In *Continuous Culture 6: Applications and New Fields*, eds. A. C. R. Dean, D. C. Ellwood, C. G. T. Evans, & J. Melling, pp. 1–30. Chichester: Ellis Horwood for the Society for Chemical Industry.

HERBERT, D., ELSWORTH, R. & TELLING, R. C. (1956). The continuous culture of bacteria: a theoretical and experimental study. *Journal of General Microbiology*, **14**, 601–22.

HOLME, T. (1957). Continuous culture studies on glycogen synthesis in *Escherichia coli* B. *Acta Chemica Scandinavica*, **11**, 762–75.

HUMPHREY, A. E. (1968). Future of large-scale fermentation for production of single-cell protein. In *Single-Cell Protein*, eds. R. I. Mateles & S. R. Tannenbaum, pp. 330–9. Cambridge Massachusetts: MIT Press.

JOHNSON, M. J. (1967). Growth of microbial cells on hydrocarbons. *Science*, **155**, 1515–19.

KUBITSCHEK, H. E. (1970). *Introduction to Research with Continuous Cultures*. Englewood Cliffs, New Jersey: Prentice-Hall.

KWOLEK, W. F. & CAVINS, J. F. (1971). Relative standard deviations in determination of amino acids. *Journal of the Association of Official Agricultural Chemists*, **54**, 1283–7.

LEVINE, D. W. & COONEY, C. L. (1973). Isolation and characterization of a thermotolerant methanol-utilizing yeast. *Applied Microbiology*, **26**, 982–90.

LINTON, J. D. & STEPHENSON, R. J. (1978). A preliminary study on growth yields in relation to the carbon and energy content of various organic growth substrates. *FEMS Microbiology Letters*, **3**, 95–8.

LUSCOMBE, B. M. (1974). The effect of dropwise addition of medium on the yield of carbon-limited cultures of *Arthrobacter globiformis*. *Journal of General Microbiology*, **83**, 197–8.

MALEK, I. & FENCL, Z. (1966). *Theoretical and Methodological Basis of Continuous Culture of Microorganisms*. Prague: Czechoslovak Academy of Sciences.

MATELES, R. I. (1968). Applications of continuous culture. In *Single-Cell Protein*, eds. R. I. Mateles & S. R. Tannenbaum, pp. 208–16. Cambridge, Massachusetts: MIT Press.

MATELES, R. I. (1971) Calculation of the oxygen required for cell production. *Biotechnology and Bioengineering*, **13**, 581–2.

MATELES, R. I. & CHIAN, S. K. (1969). Kinetics of substrate uptake in pure and mixed culture. *Environmental Science and Technology*, **3**, 569–74.

MATELES, R. I. & TANNENBAUM, S. R. (1968). *Single-Cell Protein*. Cambridge, Massachusetts: MIT Press.

MILLER, S. A. (1968). Nutritional factors in single-cell protein. In *Single-Cell Protein*, eds. R. I. Mateles & S. R. Tannenbaum, pp. 79–89. Cambridge, Massachusetts: MIT Press.

MOMOSE, H. & GREGORY, K. F. (1978). Temperature-sensitive mutants of *Saccharomyces cerevisiae* variable in the methionine content of their protein. *Applied and Environmental Microbiology*, **35**, 641–7.

MUNRO, H. N. (1968). The nature of protein needs. In *Single-Cell Protein*, eds. R. I. Mateles & S. R. Tannenbaum, pp. 27–47. Cambridge, Massachusetts: MIT Press.

NAGAI, S., NISHIZAWA, Y. & AIBA, S. (1969). Energetics of growth of *Azotobacter vinelandii* in a glucose-limited chemostat culture. *Journal of General Microbiology*, **59**, 163–9.

NYIRI, L. & LENGYEL, Z. L. (1965). Studies on automatically aerated biosynthetic processes. I. The effect of agitation and carbon dioxide on penicillin formation in automatically aerated liquid cultures. *Biotechnology and Bioengineering*, **7**, 343–54.

PAYNE, W. J. (1970). Energy yields and growth of heterotrophs. *Annual Review of Microbiology*, **24**, 17–52.

PIRT, S. J. (1965). The maintenance energy of bacteria in growing cultures. *Proceedings of the Royal Society of London*, Series B, **163**, 224–31.

PIRT, S. J. (1975). *Principles of Microbe and Cell Cultivation*, pp. 67ff. Oxford: Blackwell Scientific.

RIBBONS, D. W., HARRISON, J. F. & WADZINSKI, A. M. (1970). Metabolism of single carbon compounds. *Annual Review of Microbiology*, **24**, 135–58.

ROESLER, F. C. (1977). Introduction of nutrient medium into a fermenter. United States Patent 4 048 017.

ROKEM, J. S., GOLDBERG, I. & MATELES, R. I. (1978). Maintenance requirement for bacteria growing on C_1 compounds. *Biotechnology and Bioengineering*, **20**, 1557–64.

RYU, D. D. Y. (1967). Transient response of continuous culture. p. 193ff. Doctoral thesis, Massachusetts Institute of Technology, May, 1967.

SAHM, H. & WAGNER, F. (1972). Mikrobielle Verwertung von Methanol. Isolierung und Charakterisierung der Hefe *Candida boidinii*. *Archiv für Mikrobiologie*, **84**, 29–42.

SCHULZE, K. L. & LIPE, R. S. (1964). Relationship between substrate concentration, growth rate, and respiration rate of *Escherichia coli* in continuous culture. *Archiv für Mikrobiologie*, **48**, 1–20.

SCRIMSHAW, N. S. (1975). Single-cell protein for human consumption – an overview. In *Single-Cell Protein II*, eds. S. R. Tannenbaum & D. I. C Wang, pp. 24–45. Cambridge, Massachusetts: MIT Press.

SENIOR, P. J., BEECH, G. A., RITCHIE, G. A. F. & DAWES, E. A. (1972). The role of oxygen limitation in the formation of poly-β-hydroxybutyrate during batch and continuous culture of *Azotobacter beijerinckii*. *Biochemical Journal*, **128**, 1193–1201.

SHACKLADY, C. H. & GATUMEL, E. (1972). The nutritional value of yeast grown on alkanes. In *Proteins from Hydrocarbons*, ed. H. Gounelle de Pontanel, pp. 27–52. London, New York: Academic Press.

SHEEHAN, B. T. & JOHNSON, M. J. (1971). Production of bacterial cells from methane. *Applied Microbiology*, **21**, 511–15.

SHIBAI, H., ISHIZAKI, A., MIZUNO, H. & HIROSE, Y. (1973). Effects of oxygen and carbon dioxide on inosine fermentation. *Agricultural and Biological Chemistry*. **37**, 91–7.

SINSKEY, A. J. & TANNENBAUM, S. R. (1975). Removal of nucleic acids in SCP. In *Single-Cell Protein II*, eds. S. R. Tannenbaum & D. I. C. Wang, pp. 158–78. Cambridge, Massachusetts: MIT Press.

Society for Chemical Industry (1961). *Continuous Culture of Micro-organisms, SCI Monograph 12.* London: Society for Chemical Industry.

Stouthamer, A. H. (1977). Energetic aspects of the growth of micro-organisms. In *Microbial Energetics*, eds. B. A. Haddock & W. A. Hamilton, pp. 285–315. *27th Symposium of the Society for General Microbiology.* Cambridge University Press.

Stringer, D. A. (1975). Safety and nutrition testing of single-cell protein. In *1. Symposium Mikrobielle Proteingewinnung 1975*, ed. F. Wagner, pp. 167–72. Weinheim: Verlag Chemie.

Sukatsch, D. H. & Johnson, M. J. (1972). Bacterial cell production from hexadecane at high temperatures. *Applied Microbiology*, **23**, 543–6.

Tannenbaum, S. R. & Wang, D. I. C. (1975). *Single-Cell Protein II.* Cambridge, Massachusetts: MIT Press.

Van Dijken, J. P. & Harder, W. (1975). Growth yields of microorganisms on methanol and methane. A theoretical study. *Biotechnology and Bioengineering*, **17**, 15–30.

Van Dijken, J. P., Otto, R. & Harder, W. (1976). Growth of *Hansenula polymorpha* in a methanol-limited chemostat. *Archives of Microbiology*, **111**, 137–44.

Von Meyenburg, H. K. (1969). Energetics of the budding cycle of *Saccharomyces cerevisiae* during glucose limited aerobic growth. *Archiv für Mikrobiologie*, **66**, 289–303.

Wagner, F. (1975). *1. Symposium Mikrobielle Proteingewinnung 1975.* Weinheim: Verlag Chemie.

Wilkinson, J. F. & Munro, H. L. S. (1967). The influence of growth limiting conditions on the synthesis of possible carbon and energy storage polymers in *Bacillus megaterium*. In *Microbial Physiology and Continuous Culture*, eds. E. O. Powell, C. G. T. Evans, R. E. Strange & D. W. Tempest, pp. 173–85. London: Her Majesty's Stationery Office.

Wilkinson, T. G., Topiwala, H. H. & Hamer, G. (1974). Interactions in a mixed bacterial population growing on methane in continuous culture. *Biotechnology and Bioengineering*, **16**, 41–59.

Wimpenny, J. W. T. (1969). Oxygen and carbon dioxide as regulators of microbial growth and metabolism. In *Microbial Growth*, eds. P. M. Meadow & S. J. Pirt, pp. 161–97. *19th Symposium of the Society for General Microbiology.* Cambridge University Press.

Wodzinski, R. S. & Johnson, M. J. (1968). Yield of bacterial cells from hydrocarbons. *Applied Microbiology*, **16**, 1886–91.

Yagi, H. & Yoshida, F. (1977). Desorption of carbon dioxide from fermentation broth. *Biotechnology and Bioengineering*, **19**, 801–19.

THE PHYSIOLOGY OF
METABOLITE OVER-PRODUCTION

OENSE M. NEIJSSEL AND DAVID W. TEMPEST

Laboratorium voor Microbiologie, Universiteit van Amsterdam,
Plantage Muidergracht 14, Amsterdam, The Netherlands

INTRODUCTION

The culturing of micro-organisms with a view to producing metabolites has been, at least in a majority of cases, an empirically designed process. Usually one isolates from Nature a wide variety of organisms belonging to genera or species that are known to express a certain desired property, such as the production of an antibiotic, and these organisms are then screened for this property. If possible, one tries to acquire 'improved' strains either by direct selection or by selection after mutagen treatment. The next step involves establishing the optimal culture conditions for the process and, once these are found, a mathematical model is developed that will enable the biotechnologist to steer the process to its potential optimum. Moreover, with the advent of the recombinant DNA technique it has become possible, at least in principle, to insert the gene coding for a desired product into the DNA of a microbe and then to promote synthesis of the product by the organism. At the time of writing one report has been published on the successful use of this technique for the synthesis of the human hormone somatostatin by *Escherichia coli*. The gene coding for somatostatin was synthesized *in vitro* and was incorporated into a plasmid containing a part of the lactose operon. The different regulatory genes of the lactose operon were used for the transcription *in vivo* of this essentially eukaryotic gene. The product was a protein containing a large fragment of the β-galactosidase molecule coupled to somatostatin. This chimeric protein then could be cleaved by cyanogen bromide treatment to yield somatostatin (Itakura *et al.*, 1977).

However, despite the obvious successes achieved using the genetic approach, there are often serious setbacks, in that the so-called 'improved' strains can lose the desired property, a phenomenon generally called 'strain degeneration'. Thus, Itakura *et al.* (1977) observed that *Escherichia coli* strains containing the somatostatin gene were unstable, and that the highest yield of somatostatin was from cultures in which 30% of the cells already had deletions in the lactose operon. Clearly, 'strain degeneration' is an adequate description for this phenomenon from the point of view of the industrial microbiologist, but it neglects an appreciation of the complex interactions that exist between growth environment and micro-organism. Many producer strains are functionally crippled and suffer from one or more genetic disorders, or, in the case of an extra inserted gene or plasmid, they suffer from the extra genetic burden which they must carry. Since every environment exerts a selective pressure, the producer strain will only survive for prolonged periods of time if the growth environment does not select against it; in other words, the improved behaviour of the strain must offer a selective growth advantage. This is frequently not the case, as can be seen in the example of somatostatin synthesis where the bacterium synthesizes a protein which serves no function and has lost the capacity to synthesize a complete β-galactosidase molecule.

From these observations one may conclude that the growth environment plays a crucial role in the persistence of certain genes and in the survival of microbes containing extra or deleted genetic information. One may further conclude, therefore, that 'strain degeneration' is frequently 'strain improvement' from the organisms' point of view and that the discrepancy between the interests of the industrial microbiologist and those of the microbes can only be resolved if the former creates an environment in which the selective pressures are such that the organism with the desired property has a growth advantage over any other organism lacking this property. It is for this reason that we concentrate in this contribution on the effect of the growth environment on the behaviour of micro-organisms.

Another problem encountered with regard to metabolite production is a lack of detailed knowledge of the physiological significance of the process and, even worse, a lack of knowledge of the biochemical pathways leading to the synthesis of many metabolites. Thus, whereas the pathways of the synthesis of intermediary metabolites like amino acids, products of energy metabolism (e.g. ethanol and lactate), storage polymers (like polysaccharides) and nucleic acids are well known, the physiological significance of the synthesis of these substances in widely differing amounts is not always well understood. For example, the nature of the Pasteur effect is now known in principle but there are details still to be clarified (Krebs, 1972) and while it is known that conditions of energy excess promote the synthesis of storage polymers (Dawes & Senior, 1973), this is not invariably the case, as has been observed in potassium-limited chemostat cultures of *Klebsiella aerogenes* (Dicks & Tempest, 1967). Moreover, in industry one frequently has to use, for economic reasons, rather complex media and this leads to even greater difficulties in understanding the physiology of growth, since detailed knowledge of all the available and consumed nutrients in such media generally is lacking. Finally, there seems to be no obvious physiological reason for the synthesis of many of the so called 'secondary metabolites'. It has been proposed that these substances may help the producer strain to survive in Nature but when these strains are grown in axenic culture there is seemingly no function for the production of secondary metabolites and thus this property frequently is lost (Pirt, 1969). Again it might be argued that we still have not found the correct environmental conditions that will exert a selective pressure such that only those strains that produce secondary metabolites necessarily will survive.

It is the purpose of this contribution to try to formulate some fundamental principles regarding the physiology of metabolite over-production or 'overflow metabolism'. We shall analyse the interaction between the growth environment and the microbe and emphasize studies with chemostat cultures, since only this culture technique provides a means to manipulate the growth environment in a reproducible and unique way. However, rather than providing an exhaustive review of the available literature, we illustrate our arguments by reference to selected examples that may or may not have a potential application in industry.

Before proceeding further, it is appropriate to define some terms that will be used throughout this contribution. In industry, one generally calls every process in which metabolites are excreted a 'fermentation'. This can be misleading since the microbial physiologist will employ this term only for a selected class of microbial reactions. Thus, we will use the following definitions: *Fermentation* we define as that type of metabolism of a carbon source in which the energy is generated by substrate-level phosphorylation and in which organic molecules function as acceptors for the reducing equivalents generated during catabolism. Those processes in which the reducing equivalents generated during catabolism of the carbon source are oxidized by a respiratory chain, will be called *aerobic respiration* (where oxygen is the terminal electron acceptor) or *anaerobic respiration* (where compounds such as sulphate or nitrate provide the terminal electron acceptor). A *metabolite* is, by definition, a molecule which participates in metabolism and this term can be used for carbon dioxide, hydrogen or nitrogen gas on the one hand, or DNA, protein or enzymes on the other. We will mainly discuss the production and excretion of carbon-containing compounds of low molecular weight, and will not concentrate on the physiology of, for instance, enzyme excretion, although we believe that the same physiological principles are applicable to both groups of metabolites (Tempest & Neijssel, 1978).

EFFECT OF NUTRIENT LIMITATION

All micro-organisms currently being studied have, of course, been isolated originally from natural sources such as surface waters, soil, dairy products, faeces or from patients. However, many of these strains have been subsequently cultured for extensive periods of time in the laboratory – *Escherichia coli* K12, for instance, since 1922 (Anderson, 1975) – and therefore one can question whether these laboratory strains still are capable of expressing the same properties as they did when they grew in their natural environments. It is generally assumed that this is the case and, whereas many aspects of the behaviour of microbes in a fermenter may be the result of laboratory conditions, it is reasonable to suppose that the underlying physiological principles will be the same as those prevailing in natural ecosystems. However, a complete understanding of these principles cannot be derived without a detailed consideration of the many interactions that must occur between micro-organisms and the conditions of natural environments. It well may be the case that many aspects of the behaviour of these organisms are puzzling when one confines oneself to analyses of laboratory cultures, whereas satisfactory explanations might be forthcoming when one takes into consideration the problems that microbes face in Nature.

In Nature, conditions for growth are often unstable and rarely optimal. There may be periodic fluctuations of temperature, pH, ionic strength, water activity, oxygen tension and redox potential. But such environments rarely contain all the nutrients essential for cell synthesis in sufficient concentrations to allow growth to proceed at its potentially maximum rate. Micro-organisms react to these unstable and nutritionally extreme conditions by changing themselves in such a way that they are optimally adapted, structurally and functionally, to what has been called by Koch (1971) a 'feast and famine existence'. We have previously argued (Tempest & Neijssel, 1976, 1978) that this modification of behaviour must occur in at least four ways:

(i) induction or de-repression of the synthesis of a high-affinity uptake system for the growth-limiting nutrient;

(ii) modulation of the rates of uptake of all non-limiting nutrients in order to prevent intermediary metabolites accumulating intracellularly to toxic levels;

(iii) modulation of metabolism in order to circumvent as far as possible those 'bottlenecks' imposed by the specific growth limitation; and

(iv) modulation of the rates of synthesis of the macromolecular components of the cell in order to allow balanced growth to proceed at a submaximal rate.

In this connection it is worth mentioning the principle first formulated by Darwin (1859) that organisms strive to remain alive and create progeny. It is essential to realize that this principle, and the proposed modifications of behaviour mentioned above, do *not* imply that microbes strive towards a maximal yield with respect to every nutrient. Under some growth conditions success in remaining alive can only be attained by organisms compromising and accepting a lower than maximum yield; and when this decrease in yield is accompanied by metabolite excretion, these growth conditions are particularly interesting.

A well-known example of a metabolic compromise involving a lowered yield with respect to the carbon nutrient is, of course, a fermentation. If a heterotrophic, facultatively anaerobic bacterium like *Escherichia coli* is grown aerobically in a batch culture in a medium containing glucose and mineral salts, and in which glucose is ultimately the growth-limiting nutrient, the sole products of glucose metabolism will be new cell material and carbon dioxide. On the other hand, in anaerobic batch cultures with the same medium, fewer cells are produced, and ethanol, succinate, lactate, acetate, hydrogen and carbon dioxide are found as end-products (Blackwood, Neish & Ledingham, 1956). Although this may seem a rather trivial example, it serves to make the following points: During catabolism of the carbon source, intermediary metabolites and reducing equivalents are produced. Because of the absence of oxygen, the reducing equivalents cannot be transferred to the respiratory chain, so a metabolic bottleneck is created (see point (iii) above) since the carriers of the reducing equivalents, the pyridine and flavin nucleotides, are present in the cell only in low amounts. This accumulation of reduced nucleotides would have consequences for the metabolism of the carbon source and would result in the accumulation of intermediary metabolites (see point (ii) above); and this would place the organisms in an uncompromising condition (compare Postgate & Hunter 1963, 1964; Calcott & Postgate,

1972, 1974; Oh & Freese, 1976). Thus in order to allow carbon metabolism, and therefore growth, to proceed, the cell must dispose of its reducing equivalents and of some of the accumulated intermediary metabolites. It is under these circumstances that the cell opts for a lowered yield with respect to the carbon source: intermediary metabolites function as acceptors for the reducing equivalents and are excreted together with potentially toxic metabolites, such as acetate. The energy in this fermentative metabolism is derived mainly from substrate-level phosphorylation with a minority from electron transport in the first part of the respiratory chain with NADH as donor and the flavin cofactor of succinate dehydrogenase as acceptor. We may conclude, therefore, that if we could create environments in which it was essential for an organism to excrete metabolites in order to maintain viability and grow, metabolite production would be a stable and reproducible property of the culture and strain degeneration would not occur.

It is by no means essential to use exclusively anaerobic or oxygen-limited culture conditions to promote metabolite production as can be concluded from numerous published data (for reviews see: Pirt, 1969; Christner, 1976; Tempest & Neijssel, 1978). But an analysis of these data will show that a coherent physiological basis for these observations still has not been elucidated. It was for this reason that we undertook a systematic investigation into the utilitzation of glucose, glycerol, mannitol and lactate by variously limited chemostat cultures of *Klebsiella aerogenes* NCTC 418. Some of the results are summarized in Table 1.

As could be anticipated, with each of the carbon-limited cultures the sole products of growth were organisms and carbon dioxide. By comparison with these carbon-limited cultures, however, it was found that whenever the carbon substrate was present in excess of the minimum growth requirement, it was used less efficiently and the yield value decreased markedly (Table 2). The extent to which the carbon substrate was over-utilized depended both on the nature of the growth limitation and the identity of the carbon source. In this connection, it is clear that glycerol was taken up less extensively than the other substrates and that it was metabolized more completely. This is probably the result of the mode of regulation of glycerol assimilation in this organism. It has been found (Neijssel *et al.*, 1975) that *K. aerogenes* NCTC 418 assimilates glycerol via two alternative pathways: one involving a glycerol kinase with a high affinity for glycerol (apparent $K_m = 1–2 \times 10^{-6}$M), and the second a glycerol dehydrogenase with a much lower affinity for its

Table 1. *Substrate utilization rates and rates of product formation in variously-limited chemostat cultures* of Klebsiella aerogenes *NCTC 418*

Carbon source	Limitation	Substrate used	Cells[a]	CO$_2$	Pyr	2OG	Ac	GA	2KGA	Succ	dLac	Prt	CHO	Carbon recovery (%)
							(milliatom carbon per hour)							
Glucose	Carbon	36.8	20	15.6										97
	Sulphate	98.7	20	20.8	21.9	1.8	9.6			2.3		1.8		91
	Ammonia	107.4	20	20.2	5.2	22.5	4.0						36.0	102
	Phosphate	112.8	20	20.4			4.5	9.5	39.9				15.1	97
	Magnesium[b]	124.6	20	31.4	9.6	9.6	17.7	0.1	11.9		10.9	2.5		91
	Potassium	175.0	20	56.3	10.1	3.0	13.3	31.9	20.5			8.0		93
Glycerol	Carbon	35.9	20	15.3										98
	Sulphate	54.9	20	25.4			7.4							96
	Ammonia	45.1	20	20.2		5.7	2.0							106
	Phosphate	53.8	20	31.2										95
Mannitol	Carbon	33.1	20	13.1										100
	Sulphate	61.1	20	27.5			15.2							103
	Ammonia	64.0	20	25.0		15.8	3.3							100
	Phosphate	41.2	20	19.8			1.6	0.1	0.1					101
Lactate	Carbon	40.2	20	19.3										98
	Sulphate	60.4	20	36.5			8.5							108
	Ammonia	92.1	20	60.3	1.6	4.3	6.3							100
	Phosphate	62.0	20	39.2			2.5							100

($D = 0.17 \pm 0.01$ h^{-1}; pH 6.8; 35 °C)

Pyr = pyruvate; 2OG = 2-oxoglutarate; Ac = acetate; GA = gluconate; 2KGA = 2-ketogluconate; Succ = succinate; dLac = D-lactate; Prt = extracellular protein; CHO = extracellular polysaccharide.

[a] All values were adjusted to a cell carbon production rate of 20 milliatom carbon per hour.

[b] Because of the inherent instability of this type of culture only approximate figures are given.

Data of Neijssel & Tempest (1975) and Tempest & Neijssel (1978).

Table 2. *Yield values for carbon-substrate and oxygen, obtained with variously-limited chemostat cultures of* Klebsiella aerogenes *NCTC 418*

Substrate	Limitation	$Y_{substrate}$	Y_{oxygen}	q_{O_2}
Glucose	Carbon	0.45	20.4	4.17
	Sulphate	0.19	11.5	7.40
	Ammonia	0.16	11.5	7.40
	Phosphate	0.15	8.7	9.75
	Magnesium[a]	0.14	7.6	11.2
	Potassium	0.09	5.2	17.4
Glycerol	Carbon	0.46	18.1	4.71
	Sulphate	0.30	7.5	11.3
	Ammonia	0.36	8.3	10.2
	Phosphate	0.30	8.1	10.5
Mannitol	Carbon	0.50	16.2	5.25
	Sulphate	0.27	7.4	11.5
	Ammonia	0.26	9.2	9.26
	Phosphate	0.40	11.2	7.60
Lactate	Carbon	0.41	12.4	6.86
	Sulphate	0.20	5.0	16.9
	Ammonia	0.17	5.0	16.9
	Phosphate	0.27	6.8	12.5

($D = 0.17 \pm 0.01$ h^{-1}; pH 6.8; 35 °C)

$Y_{substrate}$ = gram organisms synthesized per gram substrate consumed.

Y_{oxygen} = gram organisms synthesized per 0.5 mole oxygen consumed.

q_{O_2} = mmole oxygen consumed per gram dry organisms per hour

[a] Because of the inherent instability of this type of culture, only approximate figures are given.

Data of Neijssel & Tempest (1975) and Tempest & Neijssel (1978).

substrate (apparent $K_m = 2$–4×10^{-2}M). Aerobic carbon-limited organisms contained only the glycerol kinase pathway whereas aerobic carbon-sufficient organisms seemingly used only the glycerol dehydrogenase pathway. This modulation of the uptake system for the carbon substrate, depending on the growth-limiting component, could not be demonstrated for the other substrates: *K. aerogenes* NCTC 418 retained a high-affinity uptake system for glucose, mannitol and lactate under all the growth conditions examined, with affinity constants in the order of 4–100 μM (Neijssel & Tempest, 1975).

The uptake kinetics, however, are not the only consideration, since, even in those cases where the uptake system seemingly was not modulated, striking differences were evident. For example, *K. aerogenes* NCTC 418 has the same high-affinity uptake system – a phosphoenolpyruvate-dependent phosphotransferase system (Tanaka & Lin, 1967; Tanaka, Lerner & Lin, 1967) – for both mannitol and glucose, irrespective of whether the substrate is in excess or is growth-limiting. Yet mannitol was invariably used more efficiently than was glucose. An examination of the pathways of metabolism of these two compounds reveals one important difference, however, in that glucose, after being taken up into the cell, is not immediately oxidized, as is mannitol. The first oxidation step in the metabolism of glucose occurs only at the level of glyceraldehyde 3-phosphate, whereas mannitol phosphate is immediately oxidized to fructose 6-phosphate. It is therefore reasonable to assume that pyridine nucleotide-linked dehydrogenases also play a key role in the regulation of substrate uptake and assimilation; a situation that can be readily envisaged, since the activity of these enzymes will depend critically on the redox state of the cell, particularly the NADH/NAD$^+$ ratio. Thus, when a dehydrogenase is among the first of the enzymes involved in the metabolism of a substrate, then more possibilities exist for regulating the rates of uptake and metabolism of this substance, sufficient to avoid a very substantial increase in the concentration of 'pool' intermediary metabolites with consequent excretion of some of them into the medium. In this connection, the switch from the high-affinity glycerol kinase pathway in glycerol-limited *K. aerogenes* to the low affinity, NAD-linked glycerol dehydrogenase pathway in glycerol-sufficient organisms may well afford extra possibilities for regulating the rate of assimilation of excess glycerol and provide additional reasons for the large differences observed between the rates of utilization of glycerol and glucose.

Nevertheless, what must be noted here is the surprising fact that, in glucose-sufficient chemostat cultures, the uptake and metabolism of glucose is not regulated to such an extent that no over-production of metabolites occurs. This is not a peculiarity of this strain of *K. aerogenes* since this phenomenon was also found with wild-type cultures of *K. aerogenes* that were freshly isolated from canal water (Neijssel, Hueting & Tempest, 1977). However, this observation is in sharp contrast with those made with batch cultures of micro-organisms, where it was commonly found that glucose uptake and metabolism were closely regulated and metabolite over-production less commonly occurred during the logarithmic growth phase. In these cultures, *if* metabolite production did occur, it was seemingly confined to the late-logarithmic or early-stationary phase of growth. Consequently the metabolites that were produced under these circumstances were named 'secondary metabolites' or 'idiolites', and the part of the growth cycle wherein such production occurred was called the 'secondary growth phase' or 'idiophase' (Bu'Lock & Powell, 1965; Martin and Demain, 1976, 1977). In contrast, the production of metabolites in chemostat cultures cannot be related to the growth phase since at all dilution rates the cells are growing exponentially. In fact, it is clearly related to the nature of the growth limitation.

Inspection of the nature of these so called 'overflow' products reveals that they are not the result of fermentative metabolism or oxygen-limited growth conditions: all products are more oxidized than the original carbon source, except for D-lactate produced by magnesium-limited cultures grown on glucose: but even here D-lactate is not a common major product in the fermentation of glucose by this organism (S. Hueting, personal communication; compare Reynolds & Werkman, 1937 and Hernandez & Johnson, 1967). The spectrum of compounds that were excreted differed (qualitatively as well as quantitatively) with the nature of the growth limitation and the identity of the carbon substrate. Thus, phosphate-limited glucose-grown cells excreted much gluconate and 2-ketogluconate whereas similarly-limited cultures grown on glycerol, mannitol or lactate excreted little or none of these two compounds. Likewise, production of pyruvate was only found when glucose served as carbon source. But one consistent factor was the excretion of 2-oxoglutarate by ammonia-limited cultures, irrespective of the carbon source.

It is important to ask, therefore, what are the physiological principles that determine the nature and quantity of overflow products in carbon-sufficient aerobic cultures. We have already mentioned that regulation of carbon-substrate uptake plays a role but clearly there must be a more specific explanation for the variety of products excreted. To provide a tentative answer to this question one has to come to a better under-standing of how a specific nutrient limitation influences carbon meta-bolism in *K. aerogenes*. Fig. 1 depicts the different mechanisms for the assimilation of ammonia, sulphate, phosphate, magnesium and potas-sium. It can be easily envisaged that if an enzyme reaction involves two or more substrates, one of which is present in a low concentration and thus limiting the rate of reaction, the effectiveness with which the reaction proceeded would be increased if the other substrate(s) were present in enzyme-saturating amounts. Thus, in the case of ammonia, high levels of ATP and glutamate (for glutamine synthetase) and of 2-oxoglutarate and NADPH (for glutamate synthase) would provide an excellent driving force for ammonia assimilation. The intracellular level of glutamate is generally high in Gram-negative bacteria because it is involved in the regulation of the intracellular osmotic pressure (Tempest, Meers & Brown, 1970). But the other compounds have to be generated by an increased rate of metabolism. The over-production of

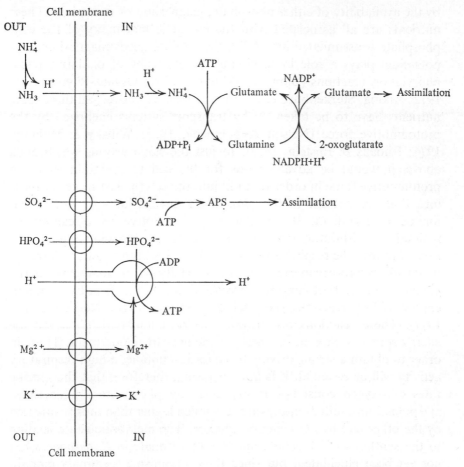

Fig. 1. Schematic representation of the high-affinity uptake systems for different growth-limiting nutrients.

2-oxoglutarate under ammonia limitation, therefore, is not really surprising. Moreover, the presence of high levels of ATP and NADPH in ammonia-limited growing cells can be inferred from their accumulation of glycogen (Tempest, Herbert & Phipps, 1967). It has been shown that synthesis of glycogen is promoted by ATP and NADPH and inhibited by AMP and oxidized pyridine nucleotides (Dawes & Senior, 1973). The synthesis of extracellular polysaccharide commonly found in ammonia-limited cultures of micro-organisms (Behrens *et al.*, 1977; Williams & Wimpenny, 1978) may be regulated in much the same way.

The synthesis of gluconate and especially 2-ketogluconate was provoked in glucose-grown cultures when the growth rate was limited by the availability of either phosphate, magnesium or potassium. These nutrients are all associated with the energetic machinery of the cell: phosphate is assimilated into ATP, the ATPase needs magnesium, and potassium plays a role in specifying the efficiency of oxidative phosphorylation reactions (Blond & Whittam, 1965; Gómez-Puyou et al., 1972; Aiking, Sterkenburg & Tempest, 1977). Moreover, all these ionic nutrients have to be taken up by transport systems energized by the protonmotive force (Nelson & Kennedy, 1972; Willsky & Malamy, 1974; Rhoads & Epstein, 1977). In just the same way as was argued above, it would be advantageous for the cell to maintain a strong protonmotive force in order to 'pull' into the cell the last traces of these ions that are present in the phosphate-, magnesium- or potassium-limited environment. But a strong protonmotive force cannot be maintained without continuous production of reducing equivalents needed to feed the respiratory chain, since it has been shown that many, if not all, micro-organisms have the capacity to dissipate the energy contained in the protonmotive force (and in ATP) as heat by means of energy-spilling reactions (Neijssel & Tempest, 1976; Neijssel et al., 1977). These reactions are believed to be stimulated whenever the energy source is present in excess of the growth requirement. Thus, in order to obtain a strong driving force for ion uptake, a high respiratory activity will be essential. It is not surprising, therefore, that the specific rates of oxygen consumption expressed by phosphate-, magnesium- and potassium-limited cultures are very much higher than those expressed by the other cultures growing on glucose. The metabolic route leading to the synthesis of gluconate and 2-ketogluconate in K. aerogenes has not yet been elucidated, but since these organisms seemingly contain no glucose dehydrogenase, whereas glucose-6-phosphate dehydrogenase could be readily detected, one must assume that these compounds are actually synthesized via glucose-6-phosphate and 6-phosphogluconate. This extra pathway of glucose metabolism provides many reducing equivalents with a minimum involvement of enzymes and phosphorylated intermediates as compared with the usual Embden–Meyerhof pathway of glucose catabolism. The 2-ketogluconate pathway of glucose metabolism therefore serves the function of saturating the respiratory chain with the necessary reducing equivalents.

In contrast to glucose, catabolism of the other carbon sources immediately generates reducing equivalents, and so the need for an extra pathway (if that were possible) does not arise. The fact that the specific rates of oxygen consumption of phosphate-limited cultures grown on, respectively, mannitol, glycerol and lactate were also not very much higher than those of ammonia-limited or sulphate-limited cultures growing on the same carbon compounds, supports this conclusion (Table 2).

Acetate was produced under almost all growth limitations. Its production, therefore, is linked more with the general condition of carbon-source excess than with a specific limitation. From the production of acetate, and sometimes pyruvate, one may conclude that the rate of glycolysis exceeds the oxidizing capacity of the citric acid cycle. It has been reported in the literature, and also found in our studies, that under conditions of glucose excess the synthesis of 2-oxoglutarate dehydrogenase is repressed (Hollywood & Doelle, 1976). Clearly, this will have consequences regarding the oxidative capacity of this cycle, and also provides an explanation for the production of quantities of 2-oxoglutarate, albeit in relatively minor amounts, under limitations other than ammonia.

The production of considerable amounts of pyruvate by sulphate-limited cultures grown on glucose can be explained by a lowered level of the sulphur-containing cofactors of the pyruvate dehydrogenase system: thiamine pyrophosphate, lipoic acid and coenzyme A. It was mentioned by Rosenberger & Elsden (1960) that thiamine deficiency led to pyruvate excretion by micro-organisms. Possibly the synthesis of pyruvate dehydrogenase suffers also from a degree of catabolite repression by glucose. Mannitol, glycerol and lactate are used at lower rates than glucose, which accounts for the fact that sulphate-limited cultures grown on these carbon sources do not produce pyruvate.

Under none of the growth conditions tested were phosphorylated products of carbon metabolism excreted. This can be explained by the fact that, in general, these compounds are not easily transported across the cell membrane. Moreover if the organisms were to excrete phosphorylated compounds in considerable amounts they would quickly deplete their 'pools' of utilizable phosphate. Clearly, it would be more advantageous for the microbe to produce an intracellular phosphatase to hydrolyse those phosphorylated metabolic intermediates that eventually would accumulate (cf. excretion of gluconate).

In order to test whether overflow metabolism was peculiar to *Klebsiella aerogenes*, we performed similar experiments with *Escherichia coli* ATCC 9001. Preliminary results indicated that these organisms, when grown in a sulphate-limited chemostat on glucose, produced large quantities of acetate but no pyruvate. When glycerol served as the carbon source in such cultures no excretion of soluble products occurred.

Glucose-limited or glycerol-limited cultures produced only cells and carbon dioxide (see also Hollywood & Doelle, 1976).

When *Bacillus subtilis* var. *niger* was grown in a chemostat culture under different limitations (carbon, ammonia, sulphate and phosphate limitation), with glucose as the carbon source, it was again found that carbon-sufficient cultures gave a lower yield with respect to the carbon source (Table 3) and increased q_{O_2} values. Under conditions of phosphate limitation, acetate was the only extracellular product, whereas sulphate-limited cultures produced both pyruvate and acetate. Significantly, no extracellular products could be found under conditions of ammonia limitation. This may be due to the impermeability of the cell membrane of *Bacillus subtilis* towards 2-oxoglutarate (Ghei & Kay, 1973).

Table 3. *Glucose utilization rates, rates of product formation, yields on glucose, and specific rates of oxygen consumption, in chemostat cultures of* Bacillus subtilis *var.* niger *WM that were growing on glucose*

Limitation	Glucose	Ammonia	Sulphate	Phosphate
Dilution rate (h^{-1})	0.10	0.13	0.13	0.16
Glucose used^a	47.0	70.6	110.3	95.8
Products formed				
Cells	20	20	20	20
CO$_2$	28	49.4	34.4	35.8
Pyruvate	0	0	24.2	0
Acetate	0	0	15.5	33.3
Protein	0	0	4.9	0
Carbon recovery (%)	102	98	90	93
$Y_{glucose}$	34.1	23.1	16.8	16.4
q_{O_2}	6.1	13.0	11.0	12.0

(pH 6.8; 35 °C)

Media and growth conditions were the same as those used for studies with *K. aerogenes*.

$Y_{glucose}$ = gram organisms synthesized per 100 glucose consumed.

q_{O_2} = mmole oxygen consumed per gram dry organisms per hour.

^{a} Rates of glucose consumption and product formation are expressed as milliatoms carbon per hour and have been normalized to a cell carbon production rate of 20 milliatoms carbon per hour.

Published data of similar experiments are very sparse, seemingly for two reasons: first, few investigators have grown aerobic or facultatively anaerobic organisms in chemostat culture under fully aerobic, carbon-sufficient growth conditions and constructed carbon balances; secondly, when there are data on excretion of metabolites by aerobic cultures of micro-organisms, the majority of these data relate to batch cultures in the late-logarithmic phase. But it is almost impossible to analyse the precise interactions between the environment and the organism in this growth phase where, as can be easily shown, changes in the environment occur at a very fast and exponential rate. Therefore we will discuss only examples in which the growth environment has been well defined with respect to nutrient composition and physicochemical parameters such as pH and temperature.

Whiting, Midgley & Dawes (1976 *a, b*) grew *Pseudomonas aeruginosa* in a chemostat under conditions of glucose limitation. Under these circumstances *P. aeruginosa* took up glucose by way of a high-affinity system ($K_m = 8$ μM). In contrast, when ammonia was the growth-limiting nutrient the high-affinity glucose transport system was repressed. Instead, glucose was oxidized extracellularly to gluconate at a high rate, by a glucose dehydrogenase reaction, and gluconate was further oxidized extracellularly to 2-ketogluconate by a gluconate dehydrogenase. Gluconate and 2-ketogluconate were taken up by the organism via a low-affinity uptake system ($K_m = 2$ mM) (Dawes, Midgley & Whiting, 1976), but considerable amounts of both metabolites remained in the extracellular fluids. This example shows the previously mentioned division into a high-affinity and a low-affinity uptake system for the carbon nutrient and the generation of a metabolic bottleneck resulting in overflow metabolism.

A further example is provided by cultures of *Aspergillus niger* which produce citrate. For citrate production to occur it is essential to limit the growth of the fungus by the availability of zinc, iron, or manganese ions (Kubicek & Röhr, 1977) but the physiological significance of citrate production is still not clear. It is well established that aconitase needs an iron–citrate complex in order to react properly (Glusker, 1968) so that an iron deficiency would create a metabolic bottleneck at the level of this enzyme. But this does not explain the effect of zinc and manganese. A comparison with cation-limited cultures of prokaryotes might yield an additional explanation. Iron-limited growth of *Escherichia coli*, and other bacteria, generally provokes the synthesis of iron-chelating compounds, called siderophores (Byers, 1974; Emery, 1974; Rosenberg & Young, 1974). It is assumed that these compounds serve the function of scavenging the last traces of iron from the environment when growth is iron-limited. In this connection, it has been proposed by Haavik (1976) that bacitracin and some other peptide antibiotics serve a similar function in *Bacillus licheniformis*. It may be, therefore, that citrate serves a chelating function for the uptake of cations by *Aspergillus niger*.

We have concentrated up to this point on a discussion of aerobic cultures, and the reasons why these cultures sometimes necessarily excrete metabolites. If one now returns to anaerobic or oxygen-limited cultures it is clear that the same considerations should be applicable. Thus, if one considers the respiratory chain of an organism that uses oxygen as the electron acceptor as the metabolic bottleneck induced by the limited availability of oxygen one sees that, in principle, the reasons for the excretion of fermentation products are comparable to those formulated for the excretion of, say, 2-oxoglutarate by ammonia-limited *K. aerogenes*. In this respect, an anaerobic culture of a facult-atively anaerobic organism (which uses the respiratory chain under aerobic circumstances) is just an extreme case of a nutrient limitation. Anaerobic cultures of organisms that carry out an anaerobic respiration also should be comparable with aerobic cultures of *K. aerogenes*. Indeed, nitrogen-limited cultures of *Desulfovibrio vulgaris* have been found to excrete 2-oxoglutarate and pyruvate (Lewis & Miller, 1975). Overflow metabolism has also been observed with photosynthetic bacteria like *Rhodopseudomonas capsulata*. When these nitrogen-fixing organisms were grown dinitrogen-limited they produced hydrogen. The production of hydrogen occurred via nitrogenase and was inhibited by molecular nitrogen or ammonium salts (Hillmer & Gest, 1977). Obviously, under dinitrogen limitation nitrogenase is continuously saturated with reducing equivalents which, in the absence of N_2, are disposed of as hydrogen.

In view of what has been discussed above it is not surprising that similar observations have been made with bacteria, e.g. the streptococci, that can only carry out a fermentation. The penetrating studies of Rosenberger & Elsden (1960) are, in this respect, worthy of further consideration. They cultured *Streptococcus faecalis* anaerobically in a defined medium in a chemostat. Table 4 shows the specific rates of glucose consumption ($q_{glucose}$) and lactate production ($q_{lactate}$), for the different growth conditions, calculated from the authors' data. As can be seen, when growth was limited by the glucose supply, the rate of glucose consumption by the organism was lower than when tryptophan was growth-limiting. Under glucose limitation there was a proportion-ality between the dilution rate and the $q_{glucose}$, but under tryptophan limitation this was not the case. These, and other data reported by the authors, indicate that under tryptophan limitation $q_{glucose}$ remained roughly constant at all growth rates. The $q_{lactate}$ increased with growth rate under both limitations but was very much higher under tryptophan limitation. It is regrettable that not many microbiologists have taken notice of the comments made by these authors, since they pointed to the important bioenergetic consequences of their findings in that the data from the tryptophan-limited cultures indicate that the rate of

Table 4. *Specific rates of glucose consumption and lactate production by chemostat cultures of* Streptococcus faecalis *that were, respectively, glucose-limited and tryptophan-limited*

Limitation	Dilution rate (h^{-1})	Specific rates of consumption and formation (mmole per gram cells per hour)	
		Glucose	Lactate
Glucose	0.22	4.9	3.8
	0.31	7.5	6.8
	0.43	9.6	17.6
Tryptophan	0.22	12.9	13.4
	0.31	16.6	20.2
	0.43	16.1	30.6

Data of Rosenberger & Elsden (1960).

glucose catabolism is *not* controlled by the rate of growth and that under carbon-sufficient growth conditions energy-spilling reactions must be present. It is this consideration which has far reaching consequences for metabolite production. Later, similar findings were reported by Yamada & Carlsson (1975) and by White *et al.* (1976). To summarize this section, the factors which are important for metabolite over-production are:

(i) carbon-sufficient growth conditions;
(ii) the nature of the carbon source;
(iii) the nature of the growth-limiting nutrient, since this will determine the metabolic bottleneck which results in a particular type of overflow product;
(iv) no extensive regulation of the uptake of the carbon source;
(v) the permeability of the cell membrane towards metabolites, since this will determine whether a particular metabolite can be excreted or not; and
(vi) the type of organism, since regulation of metabolism and membrane permeability varies from organism to organism.

But there are more environmental parameters which influence overflow metabolism and which we have not yet discussed. The following sections consider the influence which some of these factors exert.

EFFECT OF pH

The pH value of the growth environment exerts a profound influence on microbial growth and metabolism. It was shown by Gale & Epps (1942) that *Escherichia coli* and *Micrococcus lysodeikticus* synthesized different amounts of intracellular enzymes (such as catalase, urease, fumarase, formate hydrogen lyase, several deaminases and decarboxylases) when the pH of the growth medium was altered. In particular, amino acid decarboxylases were produced in response to an acidic growth environment and amino acid deaminases were synthesized in response to alkaline growth conditions. They concluded that a change in the external pH was followed by an alteration in the enzymic constitution of the cells such as to invoke a neutralization reaction, while other essential enzyme activities were maintained at a constant level.

If there is such a drastic change in enzyme level and composition in response to a change in the extracellular pH then, clearly, this must be reflected in the metabolism of the carbon source. This indeed has been found by many authors. As an example, Table 5 contains the results obtained by Gunsalus & Niven (1942) with *Streptococcus liquefaciens*

Table 5. *Effect of medium pH value on the fermentation of glucose by different organisms*

Organism	S. liquefaciens[a]			E. coli[b]		A. cloacae[c]	
pH value	5.0	7.0	9.0	6.0	7.8	5.0	7.2
(mmole produced per 100 mmole glucose fermented)							
Acetate	12.2	18.8	31.2	36.5	38.7	5.0	69.5
Acetoin	–	–	–	0.059	0.190	0	1.5
2,3-Butanediol	–	–	–	0.30	0.26	38.8	2.3
Butyrate	–	–	–	0	7.10	–	–
CO_2	–	–	–	88.0	1.75	–	–
Ethanol	7.0	14.6	22.4	49.8	50.5	61.8	67.5
Formate	15.4	33.6	52.8	2.43	86.0	–	–
Glycerol	–	–	–	1.42	0.32	–	–
Hydrogen	–	–	–	75.0	0.26	–	–
Lactate	174	146	122	79.5	70.0	0	3.6
Succinate	–	–	–	10.7	14.8	3.3	6.4
Total acid	201.6	198.4	206.0	129.13	216.6	8.3	79.5
Total neutral (including CO_2)	7.0	14.6	22.4	139.51	53.0	100.6	71.3

– = not determined.
[a] Data of Gunsalus & Niven (1942); [b] Data of Blackwood *et al.* (1956); [c] Data of Hernandez & Johnson (1967).

(Lancefield group D), Blackwood, Neish & Ledingham (1956) with *Escherichia coli*, and Hernandez & Johnson (1967) with *Aerobacter cloacae*. The conclusions of Gale & Epps (1942) are substantiated in that at acidic pH values more neutral products were formed by the members of the Enterobacteriaceae, whereas at alkaline pH values more acid was formed. It was shown by Gale & Epps (1942) that formate hydrogen lyase activity increased substantially in organisms grown at a low pH value. This is reflected in Table 5 by the higher amounts of hydrogen and carbon dioxide produced by *Escherichia coli* at pH 6.0 as compared with growth at pH 7.8. The influence of pH on the fermentation effected by *Streptococcus liquefaciens* is evident as an increase in acetate and formate production with a concomitant decrease in lactate production as the pH value was increased.

A variation in the pH value of the growth environment also had an

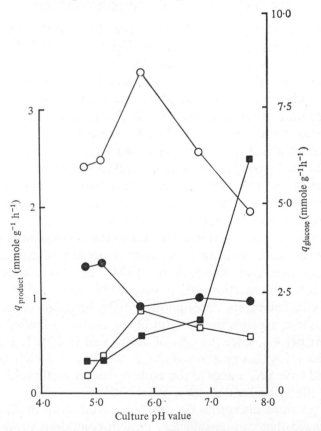

Fig. 2. Effect of culture pH value on overflow metabolism of ammonia-limited *K. aerogenes* NCTC 418 ($D = 0.1$ h^{-1}; 35 °C). Key: ○, glucose; ●, 2-oxoglutarate; □, pyruvate; ■, acetate.

effect on the overflow metabolism of *K. aerogenes* NCTC 418 growing under different limitations in an aerobic glucose-containing chemostat culture. A typical example is shown in Fig. 2. Again the specific rate of acetate production increased with the pH value but the specific rate of production of 2-oxoglutarate and pyruvate did not show the same trend; yet the total specific rate of acid production increased with pH (q_{acid} = 1.9 mmole g^{-1} h^{-1} at pH = 5.0 as compared with q_{acid} = 4 mmole g^{-1} h^{-1} at pH = 7.7). Thus the compensation mechanism also was functioning under aerobic growth conditions, and this mechanism has a strong influence on the production of different metabolites. The choice and control of the culture pH value is therefore an important parameter in the design of a production process.

EFFECT OF UNCOUPLERS OF OXIDATIVE PHOSPHORYLATION

There are several reports in the literature that acetate exerts a toxic effect on microbial growth by prolonging the lag period and/or decreasing the exponential growth rate in a batch culture (Cruess, 1943; Maesen & Lako, 1952; Tseng & Wayman, 1975). These observations led to a study on the influence of acetate on the growth of *Candida utilis* in continuous culture (Hueting & Tempest, 1977). It was found that whereas acetate-limited chemostat cultures of this yeast could be maintained in a steady state at pH values from 2.5 to 7.0, ammonia-limited cultures (growing on acetate) washed out at pH values of 4.8 and lower. Moreover, the specific rate of oxygen consumption in ammonia-limited, acetate-sufficient, chemostat cultures of *Candida utilis* increased with the extracellular acetate concentration when the dilution rate was held constant. This led to the conclusion that acetate, being a weak acid, could function as an uncoupler of oxidative phosphorylation (see also Klingenberg, 1970). In this connection, the extracellular concentration of undissociated acetic acid is crucial; when this increases, its uncoupling activity becomes more effective. This, of course, explains the washout of the ammonia-limited culture of *C. utilis* at pH 4.8, since the pK_a of acetic acid is 4.76. It is clear that other weak acids that possess sufficient solubility in a biological membrane, and have pK_a values in the same region as acetic acid, will exert a similar effect (see Ierusalimsky, 1967).

In the previous discussion on overflow metabolism of *K. aerogenes* it was argued that carbon-sufficient growth conditions promote metabolite production but that concomitantly energy-spilling reactions must be present. Therefore, an uncoupling of oxidative phosphory-

Table 6. Substrate utilization and rates of product formation in chemostat cultures of Klebsiella aerogenes NCTC 418 grown in the presence and absence of 2,4-dinitrophenol

| Limitation | DNP (mM) | D (h^{-1}) | (Milliatom carbon per hour) | | | | | | | | | | | Carbon recovery (%) |
			Cells[a]	Glucose used	CO_2	Pyr	2OG	Ac	GA	2KGA	dLac	Prt	CHO	
Carbon	0	0.08	20	40.1	17.8									94
	1	0.08	20	134	32.1			29.4			21.4	28.1		98
	0	0.11	20	38.4	15.8									93
	1	0.11	20	65.8	43.2			1.2			0			98
Ammonia	0	0.17	20	107	20.2	5.2	22.5	4.0					36.0	102
	1	0.17	20	253	17.4	28.1	17.2	8.1	11.2	143	0		0	97

Pyr = pyruvate; 2OG = 2-oxoglutarate; Ac = acetate; GA = gluconate; 2KGA = 2-ketogluconate; dLac = D-lactate; Prt = extracellular protein; CHO = extracellular polysaccharide.
[a] All values were adjusted to a cell carbon production rate of 20 milliatom carbon per hour. Data of Neijssel (1977).

lation either by acetate or other weak acids generated by overflow metabolism, or by an uncoupler added to the growth medium, should influence metabolite production. This was found to be the case, as shown in Table 6. When 1 mM 2,4-dinitrophenol (DNP) was added to the growth medium the organisms could be maintained in steady state under conditions of glucose limitation, as well as under ammonia-limiting conditions. The specific rate of oxygen consumption was increased as compared with a culture growing at the same dilution rate in the absence of DNP. This uncoupler was found to stimulate over-flow metabolism of the ammonia-limited culture: considerable amounts of 2-ketogluconate and gluconate were excreted in excess of the usual overflow products. Only the synthesis of extracellular polysaccharide was inhibited, which is in agreement with the results reported by Dietzler *et al.* (1975) for non-growing nitrogen-starved cultures of *Escherichia coli*. The extra synthesis of 2-ketogluconate and gluconate is consistent with the explanation given previously for the production of these compounds under phosphate-, potassium- or magnesium-limitation: the uncoupler induces a higher demand for reducing equivalents than is required to allow oxidative phosphorylation to proceed, albeit with a lower efficiency. Interestingly, carbon-limited cultures grown at a low dilution rate excreted D-lactate and protein. The explanation for this finding is not clear but, even under these circumstances, generation of a protonmotive force by respiration must have taken place since the amounts of D-lactate produced and oxygen consumed were not compatible with a completely fermentative mode of metabolism.

Uncouplers of oxidative phosphorylation, therefore, exert a profound influence on the metabolism of microbes. A more powerful uncoupler than DNP or acetic acid could prove to be too toxic for a micro-organism because it would prevent the generation of even a minimal protonmotive force, thereby making the uptake of ions impossible. This may also provide a rationale for the compensation mechanism (i.e. neutralization reactions) proposed by Gale & Epps (1942), for if microbes were to produce large quantities of weak organic acids, in environments with a low pH value, they would poison themselves and, in effect, commit suicide! Another consequence of the uncoupling effect of acetate is that the culture density will influence the growth and metabolism of micro-organisms because the uncoupling activity of a compound is related, among other factors, to its extracellular concentration. Thus, with any specific rate of acetate production, the denser the culture the sooner will a toxic extracellular concentration of acetate be attained (S. Hueting, personal communication). In this connection, it has been shown by Landwall & Holme (1977) that in a dialysis culture of *Escherichia coli*, culture densities of 140 to 150 g dry wt per litre could be achieved, whereas in non-dialysed cultures the maximal culture density on the same medium was about 30 to 50 g per litre. It was demonstrated that growth inhibition under the latter growth conditions was the result of the combined influence of acetate, lactate, pyruvate, succinate, propionate and isobutyrate, which were the end-products of glucose metabolism. By using the dialysis culture, accumulation of all these metabolites in the culture was prevented and growth inhibition did not take place.

EFFECT OF TEMPERATURE

Each micro-organism can grow only within certain limits of temperature. This environmental factor influences the maximum growth rate, macromolecular composition and levels of intracellular metabolites (Tempest & Hunter, 1965). One may therefore expect that metabolite over-production also will be affected by the temperature and, since in any industrial process extensive cooling must be avoided for economic reasons, it is of considerable importance to determine the optimal temperature for the production of a particular metabolite.

It is known that growth temperature has a profound effect on the metabolism of an organism. It has been shown by Chung, Cannon & Smith (1976) that a psychrotrophic strain of *Bacillus cereus* metabolized glucose by simultaneous operation of the Embden–Meyerhof pathway and the pentose phosphate pathway. As the growth temperature was decreased from 32 °C to 7 °C, glucose was metabolized with an increased participation of the pentose phosphate pathway, whereas the shift to a higher temperature of cells grown at low temperatures had the opposite effect. Cells from the late-logarithmic phase grown at 20 °C and 7 °C were able to oxidize acetate to carbon dioxide, whereas cells grown at 32 °C failed to do so. These authors found that when an amount of glucose was added to the growth medium in excess of the growth requirement, this strain of *B. cereus* produced acetate, lactate, pyruvate and minor quantities of ethanol at 32 °C. Lowering the temperature to 20 °C and further to 7 °C showed a decrease of, respectively, 20% and 75% in the level of accumulation of the total extracellular products. The proportion of ethanol increased with decreasing temperature, the proportion of pyruvate excreted remained roughly constant and the proportion of lactate, and particularly acetate, decreased sharply. However, this study suffers from a typical drawback of a common batch culture in that the pH was not controlled during growth and, particularly at 32 °C, there was a sharp decline in its value (from 7.4 to 5.3). In view of this, and the excretion of acetate (see previous sections), it remains to be shown whether the observed effects were the result of the growth temperature alone or of a combination of the effect of temperature and pH.

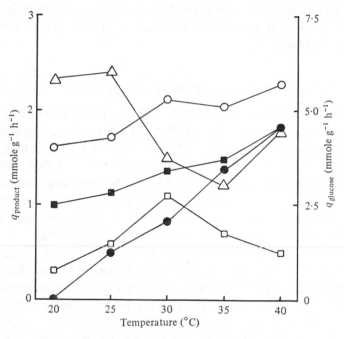

Fig. 3. Effect of temperature on overflow metabolism of ammonia-limited *K. aerogenes* NCTC 418 ($D = 0.17$ h^{-1}; pH 6.8). Key: ○, glucose; ●, 2-oxoglutarate; □, pyruvate; ■, acetate; ,△ polysaccharide expressed as glucose equivalents.

To study the effect of temperature on the overflow metabolism of *K. aerogenes* NCTC 418, this organism was grown in a chemostat under the same conditions as those used for the experiments summarized in Table 1. The growth temperature was varied from 40 °C to 20 °C at intervals of 5 °C. A typical example of the results that were found is shown in Fig. 3. In agreement with the results of Chung *et al.* (1976), it is clear that the production of acetate and 2-oxoglutarate decreased with the growth temperature. In contrast, the specific rate of extra-cellular polysaccharide production decreased when the growth temperature was lowered from 40 °C to 35 °C but at temperatures below

35 °C the production of this compound again increased. In view of the decreased production of metabolites at lower temperatures it is not surprising to find that the specific rate of glucose consumption decreased with temperature. Similar results have been found with chemostat cultures of *K. aerogenes* NCTC 418 grown on glucose and limited by the availability of phosphate or sulphate (unpublished results).

It is generally accepted that a lower temperature results in a lower rate of transport and diffusion of solutes across a biological membrane and, more frequently, in a lowered enzyme activity. Therefore, the lower rate of metabolism as expressed by the lower production rates of 2-oxoglutarate and acetate is not surprising. However, the increased rate of polysaccharide production at lower temperatures is in sharp contrast with this proposal and it is obvious that more subtle metabolic regulatory mechanisms must play a role; however, in general, it can be stated that a higher temperature stimulates metabolite production.

CONCLUSIONS

Our discussions have centred around the behaviour of *Klebsiella aerogenes* in nutrient-limited environments, such as those provided by a chemostat culture as we wished to illustrate some principles which we believe to be important in the physiology of metabolite over-production rather than to give an exhaustive review of the relevant literature. But one should not conclude that all micro-organisms will produce the same range of metabolites as *K. aerogenes* under any particular condition of nutrient limitation. The differences in the contents and properties of enzymes, membrane permeabilities, nutritional requirements (hetero-trophic versus autotrophic) and mechanisms of energy generation (photosynthesis, fermentation and respiration) will exclude such a unity in behaviour. But there is, in our opinion, some unity in the type of behaviour of microbes when growing in nutrient-limited environments. Clearly, they over-produce metabolites only when forced by particular environmental conditions to do so, and they do this in order to resolve physiological problems rather than to oblige the operator!

Regarding the physiological rationale underlying overflow meta-bolism, one might raise the philosophical question as to whether it serves to facilitate uptake of the growth-limiting nutrient, as argued above, or whether it is in fact a detoxification mechanism necessarily invoked as a consequence of the organisms' inability to regulate carbon-substrate uptake adequately. In other words, is overflow metabolism a physiological or a pathological process? The latter possibility cannot be excluded, though if such was the explanation then clearly one must accept the fact that the uptake of carbon-substrate is poorly regulated in very many organisms. Further, one might expect that if overflow metabolism was a consequence of metabolic disorganization, then those organisms that were best able to modulate their rate of carbon-substrate uptake would have a selective growth advantage over other species – but there is no evidence for this. Indeed, it has been found that during the course of experiments with chemostat cultures of *K. aerogenes*, no mutants are selected that express a higher yield value with respect to the carbon-substrate. On the contrary, it has occasionally been observed with slowly-growing ammonia-limited (glycerol-containing) cultures that mutants are selected that excrete vastly increased amounts of 2-oxoglutarate (Tempest, Herbert & Phipps, 1967). One must conclude, therefore, that overflow metabolism provides the optimal adaptation of a microbe to these types of low-nutrient environment and that also under these circumstances a maximal yield of biomass from carbon substrate is not a primary selective factor.

We have discussed the influence of some environmental parameters like pH, uncouplers of oxidative phosphorylation and temperature on overflow metabolism. But there are many more environmental parameters that influence microbial metabolism that have not been discussed; for example, effect of growth rate, osmotic pressure and attachment to surfaces. Moreover, addition to a fermentation of a suitable acceptor for reducing equivalents can alter the fermentation pathway, or addition of a specific inhibitor of some enzyme can provoke bypass reactions in the micro-organism, possibly resulting in a changed metabolism. In other words *any* manipulation of the growth environment will exert an influence on the organism.

It must be clear from the data that have been presented that metabolites can be, and are, over-produced by growing cultures. Thus, it may be that the 'secondary metabolites' from a batch culture serve an important function for the producer organism when it is faced with the environment prevailing in the 'secondary growth phase'. Although the relevance of this function to the overall physiology of the cell is still obscure there is yet no good reason for calling these metabolites 'secondary' since, if they serve a function, they are of primary importance to the cell. An investigation into the function of these metabolites also will provide a better strategy for creating the optimal culture conditions for a maximal production. Thus, only if genetic considerations are combined with environmental considerations will one be able to devise an optimal culturing system for the production of desired compounds.

REFERENCES

AIKING, H., STERKENBURG, A. & TEMPEST, D. W. (1977). Influence of specific growth limitation and dilution rate on the phosphorylation efficiency and cytochrome content of mitochondria of Candida utilis NCYC 321. Archives of Microbiology, 113, 65–72.

ANDERSON, E. S. (1975). Viability of, and transfer of a plasmid from, E. coli K12 in the human intestine. Nature, London, 255, 502–4.

BEHRENS, U., MAKARSKIJ, A., AMBROSIUS, J. & FRANKE, R. (1977). Stickstoffverwertung und Xanthanbildung bei Xanthomonas campestris. Zeitschrift für Allgemeine Mikrobiologie, 17, 339–45.

BLACKWOOD, A. C., NEISH, A. C. & LEDINGHAM, G. A. (1956). Dissimilation of glucose at controlled pH values by pigmented and non-pigmented strains of Escherichia coli. Journal of Bacteriology, 72, 497–9.

BLOND, D. M. & WHITTAM, R. (1965). Effects of sodium and potassium ions on oxidative phosphorylation in relation to respiratory control by a cell-membrane adenosinetriphosphatase. Biochemical Journal, 97, 523–31.

BU'LOCK, J. D. & POWELL, A. J. (1965). Secondary metabolism: an explanation in terms of induced enzyme mechanisms. Experientia, 21, 55–6.

BYERS, B. R. (1974). Iron transport in Gram-positive and acid-fast bacilli. In Microbial Iron Metabolism, A Comprehensive Treatise, ed. J. B. Neilands, pp. 83–105. London, New York: Academic Press.

CALCOTT, P. H. & POSTGATE, J. R. (1972). On substrate accelerated death of Klebsiella aerogenes. Journal of General Microbiology, 70, 115–22.

CALCOTT, P. H. & POSTGATE, J. R. (1974). The effects of β-galactosidase activity and cyclic AMP on lactose accelerated death. Journal of General Microbiology, 85, 85–90.

CHRISTNER, A. (1976). Spezifische Produktbildungsraten von mikrobiellen Produktbildnern. Zeitschrift für Allgemeine Mikrobiologie, 16, 157–72.

CHUNG, B. H., CANNON, R. Y. & SMITH, R. C. (1976). Influence of growth temperature on glucose metabolism of a psychrotrophic strain of Bacillus cereus. Applied and Environmental Microbiology, 31, 39–45.

CRUESS, W. V. (1943). Microorganisms and enzymes in wine making. Advances in Enzymology, 3, 349–86.

DARWIN, C. (1859). The Origin of Species by means of Natural Selection, ed. John Murray. Harmondsworth: Penguin Books.

DAWES, E. A., MIDGLEY, M. & WHITING, P. H. (1976). Control of transport systems for glucose, gluconate and 2-oxo-gluconate, and of glucose metabolism in Pseudomonas aeruginosa. In Continuous Culture 6: Applications and New Fields, eds. A. C. R. Dean, D. C. Ellwood, C. G. T. Evans & J. Melling, pp. 195–207. Chichester: Ellis Horwood for the Society for Chemical Industry.

DAWES, E. A. & SENIOR, P. J. (1973). The role and regulation of energy reserve polymers in micro-organisms. Advances in Microbial Physiology, 10, 135–266.

DICKS, J. W. & TEMPEST, D. W. (1967). Potassium–ammonium antagonism in polysaccharide synthesis by Aerobacter aerogenes NCTC 418. Biochimica et Biophysica Acta, 136, 176–9.

DIETZLER, D. N., LECKIE, M. P., MAGNANI, J. L., SUGHRUE, M. J. & BERGSTEIN, P. E. (1975). Evidence for the coordinate control of glycogen synthesis, glucose utilization, and glycolysis in Escherichia coli. II Quantitative correlation of the inhibition of glycogen synthesis and the stimulation of glucose utilization by 2,4-dinitrophenol with the effects on the cellular levels of glucose 6-phosphate, fructose 1,6-diphosphate and total adenylates. Journal of Biological Chemistry, 250, 7194–203.

EMERY, T. (1974). Biosynthesis and mechanism of action of hydroxamate-type siderochromes. In *Microbial Iron Metabolism, A Comprehensive Treatise*, ed. J. B. Neilands, pp. 107–23. London. New York: Academic Press.

GALE, E. F. & EPPS, H. M. R. (1942). The effect of the pH of the medium during growth on the enzymic activities of bacteria (*Escherichia coli* and *Micrococcus lysodeikticus*) and the biological significance of the changes produced. *Biochemical Journal* **36**, 600–18.

GHEI, O. K. & KAY, W. W. (1973). Properties of an inducible C₄-dicarboxylic acid transport system in *Bacillus subtilis*. *Journal of Bacteriology*, **114**, 65–79.

GLUSKER, J. P. (1968). Mechanism of aconitase action deduced from crystallographic studies of its substrates. *Journal of Molecular Biology*, **38**, 149–62.

GÓMEZ-PUYOU, A., SANDOVAL, F., TUENA DE GÓMEZ-PUYOU, M., PENA, A. & CHÁVEZ, E. (1972). Coupling of oxidative phosphorylation by monovalent cations. *Biochemistry*, **11**, 97–102.

GUNSALUS, I. C. & NIVEN, C. F. (1942). The effect of pH on the lactic acid fermentation. *Journal of Biological Chemistry*, **145**, 131–6.

HAAVIK, H. I. (1976). Possible functions of peptide antibiotics during growth of producer organisms: bacitracin and metal(II) ion transport. *Acta Pathologica et Microbiologica Scandinavica*, Section B, **84**, 117–24.

HERNANDEZ, E. & JOHNSON, M. J. (1967). Anaerobic growth yields of *Aerobacter cloacae* and *Escherichia coli*. *Journal of Bacteriology*, **94**, 991–5.

HILLMER, P. & GEST, H. (1977). H₂ metabolism in the photosynthetic bacterium *Rhodopseudomonas capsulata:* H₂ production by growing cultures. *Journal of Bacteriology*, **129**, 724–31.

HOLLYWOOD, N. & DOELLE, H. W. (1976) Effect of specific growth rate and glucose concentration on growth and metabolism of *Escherichia coli* K12. *Microbios*, **17**, 23–33.

HUETING, S. & TEMPEST, D. W. (1977). Influence of acetate on the growth of *Candida utilis* in continuous culture. *Archives of Microbiology*, **115**, 73–8.

IERUSALIMSKY, N. D. (1967). Bottle-necks in metabolism as growth rate controlling factors. In *Microbial Physiology and Continuous Culture*, eds. E. O. Powell, C. G. T. Evans, R. E. Strange & D. W. Tempest, pp. 23–33. London: Her Majesty's Stationery Office.

ITAKURA, K., HIROSE, T., CREA, R., RIGGS, A. D., HEYNEKER, H. L., BOLIVAR, F. & BOYER, H. W. (1977). Expression in *Escherichia coli* of a chemically synthesized gene for the hormone somatostatin. *Science*, **198**, 1056–63.

KLINGENBERG, M. (1970). Metabolite transport in mitochondria: an example for intracellular membrane function. In *Essays in Biochemistry*, **6**, eds. F. Dickens & P. N. Campbell, pp. 119–159. London, New York: Academic Press.

KOCH, A. L. (1971). The adaptive responses of *Escherichia coli* to a feast and famine existence. *Advances in Microbial Physiology*, **6**, 147–217.

KREBS, H. A. (1972). The Pasteur effect and the relations between respiration and fermentation. In *Essays in Biochemistry*, **8**, eds. P. N. Campbell & F. Dickens, pp. 1–34. London, New York: Academic Press.

KUBICEK, C. P. & RÖHR, M. (1977). Influence of manganese on enzyme synthesis and citric acid accumulation in *Aspergillus niger*. *European Journal of Applied Microbiology*, **4**, 167–75.

LANDWALL, P. & HOLME, T. (1977). Removal of inhibitors of bacterial growth by dialysis culture. *Journal of General Microbiology*, **103**, 345–52.

LEWIS, A. J. & MILLER, J. D. A. (1975). Keto acid metabolism in *Desulfovibrio*. *Journal of General Microbiology*, **90**, 286–92.

MAESEN, Th. J. M. & LAKO, E. (1952). The influence of acetate on the fermentation of baker's yeast. *Biochimica et Biophysica Acta*, **9**, 106–7.

MARTIN, J. F. & DEMAIN, A. L. (1976). Control by phosphate of candicidin production. *Biochemical and Biophysical Research Communications*, **71**, 1103–9.

MARTIN, J. F. & DEMAIN, A. L. (1977). Effect of exogenous nucleotides on the candicidin fermentation. *Canadian Journal of Microbiology*, **23**, 1334–9.

NEIJSSEL, O. M. (1977). The effect of 2,4-dinitrophenol on the growth of *Klebsiella aerogenes* NCTC 418 in aerobic chemostat cultures. *FEMS Microbiology Letters*, **1**, 47–50.

NEIJSSEL, O. M. & TEMPEST, D. W. (1975). The regulation of carbohydrate metabolism in *Klebsiella aerogenes* NCTC 418 organisms, growing in chemostat culture. *Archives of Microbiology*, **106**, 251–8.

NEIJSSEL, O. M. & TEMPEST, D. W. (1976). The role of energy-spilling reactions in the growth of *Klebsiella aerogenes* NCTC 418 in aerobic chemostat culture. *Archives of Microbiology*, **110**, 305–11.

NEIJSSEL, O. M., HUETING, S., CRABBENHAM, K. J. & TEMPEST, D. W. (1975). Dual pathways of glycerol assimilation in *Klebsiella aerogenes* NCIB 418. Their regulation and possible functional significance. *Archives of Microbiology*, **104**, 83–7.

NEIJSSEL, O. M., HUETING, S. & TEMPEST, D. W. (1977). Glucose transport capacity is not the rate-limiting step in the growth of some wild-type strains of *Escherichia coli* and *Klebsiella aerogenes* in chemostat culture. *FEMS Microbiology Letters*, **2**, 1–3.

NELSON, D. L. & KENNEDY, E. P. (1972). Transport of magnesium by a repressible and a nonrepressible system in *Escherichia coli*. *Proceedings of the National Academy of Sciences, USA*, **69**, 1097–3.

OH, Y. K. & FREESE, E. (1976). Manganese requirement of phosphoglycerate phosphomutase and its consequences for growth and sporulation of *Bacillus subtilis*. *Journal of Bacteriology*, **127**, 739–46.

PIRT, S. J. (1969). Microbial growth and product formation. In *Microbial Growth*, eds. P. Meadow & S. J. Pirt, pp. 199–211. *19th Symposium of the Society for General Microbiology*, Cambridge University Press.

POSTGATE, J. R. & HUNTER, J. R. (1963). The survival of starved bacteria. *Journal of Applied Bacteriology*, **26**, 295–306.

POSTGATE, J. R. & HUNTER, J. R. (1964). Accelerated death of *Aerobacter aerogenes* starved in the presence of growth-limiting substrates. *Journal of General Microbiology*, **34**, 459–73.

REYNOLDS, H. & WERKMAN, C. H. (1937). The intermediate dissimilation of glucose by *Aerobacter indologenes*. *Journal of Bacteriology*, **33**, 603–14.

RHOADS, D. B. & EPSTEIN, W. (1977). Energy coupling to net K^+ transport in *Escherichia coli*. *Journal of Biological Chemistry*, **252**, 1394–1401.

ROSENBERG, H. & YOUNG, I. G. (1974). Iron transport in the enteric bacteria. In *Microbial Iron Metabolism, A Comprehensive Treatise*, ed. J. B. Neilands, pp. 67–82. London, New York: Academic Press.

ROSENBERGER, R. F. & ELSDEN, S. R. (1960). The yields of *Streptococcus faecalis* grown in continuous culture. *Journal of General Microbiology*, **22**, 726–39.

TANAKA, S. & LIN, E. C. C. (1967). Two classes of pleiotropic mutants of *Aerobacter aerogenes* lacking components of a phosphoenolpyruvate-dependent phosphotransferase system. *Proceedings of the National Academy of Sciences, USA*, **57**, 913–19.

TANAKA, S., LERNER, S. A. & LIN, E. C. C. (1967). Replacement of a phosphoenolpyruvate-dependent phosphotransferase by a nicotinamide adenine dinucleotide-linked dehydrogenase for the utilization of mannitol. *Journal of Bacteriology*, **93**, 642–8.

TEMPEST, D. W., HERBERT, D. & PHIPPS, P. J. (1967). Studies on the growth of

Aerobacter aerogenes at low dilution rates in a chemostat. In *Microbial Physiology and Continuous Culture*, eds. E. O. Powell, C. G. T. Evans, R. E. Strange & D. W. Tempest, pp. 240–54. London: Her Majesty's Stationery Office.

TEMPEST, D. W. & HUNTER, J. R. (1965). The influence of temperature and pH value on the macromolecular composition of magnesium-limited and glycerol-limited *Aerobacter aerogenes* growing in a chemostat. *Journal of General Microbiology*, **41**, 267–73.

TEMPEST, D. W., MEERS, J. L. & BROWN, C. M. (1970). Influence of environment on the content and composition of microbial free amino acid pools. *Journal of General Microbiology*, **64**, 171–85.

TEMPEST, D. W. & NEIJSSEL, O. M. (1976). Microbial adaptation to low-nutrient environments. In *Continuous Culture 6: Applications and New Fields*, eds. A. C. R. Dean, D. C. Ellwood, C. G. T. Evans & J. Melling, pp. 283–96. Chichester: Ellis Horwood for the Society for Chemical Industry.

TEMPEST, D. W. & NEIJSSEL, O. M. (1978). Eco-physiological aspects of microbial growth in aerobic nutrient-limited environments. *Advances in Microbial Ecology*, **2** (in press).

TSENG, M. M. C. & WAYMAN, M. (1975). Kinetics of yeast growth: inhibition threshold substrate concentrations. *Canadian Journal of Microbiology*, **21**, 994–1003.

WHITE, G. E., COONEY, C. L., SINSKEY, A. J. & MILLER, S. A. (1976). Continuous culture studies on the growth and physiology of *Streptococcus mutans*. *Journal of Dental Research*, **55**, 239–43.

WHITING, P. H., MIDGLEY, M. & DAWES, E. A. (1976a). The regulation of transport of glucose, gluconate and 2-oxogluconate and of glucose catabolism in *Pseudomonas aeruginosa*. *Biochemical Journal*, **154**, 659–68.

WHITING, P. H., MIDGLEY, M. & DAWES, E. A. (1976b). The role of glucose limitation in the regulation of the transport of glucose, gluconate and 2-oxogluconate, and of glucose metabolism in *Pseudomonas aeruginosa*. *Journal of General Microbiology*, **92**, 304–10.

WILLIAMS, A. G. & WIMPENNY, J. W. T. (1978). Exopolysaccharide production by *Pseudomonas* NCIB 11264 grown in continuous culture. *Journal of General Microbiology*, **104**, 47–57.

WILLSKY, G. R. & MALAMY, M. H. (1974). The loss of the phoS periplasmic protein leads to a change in the specificity of a constitutive inorganic phosphate transport system in *Escherichia coli*. *Biochemical and Biophysical Research Communications*, **60**, 226–33.

YAMADA, T. & CARLSSON, J. (1975). Regulation of lactate dehydrogenase and change of fermentation products in streptococci. *Journal of Bacteriology*, **124**, 55–61.

THE INFLUENCE OF MICROBIAL PHYSIOLOGY ON REACTOR DESIGN

ROBERT K. FINN* AND ARMIN FIECHTER†

*School of Chemical Engineering, Cornell University, Ithaca,
NY 14853, USA
†Mikrobiologisches Institut, Eidgen. Techn. Hochschule,
Weinbergstr. 38, CH-8006, Zurich, Switzerland

INTRODUCTION

A major strength of the fermentation industry has always been the simplicity and versatility of its reactors. A standard stainless steel tank, baffled, aerated and with a set of turbine agitators, can turn out a variety of pharmaceutical products. Such versatility is rare in the chemical industry. A factory making tetraethyl lead, for example, may have to be dismantled when market conditions change but old penicillin plants never die! One simply changes the organism, the medium, and some of the recovery equipment so as to manufacture vitamins or amino acids instead of antibiotics. In such a situation, the design of special fermenters to fit the physiology of particular organisms can hardly be justified.

Nowadays the picture is changing. Production of cheap proteins for animal feed by continuous fermentation does favour radical changes in reactor design. Moreover, some of the new processes use water-insoluble substrates such as petroleum factions, agricultural wastes or hydrogen gas. The production of microbial polysaccharides (see Sutherland & Ellwood, this volume pp. 107–50) involves special problems in agitation because of the unusually high viscosity encountered. Anaerobic processes, like the ethanol fermentation, are making a comeback (see Bu'Lock, this volume pp. 309–25). The cultivation of plant or animal cells also requires some modification of the standard fermenter. An additional impetus for new designs is the realization that the industry is not only diversifying but expanding. Many new fermentation plants will be built, especially in tropical regions, during the remaining years of this century. If ever fresh ideas are to be tried, now is the time.

Much of the innovation in design is directed toward reducing costs. Nevertheless, it is inappropriate in this symposium to catalogue new devices on such a basis, or to argue the economy of bubble columns as against stirred tanks (Schügerl, Lücke & Oels, 1977; Sigurdson &

Robinson, 1977). Rather, attention will be given to the problems of microbial growth and product formation as influencing not only new hardware but also new concepts in overall process design.

There is still need to exploit the full potential of conventional fermenters. To this end, increasing use is being made of dissolved oxygen electrodes (Vincent, 1974) and other probes that relate fermenter performance to biological function. Harrison & Harmes (1972) have provided a useful summary of culture parameters that are conveniently logged by computer. The monitoring of intracellular level of NADH (Harrison & Chance, 1970), while not yet quite routine, promises to afford a more accurate measure of the true redox state of the culture than can be provided by external measurements using noble-metal electrodes. An especially useful index of metabolism is the respiratory quotient (RQ), readily calculated from the metered gas flow rate and analyses for O_2 and CO_2 in the gases leaving the fermenter. It has been used to maximize the yield of baker's yeast (Nyiri, Toth, & Charles, 1975; Nagai, Nishizawa, & Yamagata, 1976). A more sophisticated use of the computer involves component balancing to estimate indirectly the concentration of biomass and growth rate (Humphrey, 1977; Wang, Cooney, & Wang, 1977). Perhaps a simple measurement of ethanol in the off-gases would provide the needed signal for substrate dosage to optimize the cell yield for baker's yeast. In other situations though, RQ could be an indispensable control parameter for following the shifts in metabolism. Formation of exopolysaccharide as a secondary metabolite from methanol is a case in point (Tam & Finn, 1977).

Secondary metabolite formation is much more sensitive to small changes in temperature, pH, or phosphate concentration than growth itself (Demain, 1968; Weinberg, 1974). Because these parameters are easily measured and controlled it is rather surprising that so few processes that make use of a shifting environment have evolved. The influence of temperature on metabolism has been well reviewed (Farrell & Rose, 1967; Hunter & Rose, 1972) but a few examples are worth reciting because the effects can be rather dramatic. Growing cultures of the red bacillus *Serratia marcesens* at 38 °C are colourless, but if one shifts washed cells to 27 °C and promptly adds certain amino acids, the bacteria rapidly produce the pigment prodigiosin in an amount equal to a third of their protein content (Qadri & Williams, 1972). With *Pseudomonas aeruginosa* at 5 °C, glucose was not oxidized beyond 2-ketogluconate, and fumarate was not oxidized beyond pyruvate (Campbell, Gronlund, & Menu, 1968). At 30 °C, oxidation was more complete and these intermediates did not accumulate. To accomplish such shifts to a lower temperature would be troublesome on a large scale but technically feasible. However, there would necessarily be time lags in attaining new temperatures; these would have to be reconciled also with metabolic lags of the sort investigated by Topiwala & Sinclair (1971) in their study of transient responses during continuous culture.

Still another way to make full use of conventional fermenters is by nutrient feeding, an established practice in much of the industry. By controlling the gaseous atmosphere and judiciously adding mineral nutrients in response to the exponential growth of cells, Repaske & Mayer (1976) were able to attain a density of 25 g dry weight per litre of *Alcaligenes eutrophus*. The remarkable feature of their work was not so much the density itself as the fact that an exponential increase in cells could be extended right up to the end of growth, without resorting to special apparatus like dialysis culture. The quantitative as well as qualitative growth requirements must be known in order to accomplish this (Repaske & Repaske, 1976). Their study revealed not only the value of automatic rather than manual controls, but also the fact that knowledge of the physiology is usually lacking. The accumulation of secondary metabolites by a mutant, for example, is much more complex than the growth of a non-fastidious autotroph which forms no toxic by-products.

AERATION AND AGITATION

High and low oxygen levels

In a symposium of the Society ten years ago, Wimpenny (1969) thoroughly discussed the physiological effects of oxygen and carbon dioxide. Since there have now been a number of other reviews (Harrison, 1972, 1973, 1976; Mukhopadhyay & Ghose, 1976), including a recent one on the measurement and significance of the redox potential (Kjaergaard, 1977), only a few more examples from the technical side need be cited in this review.

The respiratory enzymes of most organisms are saturated above a dissolved oxygen level equivalent to 8–10 mmHg; this corresponds to only 5% of the air-saturation value of 160 mmHg. On the other hand, higher levels may be necessary to promote oxidative reactions. Flickinger & Perlman (1977) found that by using oxygen-enriched air to maintain a dissolved level equivalent of 25% air saturation, the maximum yield of dihydroxyacetone from glycerol in a *Gluconobacter* fermentation was almost double the yield from an oxygen-limited fermentation. No change in total cell growth occurred as a result of the enhanced level of oxygen. In fact during the period of oxygen enrichment, the specific growth rate of the *Gluconobacter* declined from 0.23 to 0.06 h^{-1} and cell yields were even somewhat lower than when oxygen was restricted. Similar observations had been made by Bu'Lock & Yuen (1970) regarding effects of oxygen on yields of antibiotics relative to biomass of the mould *Trichoderma viride*. The explanation was offered that higher concentrations of cytochrome P-450, a component of mixed-function oxidases, were formed when oxygen levels exceeded those needed for the normal respiratory enzymes. Actually, in yeast, the P-450 component is produced more abundantly in semi-anaerobic cells (Lindenmayer & Smith, 1964) so that the true cause for altered metabolism may lie in the flavin enzymes, as proposed earlier (Terui & Konno, 1960).

In any case, a suggestion is often made that enriching the air with pure oxygen during a short critical period of the fermentation might pay off substantially. To accomplish this economically, the gases would probably have to be recycled through gas-inducing impellers (Topiwala & Hamer, 1973; Zlokarnik, 1978). The task of oxygen enrichment is not so easy as in activated sludge plants, however. Because sewage is such a dilute nutrient, the volumetric rate of CO_2 production is low compared to the liquid throughput. Consequently, enough of the CO_2 is removed in the effluent liquors to prevent its build-up as a diluent in the gas phase. A waste treatment plant can therefore use oxygen more efficiently than even a multistage fermentation. Closed gas systems, however, are proposed for methane and hydrogen fermentations (Bongers, 1970).

High dissolved oxygen (DO) levels can also be achieved by using air under pressure or by high-energy dispersion of air into the liquid. The ICI pressure-cycle fermenter (Fig. 1 *left*) operates on an air-lift principle, but it is tall enough for oxygen solubility to be significantly enhanced at the bottom where air enters. A plunging jet design (Vogelbusch IZ fermenter, Fig. 1, *right*) can give an oxygen transfer rate of 8 to 12 g O_2 $l^{-1} h^{-1}$ (Schreier, 1975), but at such high aeration capacities the oxygen economy is only 1.8–2.0 kg O_2 kWh^{-1}. The recirculation rate is almost 2 tank volumes per minute, made possible by a specially designed centrifugal pump that incorporates an air bleed. All loop reactors of the above types expose the micro-organisms to fluctuating oxygen concentrations, and the physiological consequences of such fluctuations are now being studied (Katinger, 1976).

At the other end of the scale, very low levels of DO are known to cause complex responses in facultative bacteria (Harrison & Pirt, 1967).

Many of the interesting physiological effects occur at or below the limit of sensitivity of the usual oxygen electrodes, i.e. below 1% of air saturation. The redox potential, E_h, can be used to get at least some measure of DO under such circumstances (Wimpenny, 1976). Akashi *et al.* (1978) have used careful E_h measurements to study the respiration and leucine production of *Brevibacterium*. A critical DO concentration,

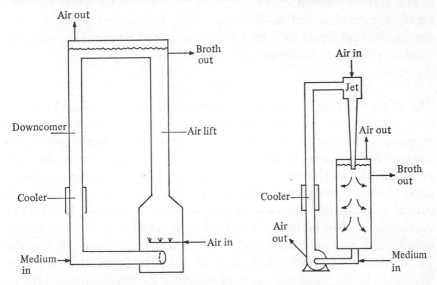

Fig. 1. *Left*, air-lift pressure loop design of Imperial Chemical Industries, Ltd; *right*, plunging jet design of Vogelbusch IZ fermenter.

below which the respiratory demand of the cells was not satisfied, corresponded to −180 mV; but the cells accumulated the largest amount of leucine at −200 mV where the respiratory demand was only 85% satisfied. Completely anaerobic cells formed very little of the amino acid. Oxygen concentration is also a controlling parameter in the production of some exoenzymes and enterotoxins. Thus neither highly aerobic nor anaerobic cultural conditions favoured production of asparaginase by *Escherichia coli* (Boeck *et al.*, 1970). An optimum process involved aerobic growth followed by anaerobic ageing in the presence of glucose. It has also been found that toxin production by *Bacillus cereus* was greatest when the DO was barely detectable (Carpenter, Spira & Silverman, 1976). No measurements of E_h were reported on this system. For toxin production by *Staphylococcus*, a somewhat higher DO level of 10% air saturation was about optimum. Cell formation rather than toxin formation was favoured at still higher levels, corresponding to 50% saturation (Carpenter & Silverman,

1974). In a conventional batch fermenter with provision for constant aeration rate, rather than constant DO level, the appearance of a burst of toxin formation in late-log or early-stationary phase might be misread as a feature of the 'physiological age' of the culture, when in fact the root cause was an uncontrolled variable. Chemostat studies are called for to dissociate the effects of growth rate from other variables.

The role of carbon dioxide cannot be overlooked when studying aeration variables. Pirt & Mancini (1975) found that penicillin production was inhibited 50% by a CO_2 pressure of 60 mmHg under constant conditions of adequate dissolved oxygen.

Shear

The effect of shear on micro-organisms is difficult to describe quantitatively, in part because fragility of the tissue is such a variable property. Plant and animal tissues are very sensitive. Some agitation is essential for maintaining suspension culture but large, thin-walled cells are easily broken (Mandels, 1972). Rapid shaking or stirring of tobacco cells, or even too-vigorous pulsing with air, reduced the cell yield, and caused bulking and foaming (Kato, Shimizu & Nagai, 1975). Only gentle stirring can be used for animal cells, whether in free suspension or attached to microcarrier beads as a film (van Wezel & van der Velden-de-Groot, 1978). Adhesion of bacteria to particle surfaces, important in the biodegradation of cellulose for example, can also be disrupted by excessive stirring (Ng, Weimer & Zeikus, 1977; Latham et al., 1978). Formation of flocs and their stability, important in such diverse areas as brewing and waste treatment, are also affected by mechanical shear. High shear, as might occur in centrifugal pumps, caused release of surface polymers from bacterial cells, and these in turn affected the flocculating properties of the bacteria (McGregor & Finn, 1969). Of special significance are the effects of shear on the morphology and growth of filamentous organisms, both moulds and actinomycetes. On the one hand, shear can damage the filaments causing leakage of low molecular weight nucleotides and reduced cell growth (Tanaka & Ueda, 1975); on the other hand, it is only through shearing action that viscous broths can be adequately aerated (Wang & Fewkes, 1977) or mould fermentations can be maintained in a pulpy condition without the formation of large, metabolically inactive pellets.

Attempts to characterize the shear fields accurately are frustrated by the complexities of turbulence and the further fact that a stirred tank has a region of high shear near the impeller as well as moderate shear throughout. Taguchi (1971) reviewed many of the shear problems relating to fermentation fluids. His own work with mould pellets of *Aspergillus niger* described two processes related to shear. Abrasion or attrition from the pellet surface depended very strongly on both the diameter of the pellet, D_p, and tip speed of the agitator ($N D_i$), where N is stirrer speed and D_i is diameter of the impeller

$$- \frac{dD_p}{dt} \propto (N D_i)^{5 \cdot 5} D_p^{5 \cdot 7}$$

The second process was outright rupture of pellets caused by the extreme velocity fluctuations occurring in the region near the impeller. The disruption could be described by an exponential decay in the number of whole pellets with time. The first-order rate constant for rupture depended on the frequency of circulation through the impeller, $ND_i^3 V^{-1}$, where V is the volume of liquid in the tank. A similar relationship was noted for disruption of the more fragile protozoan cells of *Tetrahymena pyriformis;* the first-order rate constant was proportional to $N \times ND_i^3 V^{-1}$ (Kennedy & Finn, unpublished data). The dependency of disruption on N was ascribed to a relationship for average rate of shear in the impeller zone (Metzner *et al.*, 1961). Other correlations relate shear damage directly to the first power of tip speed, ND_i (Tanaka, Takahashi & Ueda, 1975). Such apparent discrepancies need to be resolved, but it is possible that each model system chosen to describe shear effects has its own characteristic sensitivities.

Out in the bulk fluid of a stirred tank, away from the region of high shear near the impeller, the turbulent shear stress is proportional to $N^2 D_i^2$. Wang & Fewkes (1977) have suggested that in this domain stagnant packets of mycelium may be starved for oxygen unless adequate shearing action is provided. They found, for *Streptomyces* broths, that the specific oxygen uptake rate, Q_{O_2}, as measured directly in the fermenter, varied with dissolved oxygen concentration, as shown schematically in Fig. 2. Even at oxygen concentrations approaching full saturation with air, the *apparent* Q_{O_2} continued to increase. Different lines with different slopes and intercepts were obtained as the size and speed of the impellers were varied. The extrapolated value of DO (for which Q_{O_2} became zero) could be considered a 'pseudocritical DO', C_c^* in Fig. 2. The values for C_2^* were typically 8–30% of air saturation, well above any true critical oxygen concentration for the respiration as would be measured normally with a dilute cell suspension in a Warburg apparatus. This phenomenon has been observed before and can be explained by lack of respiration in the stagnant packets of mycelium. The significant

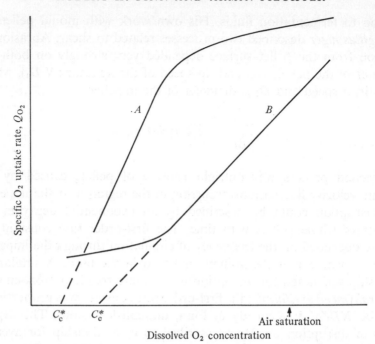

Fig. 2. Schematic diagram of apparent respiration rate for a *Streptomyces* culture, as measured directly in the fermenter. Lines with different slopes and intercepts resulted from the use of different impeller sizes and speeds. The shear : flow ratio was greater for *A* than for *B*. (After Wang & Fewkes, 1977.)

observation was made (Wang & Fewkes, 1977) that the pseudocritical dissolved oxygen decreased with increasing shear:flow ratio for various turbine impellers:

$$\frac{\text{Impeller shear}}{\text{Impeller flow}} \propto \frac{N^2 D_1^2}{N D_1^3} \propto \frac{N}{D_1}$$

This result agrees with earlier observations that small, high-speed stirrers are effective for dispersing air in actinomycete fermentations (Steel & Maxon, 1962).

The rheology of non-Newtonian fermentation broths is now being studied in several laboratories, with an emphasis on better definition of the properties and morphology of the mycelium. Progress in this direction has been reviewed recently by Charles (1977). For example, the inability of a rotating cylindrical bob to measure a meaningful viscosity in a suspension of coarse particles is now recognized. Production of microbial polysaccharides presents special problems because at the end of fermentation the broths attain very high viscosities, amounting to several thousands of centipoises. Aeration and agitation are possible only because of the high degree of shear thinning exhibited by the broth. Even so, the microbe challenges the engineer in these situations. Among the problems for him to solve are provision of uniform distribution of shear throughout the tank (so as to avoid stagnant regions) and allowing for the disengagement of very small bubbles. No detailed reports on the design of fermenters for polysaccharide production have been published. Even some of the basic parameters, such as the rate of rise of a bubble, are not well understood (Tam & Finn, 1974). An additional difficulty, encountered also in growth of filamentous organisms, is that physical properties change so drastically with time. Because of contamination and culture degeneration, multistage continuous operation may be impractical. The possibility though of blowing a batch culture into a finishing tank, equipped with agitators of different design, may merit consideration.

Mixing

In addition to an effect of shear on the aeration of *Streptomyces* broth, Wang & Fewkes (1977) also noted effects attributable to the blending capacities of various impellers. Although their conclusions were tentative, mixing time appeared to be a useful correlating parameter.

Inadequate mixing can also cause difficulty in the cultivation of yeasts on insoluble substrates like alkanes. Flocs form which consist of cells and alkane droplets, with gas bubble inclusions. These flocs rise to the surface and tend to agglomerate there, even at high speeds of a flat-bladed turbine. As a consequence the culture becomes starved for substrate despite an apparent excess. Fig. 3 shows that for batch cultivation of *Candida tropicalis* on hexadecane (Blanch & Einsele, 1973), growth was exponential for only a short period. After the peak in respiration, there was an extended period of linear growth during which hexadecane was only slowly available. Experiments in a chemostat with the same system showed premature washout of cells, i.e. washout occurred at a value of the dilution rate well below that corresponding to maximum specific growth rate. By using a fermenter with rapid recirculation, of the type shown in Fig. 4, such anomalies were less severe because separation of the substrate was avoided.

The degree of mixing was found to be important in chemostat operation at low dilution rates (Hansford & Humphrey, 1966). Slow addition of concentrated glucose solution can surfeit a small fraction of the cells and starve the rest. The adding of substrate onto the free surface of a stirred tank is especially to be avoided, but Hansford & Humphrey (1966) showed that multiple feed was superior to a single

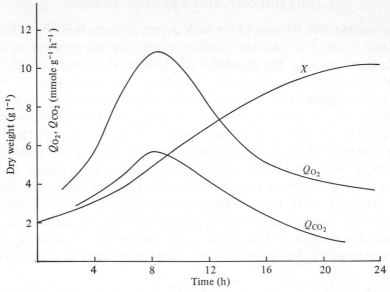

Fig. 3. Growth curve of *C. tropicalis* on hexadecane in batch culture with a flat-bladed turbine in a standard 14 l fermenter. X = dry weight; Q_{O_2}, Q_{CO_2} = specific rates of O_2 uptake and CO_2 production. (Redrawn from Blanch & Einsele, 1973, with permission.)

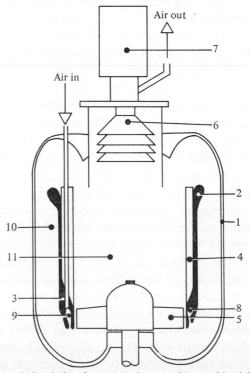

Fig. 4. Cross-section of circulating fermenter that provides rapid mixing. It can be completely filled with medium. Key: 1, vessel wall; 2, 3, draught tube; 4, baffle; 5, stirrer; 6, 7, foam breaker; 8, 9, air sparger; 10, outer space; 11, inner space.

feed even when the latter was near the tip of the impeller. Fig. 5 shows that there may be 'mixing effects' even for turbine agitation at 900 rpm in a bench-scale chemostat. Dropwise addition of glucose solution on

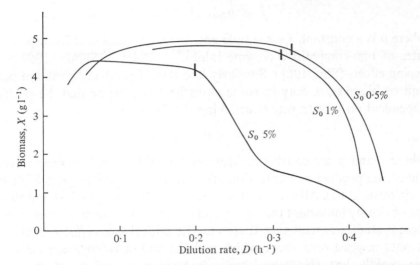

Fig. 5. Continuous cultivation of *S. cerevisiae* (glucose sensitive). Biomass concentration X as a function of D. Critical dilution rate D_R for purely respiratory turnover is marked by a vertical bar. D_R depends on substrate concentration S_0 in the inflowing medium (0.5%, 1% and 5%), indicating limitations of mixing effectiveness. D_R corresponds to maximum productivity (XD). Values for X are calculated for a common value of $S_0 = 1\%$. Yields at dilution rates above D_R are reduced because of repression of respiration. (Fiechter, 1975, with permission.)

to the surface, especially a concentrated solution, resulted in local repression of respiratory enzymes, a decrease in biomass formation, and therefore premature washout. The completely filled reactor (Fig. 4) not only provided more rapid and complete mixing than a stirred tank, but also eliminated the problem of wall growth when cultivating moulds or actinomycetes (A. Keller & A. Fiechter, unpublished).

GROWTH AND PRODUCT FORMATION

Removal of inhibitory products

A major disadvantage of fermentation as compared to other chemical processes is that the desired products are usually formed in dilute aqueous solution. As a consequence, recovery costs often outweigh production costs. If one attempts to ferment a concentrated mash even slow feeding will fail, because the accumulation of concentrated products will inhibit growth and metabolism.

When growth rate declines not because the medium is exhausted but because toxic products accumulate, the underlying mechanisms are usually complex. A simple linear decline in specific growth rate, μ, as a function of product concentration, P, will often fit the data:

$$\mu = \mu_{max} (1-aP),$$

where a is a constant. Less simple equations are fashioned from postulates of non-competitive enzyme inhibition (Aiba & Shoda, 1969) or ageing effects (Shu, 1961). Similarly, the rate of product formation per unit of biomass, ν, may in some cases be constant, or may be partly dependent on growth rate (Luedeking, 1967).

$$\nu = \alpha \mu + \beta$$

where α and β are constants. More commonly β itself decreases with increasing product concentration after growth ceases (Holzberg, Finn & Steinkraus, 1967). All such relationships are still very empirical, but they are obviously important for the correct design of a fermentation process. Recent studies suggest that there may be critical *intracellular* levels of alcohol in yeast cells above which viability and dehydrogenase activity are rapidly lost (Nagodawithana & Steinkraus, 1976). More such physiological studies are needed; mathematical models of gross behaviour may help a bit to suggest which parameters could be significant (Aborhey & Williamson, 1977) but of course quite diverse models can give equally satisfactory fits to the data.

Dialysis culture

In its simplest form the system consists of a dialysis sac containing the culture immersed in a nutrient reservoir. The culture is not only fed by passage of substrates through the membrane, but metabolic products are also removed. Gerhardt and his students have improved on the scheme over the years (review by Schultz & Gerhardt, 1969). Remarkably high cell densities can be achieved in this manner. The recent designs use direct feed of nutrient solutions to the fermenter so that the sole function of dialysis is to remove inhibitory products. This is accomplished by continuously circulating a portion of the fermenter contents through an external plate-and-frame membrane unit where indirect contact with water is made. A continuous dialysis fermentation system was used to convert the lactose from milk whey into ammonium lactate suitable for ruminant feed or for chemical purposes (Coulman, Stieber & Gerhardt, 1977; Stieber, Coulman & Gerhardt, 1977). Simple kinetic expressions, of the sort outlined above, were successful in modelling performance of the system and in predicting the effects of changes in some operating variables. As the culture of *Lactobacillus* adapted to the whey substrate over many days its maximum specific growth rate increased. Concentrated whey substrate could be fermented because dialysis removed the acidic end-product, and the resulting productivity of the equipment was superior to either a batch process or a normal continuous process. Landwall & Holme (1977) used batch dialysis culture, with limited glucose feeding, to demonstrate the continued growth of *E. coli* B at high cell yields. Even in the absence of inhibition, the molar growth yields were low (due to fermentation rather than respiration) unless oxygen was in excess and glucose was kept at low levels. In non-dialysis culture, growth inhibition was traced not only to acetic acid but to the combined influence of all the metabolic acids.

Other membrane processes, using electrodialysis for example, have not been used much in fermentation, but these should also show significant advantages if problems of membrane fouling can be over-overcome. One awaits selective membranes for particular fermentation products like butanol.

Vacuum fermentation

Inhibition due to volatile products like ethanol can be overcome by conducting the fermentation under reduced pressure and distilling off the alcohol as it is formed (Cysewski & Wilke, 1977; Ramalingham & Finn, 1977). When combined with continuous fermentation and recycle of a yeast cream to maintain high cell densities, very rapid fermentations are possible. Cysewski & Wilke (1977) achieved 82 g l^{-1} h^{-1} of ethanol using a 33% glucose feed at 35 °C and 50 mmHg. Because of the very low pressure needed to distil without thermal damage to the yeast, recovery of alcohol from the vapours can be expensive, especially since the carbon dioxide must ultimately be compressed up to atmospheric pressure for discharge. One scheme for accomplishing this is shown in Fig. 6 (Cysewski & Wilke, 1978). A vapour recompression cycle, operating at an intermediate pressure of 340 mmHg, is used to supply heat for distilling in the fermenter. The non-condensable gases are further compressed to atmospheric pressure and the beer is centrifuged at atmospheric pressure before distillation to 95% (w/v) ethanol. The economic advantage of being able to ferment more concentrated sub-strates lies partly in having less water to evaporate from the fermen-tation residues; these are normally sold as dry feed. The continuous vacuum process with cell recycle is economically much more attractive than conventional batch fermentation, according to Cysewski & Wilke

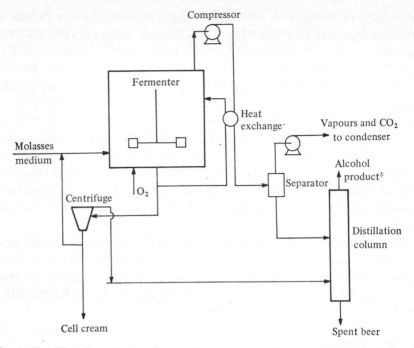

Fig. 6. Simplified flow diagram for continuous vacuum fermentation with cell recycle. Oxygen addition is needed under such highly anaerobic conditions. (After Cysewski & Wilke 1978.)

(1978). The major economies, though, come from continuous operation and cell recycle rather than from the use of vacuum. To the extent that inhibitory salts and caramelized substances from full-strength molasses can be tolerated, it should be possible to recycle the fermenter residues as make-up dilution water for the incoming substrate (Tajima & Yoshizuma, 1975; Topiwala & Khosrovi, 1978). Normal practice is to dilute molasses 1:3 before fermentation, so as to have a sugar content of 15–18% at the start, and thus an alcohol content of 7–9% (w/v) at the end. Any higher final alcohol level would slow down the batch fermentation too much. The amount of steam needed for distillation is excessive below 5% (w/v) alcohol but above about 8% (w/v) the steam requirement begins to level off, on account of the shape of the distillation diagram.

Commendable as it may be to speed up the fermentation and thereby to reduce the size of equipment, a number of physiological as well as engineering problems arise in a vacuum process. Special nutritional needs must be met because of the highly anaerobic conditions and elevated growth temperature. Addition of oxygen to the evacuated mash can allow the cells to synthesize ergosterol and fatty acids, or the recycled cell cream can also be aerated for a time. As one seeks yeast strains with higher temperature tolerance (so as to operate at higher pressures), the nutritional needs become even more stringent; 40 °C is probably an upper limit (Loginova, Gerasimova, & Seregina, 1962). Under vacuum conditions the yield of yeast cells is less than for a normal fermentation. Energy must apparently be diverted for cell maintenance and this represents an economic loss, since the production and sale of by-product yeast is important for any alcohol process.

One way to avoid the need for very low pressures in vacuum fermentation is to use an entrainer like hexane to remove alcohol and water from the fermenting mash. Preliminary evaluation of such a process (Finn & Boyajian, 1976) suggested it might be more economical than a conventional batch fermentation. Later laboratory studies and careful calculation of distillation data (S. Feldman & R. K. Finn, unpublished) showed that such an 'azeoferm' process would offer little advantage over straightforward vacuum distillation without the entrainer. Anaerobically, or with limited air supply, yeast can ferment in the presence of hexane even though aerobic growth is severely inhibited (Finn, 1966). Nevertheless, hexane losses and other complications make the process less attractive than it first seemed.

Solvent extraction and other methods

To remove ethanol from fermenting beer by solvent extraction will prove difficult because the solvents must have some polar character and yet be relatively insoluble in water. Such solvents (amyl alcohol or chlorobenzene) are more highly toxic to the yeast than ethanol itself (Blennemann *et al.* 1963). Lipid cell products, however, can be extracted by higher paraffin hydrocarbons without damage to the cell. One can conduct an emulsion fermentation. Thus prodigiosin was extracted into kerosene so as to produce colourless cells, but no higher yields of pigment resulted (Finn, 1966).

If only small amounts of a toxic by-product are formed, ion-exchange resins can conveniently be used for removal. Harwood & Pirt (1972) adsorbed formaldehyde from a culture of methane-oxidizing bacteria in this manner. Use of adsorbents is generally too expensive, however, for macroproducts of fermentation.

Alternatives for alcohol fermentation

Selection of yeast strains for ability to flocculate and for tolerance to alcohol (Rose, 1976) will certainly improve any alcohol process. It may even be practical to add certain proteolipid supplements to further enhance alcohol tolerance of the yeast (Hayashida, Feng, & Hongo, 1975). Flocculant strains can be used in tower fermenters of the type now being used for beer manufacture in Britain (Greenshields & Smith, 1974) or in fixed film anaerobic columns (Griffith & Compere, 1976). The latter type has been used in a pilot plant for ethanol from sulphite waste liquor (Amberg *et al.* 1968).

Kinetics as a basis for design

Fed batch culture

The various practical aspects of continuous culture should properly be reviewed here, but the task is too great. Fortunately Melling (1977) has provided a recent survey of the interplay between physiology and design for chemostats. It is sometimes observed that sudden shifts in nutrient level or feed rate can result in transient overproduction of some fermentation product. For example, a step-down in dilution rate enhanced the production of β1–3 glucanase over the steady-state level by about 20% (Lilley, Rowley & Bull, 1974). A similar overshoot in formation of acetate kinase was observed by H. M. Koplove & C. L. Cooney (1978). Still other examples have been cited by Harrison & Topiwala (1974). Pirt (1975a) has pointed out that by feeding a batch culture with a sufficiently dilute nutrient solution so that the volume increases, one can simulate such transient performance of a chemostat. If part of the culture is then withdrawn at intervals, a 'repeated fed batch' system can be maintained. Dunn & Mor (1975) have extended the analysis and suggested further applications looking toward overproduction of metabolic products.

Multistage continuous reactors

A number of new fermenters have as their objective the maintenance of plug flow or of a staged pattern which approaches plug flow. These include tubular and column designs with or without partitions. The effort is largely based on an attempt to optimize the system at high conversions by a combination of completely mixed reactor followed by a plug flow reactor. Bischoff (1966) first applied this principle to fermentation, but its advocates often apply it blindly even when there is negligible advantage to be gained. If, for example, a non-growth-associated product is formed, or if biomass production declines gradually with time because a toxic product is accumulating then the system described by Bischoff (1966) has merit (Fig. 7a, b). If growth is governed

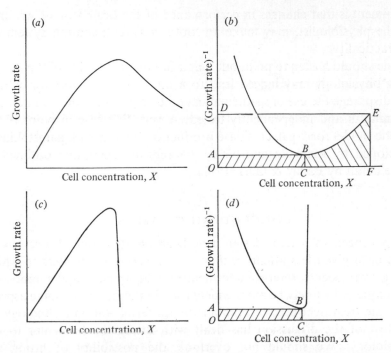

Fig. 7. (a) Growth kinetics showing an autocatalytic rise to maximum rate followed by gradual decline due to inhibition. (b) Reciprocal rate plotted to show sizes of reactors necessary to achieve a cell concentration equal to F. The area $DEFO$ represents the size of a single back-mixed reactor. The same value F can be achieved with two reactors if a back-mixed reactor, $ABCO$, is followed by a plug flow reactor, $BEFC$; in that case the total reactor volume will be less. (See Bailey & Ollis, 1977.) (c) Growth kinetics showing an autocatalytic rise to a maximum followed by precipitous decline to zero rate according to the Monod equation. (d) Reciprocal rate plotted to show that despite the first-order nature of the Monod equation, there is little benefit to be gained beyond a simple back-mixed reactor.

by the Monod equation, however, the plot of growth rate against biomass concentration, X, bends over very sharply (Fig. 7c, d). In such a case there is no merit in using anything more complicated than a single stirred tank. The significant parameter is the numerical value of K_s, the Monod saturation constant. If this is very small compared to the substrate concentration in the feed stream, then there is little advantage to be gained from a combined plug flow system. As Pirt (1975b) points out, almost all K_s values are about 10^{-5} M. Values as high as 10^{-2} M have been reported for growth of yeast and bacteria on methanol (Pilat & Prokop, 1975; Tam & Finn, 1977) or for growth of *Aspergillus niger* on glucose (Kobayashi, Van Dedem & Moo-Young, 1973). Even these values, however, are so low as to afford the plug flow reactor only marginal advantage. One of the hazards facing any designer of equipment is that changes in performance of the organism, as wrought by the physiologist, may make an 'optimum' fermentation system an impracticality.

One should hasten to point out that a full understanding of the underlying physiology may indeed lead to a multistage system. Nor should one disparage the use of mathematical models provided the numbers are meaningful and independently tested. A very fine case in point is the kinetic model for β-galactosidase production by a mould, painstakingly developed by Imanaka and his co-workers in Japan, and outlined in some detail by Bailey & Ollis (1977).

CONCLUDING COMMENT

Many aspects of reactor design have been omitted from this review or have been given less attention than they deserve. Fermenters in which film growth occurs, semi-solid fermentation equipment, and composting techniques will surely be more widely used in the future, as some aspects of waste treatment become combined with fermentation technology.

Most of the discussion has dealt with fitting the fermenter to the physiology. We should not overlook the possibility of fitting the physiology to the fermenter. After all the rumen has turned out to be a pretty good device for bioconversion of renewable resources. The point was cogently made by Professor M. J. Johnson (1971) that environmental control rather than pure culture technique may be a direction for the future. His own work on production of biomass from methane exemplified that approach (Sheehan & Johnson, 1971).

Acknowledgement

Appreciation is extended to the Guggenheim Foundation, New York, for fellowship assistance to permit the authors to work together in Zurich during 1975 and 1976.

REFERENCES

ABORHEY, S. & WILLIAMSON, D. (1977). Modelling of lactic acid production by *Streptococcus cremoris* HP. *Journal of General and Applied Microbiology*, **23**, 7–21.

AIBA, S. & SHODA, M. (1969). Reassessment of the product inhibition in alcohol fermentation. *Journal of Fermentation Technology, Japan*, **47**, 790–4.

AKASHI, K., IKEDA, S., SHIBAI, H., KOBAYASHI, K. & HIROSE, Y. (1978). Determination of redox potential levels critical for cell respiration and suitable for L-leucine production. *Biotechnology and Bioengineering*, **20**, 27–41.

AMBERG, H. R., ASPITARTE, T. R., CORMACK, J. F. & MUGG, J. B. (1968). *US Patent* 3 402 103, September 17, assigned to Crown Zellerbach Co.

BAILEY, J. E. & OLLIS, D. F. (1977). In *Biochemical Engineering Fundamentals*, pp. 554–68. New York: McGraw-Hill.

BISCHOFF, K. B. (1966). Optimal continuous fermentation reactor design. *Canadian Journal of Chemical Engineering*, **44**, 281–4.

BLANCH, H. W. & EINSELE, A. (1973). The kinetics of yeast growth on pure hydrocarbon. *Biotechnology and Bioengineering*, **15**, 861–77.

BLENNEMANN, H., JANOCHA, S., KELLER, H. & NETTER, H. (1963). Die Wirkung organischer Lösungsmittel auf die Funktion der Hefe-Zellmembran. *Hoppe-Seylers Zeitschrift für physiologische Chemie*, **331**, 164–79.

BOECK, L. D., SQUIRES, R. W., WILSON, M. W. & HO, P. P. K. (1970). Effect of glucose and low oxygen tension on L-asparaginase production by a strain of *Escherichia coli B*. *Applied Microbiology*, **20**, 964–9.

BONGERS, L. (1969). Some aspects of continuous culture of hydrogen bacteria. *Developments in Industrial Microbiology*, **11**, 241–55.

BU'LOCK, J. D. & TSE HING YUEN, T. L. S. (1970). Oxygen requirements for secondary metabolism in *Trichoderma viride* and the effect of barbiturate. *Phytochemistry*, **10**, 1835–6.

CAMPBELL, J. J. R., GRONLUND, A. & MENU, P. L. (1968). Low temperature as a means of indicating metabolic pathways. *Journal of Bacteriology*, **95**, 718–19.

CARPENTER, D. F. & SILVERMAN, G. J. (1974). Staphylococcal enterotoxin B and nuclease production under controlled dissolved oxygen conditions. *Applied Microbiology*, **28**, 628–37.

CARPENTER, D. F., SPIRA, W. M. & SILVERMAN, G. J. (1976). Effects of dissolved oxygen tension on bacterial enterotoxin production. *Developments in Industrial Microbiology*, **17**, 363–74.

CHARLES, M. (1977). Technical aspects of the rheological properties of microbial cultures. *Advances in Biochemical Engineering*, **8**, 1–62.

COULMAN, G. A., STIEBER, R. W. & GERHARDT, P. (1977). Dialysis continuous process for ammonium lactate fermentation of whey: mathematical model and computer simulation. *Applied and Environmental Microbiology*, **34**, 725–32.

CYSEWSKI, G. R. & WILKE, C. R. (1977). Rapid ethanol fermentation using vacuum and cell recycle. *Biotechnology and Bioengineering*, **19**, 1125–43.

CYSEWSKI, G. R. & WILKE, C. R. (1978). Process design and economic studies of alternative fermentation methods for the production of ethanol. *Biotechnology and Bioengineering*, **20**, 1421–4.

DEMAIN, A. L. (1968). Regulatory mechanisms and the industrial production of microbial metabolites. *Lloydia*, **31**, 395–418.

DUNN, I. J. & MOR, J. R. (1975). Variable-volume continuous cultivation. *Biotechnology and Bioengineering*, **17**, 1805–22.

FARRELL, J. & ROSE, A. H. (1967). Temperature effects on microorganisms. *Annual Review of Microbiology*, **21**, 101–20.

FIECHTER, A. (1975). Continuous cultivation of yeasts. In *Methods in Cell Biology*, vol. XI, ed. D. M. Prescott, pp. 97–130. London, New York: Academic Press.

FINN, R. K. (1966). Inhibitory cell products: their formation and some new methods of removal. *Journal of Fermentation Technology, Japan*, **44**, 305–10.

FINN, R. K. & BOYAJIAN, R. (1976). Preliminary economic evaluation of the low-temperature distillation of alcohol during fermentation. In *Abstracts of Papers, Fifth International Fermentation Symposium*, ed. H. Dellweg, p. 48. Berlin: Westkreuz-Druckerei und Verlag.

FLICKINGER, M. C. & PERLMAN, D. (1977). Application of oxygen-enriched aeration in the conversion of glycerol to dihydroxyacetone by *Gluconobacter melanogenus* IFO 3293. *Applied and Environmental Microbiology*, **33**, 706–12.

GREENSHIELDS, R. N. & SMITH, E. L. (1974). The tubular reactor in fermentation. *Process Biochemistry*, **9**(4), 11–17; 28.

GRIFFITH, W. L. & COMPERE, A. L. (1976). A new method for coating fermentation tower packing so as to facilitate microorganism attachment. *Development in Industrial Microbiology*, **17**, 241–6.

HANSFORD, G. S. & HUMPHREY, A. E. (1966). The effect of equipment scale and degree of mixing on continuous fermentation yield at low dilution rates. *Biotechnology and Bioengineering*, **8**, 85–96.

HARRISON, D. E. F. (1972). Physiological effects of dissolved oxygen tension and redox potential on growing populations of microorganisms. *Journal Applied Chemistry and Biotechnology*, **22**, 417–40.

HARRISON, D. E. F. (1973). Growth, oxygen, and respiration. *CRC Critical Reviews in Microbiology*, **2**, 185–228.

HARRISON, D. E. F. (1976). The regulation of respiration rate in growing bacteria. *Advances in Microbial Physiology*, **14**, 243–343.

HARRISON, D. E. F. & CHANCE, B. (1970). Fluorometric technique for monitoring changes in the level of reduced nicotinamide nucleotides in continuous cultures of microorganisms. *Applied Microbiology*, **19**, 446–50.

HARRISON, D. E. F. & HARMES, C. S. (1972). Control of culture systems for ultimate process optimization. *Process Biochemistry*, **7**(4), 13–16.

HARRISON, D. E. F. & PIRT, S. J. (1967). The influence of dissolved oxygen concentration on the respiration and glucose metabolism of *Klebsiella aerogenes* during growth. *Journal of General Microbiology*, **64**, 193–211.

HARRISON, D. E. F. & TOPIWALA, H. H. (1974). Transient and oscillatory states of continuous culture. *Advances in Biochemical Engineering*, **3**, 167–219.

HARWOOD, J. H. & PIRT, S. J. (1972). Quantitative aspects of growth of the methane oxidizing bacterium *Methylococcus capsulatus* on methane in shake flask and continuous chemostat culture. *Journal of Applied Bacteriology*, **35**, 597–607.

HAYASHIDA, S., FENG, D. D. & HONGO, M. (1975). Mechanism of formation of high concentration of alcohol in sake brewing. Physiological properties of yeast cells grown in the proteolipid-supplemented medium. *Agricultural and Biological Chemistry*, **39**, 1025–31.

HOLZBERG, I., FINN, R. K. & STEINKRAUS, K. H. (1967). A kinetic study of the alcoholic fermentation of grape juice. *Biotechnology and Bioengineering*, **9**, 413–27.

HUMPHREY, A. E. (1977). Computer-assisted fermentation developments. *Developments in Industrial Microbiology*, **18**, 58–70.

HUNTER, K. & ROSE, A. H. (1972). Influence of growth temperature on composition and physiology of microorganisms. *Journal of Applied Chemistry and Biotechnology*, **22**, 527–40.

JOHNSON, M. J. (1971). Fermentation – yesterday and tomorrow. *Chemtech*, **1**, 338–41.

KATINGER, H. W. D. (1976). Physiological response of *Candida tropicalis* grown on *n*-paraffin to mixing in a tubular closed loop fermentor. *European Journal of Applied Microbiology*, **3**, 103–14.

KATO, A., SHIMIZU, Y. & NAGAI, S. (1975). Effect of initial $k_{L}a$ on the growth of tobacco cells in batch culture. *Journal of Fermentation Technology, Japan*, **53**, 744–51.

KJAERGAARD, L. (1977). The redox potential: its use and control in biotechnology. *Advances in Biochemical Engineering*, **7**, 131–50.

KOBAYASHI, T., VAN DEDEM, G. & MOO-YOUNG, M. (1973). Oxygen transfer into mycelial pellets. *Biotechnology and Bioengineering*, **15**, 27–45.

KOPLOVE, H. M. & COONEY, C. (1978). Acetate kinase production by *Escherichia coli* during steady-state and transient growth in continuous culture. *Journal of Bacteriology*, **134**, 992–1001.

LANDWALL, P. & HOLME, T. (1977). Removal of inhibitors of bacterial growth by dialysis culture. *Journal of General Microbiology*, **103**, 345–52.

LATHAM, M. J., BROOKER, B. E., PETTIPHER, G. L. & HARRIS, P. J. (1978). *Ruminococcus flavifaciens* cell coat and adhesion to cotton cellulose and to cell walls in leaves of perennial ryegrass. *Applied and Environmental Microbiology*, **35**, 156–65.

LILLEY, G., ROWLEY, B. I. & BULL, A. T. (1974). Exocellular β-1,3 glucanase synthesis by continuous-flow cultures of a thermophilic streptomycete. *Journal of Applied Chemistry and Biotechnology*, **24**, 677–86.

LINDENMAYER, A. & SMITH, L. (1964). Cytochromes and other pigments of baker's yeast grown aerobically and anaerobically. *Biochimica et Biophysica Acta*, **93**, 445–61.

LOGINOVA, L. G., GERASIMOVA, N. F. & SEREGINA, L. M. (1962). Requirement of thermo-tolerant yeasts for supplementary growth factors. *Mikrobiologiya (English translation)*, **31**, 21–5.

LUEDEKING, R. (1967). Fermentation process kinetics. In *Biochemical and Biological Engineering Science*, vol. 1, ed. N. Blakebrough, pp. 181–243. London, New York: Academic Press.

McGREGOR, W. C. & FINN, R. K. (1969). Factors affecting the flocculation of bacteria by chemical additives. *Biotechnology and Bioengineering*, **9**, 127–38.

MANDELS, M. (1972). The culture of plant cells. *Advances in Biochemical Engineering*, **2**, 201–15.

MELLING, J. (1977). Regulation of enzyme synthesis in continuous culture. In *Enzyme and Fermentation Biotechnology*, vol. 1. A. Wiseman, pp. 10–42. Chichester: Ellis Horwood.

METZNER, A. B., FEEHS, R. H., RAMOS, H. L., OTTO, R. E. & TUTHILL, J. D. (1961). Agitation of viscous Newtonian and non-Newtonian fluids. *American Institute of Chemical Engineers*, **7**, 3–9.

MUKHOPADHYAY, S. N. & GHOSE, T. K. (1976). Oxygen participation in fermentation: part 1, oxygen microorganism interactions. *Process Biochemistry*, **11**, 19–27; 40.

NAGAI, S., NISHIZAWA, Y. & YAMAGATA, T. (1976). RQ control fed batch culture to enhance the productivity in baker's yeast cultivation. In *Abstracts of Papers, Fifth International Fermentation Symposium*, ed. H. Dellweg, p. 30. Berlin: Westkreuz-Druckerei und Verlag.

NAGODAWITHANA, T. W. & STEINKRAUS, K. H. (1976). Influence of the rate of ethanol production and accumulation on the viability of *Saccharomyces cerevisiae* in 'rapid fermentation'. *Applied and Environmental Microbiology*, **31**, 158–62.

NG, T. K., WEIMER, P. J. & ZEIKUS, J. G. (1977). Cellulolytic and physiological properties of *Cl. thermocellum*. *Archives for Microbiology*, **114**, 1–7.

NYIRI, L. K., TOTH, G. M. & CHARLES, M. (1975). Measurement of gas exchange conditions in fermentation processes. *Biotechnology and Bioenineering*, **17**, 1663–78.

PILAT, P. & PROKOP, A. (1975). The effect of methanol, formaldehyde, and formic acid on growth of *Candida boidinii* Bh. *Biotechnology and Bioengineering*, **17**, 1717–28.

PIRT, S. J. (1975a). Batch cultures with substrate feeds. In *Principles of Microbe and Cell Cultivation*, pp. 211–18. Oxford: Blackwell Scientific.

PIRT, S. J. (1975b). In *Principles of Microbe and Cell Cultivation*, pp. 10–12. Oxford: Blackwell Scientific.

PIRT, S. J. & MANCINI, B. (1975). Inhibition of penicillin production by carbon dioxide. *Journal of Applied Chemistry and Biotechnology*, 25, 781–3.

QADRI, S. M. H. & WILLIAMS, R. P. (1972). Induction of prodigiosin biosynthesis after shift-down in temperature of nonproliferating cells of *Serratia marcescens*. *Applied Microbiology*, 23, 704–9.

RAMALINGHAM, A. & FINN, R. K. (1977). The vacuferm process: a new approach to fermentation alcohol. *Biotechnology* and *Bioengineering*, 19, 583–9.

REPASKE, R. & MAYER, R. (1976). Dense autotrophic cultures of *Alcaligenes eutrophus*. *Applied and Environmental Microbiology*, 32, 592–7.

REPASKE, R. & REPASKE, A. C. (1976). Quantitative requirements for exponential growth of *Alcaligenes eutrophus*. *Applied and Environmental Microbiology*, 32, 585–91.

ROSE, D. (1976). Yeasts for molasses alcohol. *Process Biochemistry*, 11(3), 10–12; 36.

SCHREIER, K. (1975). Neuer Hochleistungsfermenter nach dem Tauchstrahlverfahren. *Chemische Zeitung*, 99, 328–33.

SCHÜGERL, K., LÜCKE, J. & OELS, U. (1977). Bubble column bioreactors: tower bioreactors without mechanical agitation. *Advances in Biochemical Engineering*, 7, 1–84.

SCHULTZ, J. S. & GERHARDT, P. (1969). Dialysis culture of microorganisms: design, theory, and results. *Bacteriology Reviews*, 33, 1–47.

SHEEHAN, B. T. & JOHNSON, M. J. (1971). Production of bacterial cells from methane. *Applied Microbiology*, 21, 511–15.

SHU, P. (1961). Mathematical models for the product accumulation in microbial processes. *Journal of Biochemical and Microbiological Technology and Engineering*, 3, 95–109.

SIGURDSON, S. P. & ROBINSON, C. W. (1977). A comparison of oxygen transfer power requirements and economy in nonmechanically and mechanically agitated fermenters. *Developments in Industrial Microbiology*, 18, 529–47.

STEEL, R. & MAXON, W. D. (1962). Some effects of turbine size on novobiocin fermentation. *Biotechnology* and *Bioengineering*, 4, 231–40.

STIEBER, R. W., COULMAN, G. A. & GERHARDT, P. (1977). Dialysis continuous process for ammonium lactate fermentation of whey: experimental tests. *Applied and Environmental Microbiology*, 34, 733–9.

TAGUCHI, H. (1971). The nature of fermentation fluids. *Advances in Biochemical Engineering*, 1, 1–30.

TAJIMA, K. & YOSHIZUMA, H. (1975). Mechanisms of abnormal fermentation of distiller's yeast in salted media (such as molasses media) from the point of NAD(P) redox balances. *Journal of Fermentation Technology, Japan*, 53, 841–53.

TAM, K. T. & FINN, R. K. (1974). Residence time of sphere or air bubble in sheared non-Newtonian fluids. *Nature, London*, 252, 572–4.

TAM, K. T. & FINN, R. K. (1977). Polysaccharide formation by a *Methylomonas*. In *Extracellular Microbial Polysaccharides*, eds. P. A. Sandford & A. Laskin, pp. 58–80. Symposium Series, No. 45. Washington: American Chemical Society.

TANAKA, H., TAKAHASHI, J. & UEDA, K. (1975). A standard for the intensity of agitation shock on mycelia. *Journal of Fermentation Technology, Japan*, 53, 18–26.

TANAKA, H. & UEDA, K. (1975). Kinetics of mycelial growth accompanied by leakage of intracellular nucleotides caused by agitation. *Journal of Fermentation Technology, Japan*, 53, 27–34.

TERUI, G. & KONNO, N. (1960). Analysis of the behaviours of industrial microbes toward oxygen: II Respiration of gaseous and dissolved oxygen by *Aspergillus*. *Technology Reports of Osaka University*, **10**, 889–903.

TOPIWALA, H. H. & HAMER, G. (1973). A study of gas transfer in fermenters. *Biotechnology and Bioengineering Symposium*, **4**, 547–57.

TOPIWALA, H. H. & KHOSROVI, B. (1978). Water recycle in biomass production processes. *Biotechnology and Bioengineering*, **17**, 73–85.

TOPIWALA, H. & SINCLAIR, C. G. (1971). Temperature relationships in continuous culture. *Biotechnology and Bioengineering*, 813.

VAN WEZEL, A. L. & VAN DER VELDEN-DE-GROOT, C. A. M. (1978). Large scale cultivation of animal cells in microcarrier culture. *Process Biochemistry*, **13**(3), 6–8; 28.

VINCENT, A. (1974). Control of dissolved oxygen. *Process Biochemistry*, **9**(6), 30–2.

WANG, D. I. C. & FEWKES, R. C. J. (1977). Effect of operating variables and geometric parameters on the behaviour of non-Newtonian, mycelial, antibiotic fermentations. *Developments in Industrial Microbiology*, **18**, 39.

WANG, H. Y., COONEY, C. L. & WANG, D. I. C. (1977). Computer-aided baker's yeast fermentations. *Biotechnology and Bioengineering*, **19**, 69–86.

WEINBERG, E. D. (1974). Secondary metabolism: control by temperature and inorganic phosphate. *Developments in Industrial Microbiology*, **15**, 70–81.

WIMPENNY, J. W. T. (1969). Oxygen and carbon dioxide as regulators of microbial growth and metabolism. In *Microbial Growth*, eds. P. Meadow & S. J. Pirt, pp. 161–97. *19th Symposium of the Society for General Microbiology*. Cambridge University Press.

WIMPENNY, J. W. T. (1970). Can culture redox potential be a useful indicator of oxygen metabolism by microorganisms? *Journal of Applied Chemistry and Biotechnology*, **26**, 48–9.

ZLOKARNIK, M. (1978). Sorption characteristics for gas-liquid contacting in mixing vessels. *Advances in Biochemical Engineering*, **8**, 133–51.

MICROBIAL EXOPOLYSACCHARIDES – INDUSTRIAL POLYMERS OF CURRENT AND FUTURE POTENTIAL

IAN W. SUTHERLAND AND DEREK C. ELLWOOD*

Department of Microbiology, University of Edinburgh, West Mains Road, Edinburgh EH9 3JG, UK

*Microbiological Research Establishment, Porton Down, Salisbury SP4 0JG, UK

Exopolysaccharides – polysaccharides found outside the microbial cell wall and membrane – are a common product of microbial cells. Despite their widespread occurrence, relatively few have been studied in detail. There are clearly two groups of these polymers: those composed of a single structural unit, *homopolysaccharides*; and those which are constructed from two or more monomers, *heteropolysaccharides*. Representative types of both groups have been studied, their structures determined and applications in industry found. The relative ease with which carbohydrate structures can be determined through the application of combined gas–liquid chromatography and mass spectroscopy techniques now means that a far greater range of microbial isolates can be screened to the stage of structural studies on the polysaccharides, perhaps even before optimal culture conditions have been determined. In this review, we shall attempt to place in perspective the current state of the art for polysaccharides of industrial value and what we consider to be the future for such polymers, as well as considering how to obtain further polysaccharides for industrial use.

Exopolysaccharides are found in two different forms: attached to the microbial cells which synthesize them, as discrete physical structures termed *capsules*; or secreted from the cell into the surrounding environment in the form of *soluble slime*. The two forms can readily be distinguished through negative-staining techniques, among the most useful of which is the India Ink procedure developed by Duguid (1951). At a finer level, several workers have recently used improved electron microscopic techniques to study the nature of polysaccharides and the way in which they are excreted from the cell. Such approaches are still in the development stage and it is not yet possible to relate the nature of the fibres visualized in the electron microscope (e.g. Brooker, 1976,

1979; Bayer & Thurow, 1977) with the chemical and physical properties of the polysaccharides.

CHEMICAL COMPOSITION OF EXOPOLYSACCHARIDES

(1) *Homopolysaccharides*   Extracellular homopolysaccharides form two distinct groups, depending on their site of synthesis. *Dextrans* and *levans* differ from other homopolysaccharides in that they are formed from a specific substrate, sucrose, by essentially extracellular biosynthetic processes which probably also involve a suitable acceptor molecule. Although the polyfructans (levans) have not found industrial applications, dextrans are widely used in the pharmaceutical and other industries (see below).

Of the other group of homopolysaccharides synthesized by cell-bound enzymes, few have received attention for possible commercial applications. An exception may be cellulose from *Acetobacter xylinum* and related species; this can be readily labelled through growth of the bacteria on [^{14}C]glucose to provide labelled polymer. This is, however, an extremely specialized application. We have not yet reached the stage where shortage of plant cellulose forces us to prepare the polymer from bacteria instead!

(2) *Heteropolysaccharides*   Most microbial exopolysaccharides are heteropolymers composed of neutral sugars and, commonly, uronic acids. Some may contain aminosugars in place of uronic acids or along with them. In addition, many contain acetyl groups or, more rarely, other acyl groups such as formate and succinate. The other common component of bacterial heteropolysaccharides is pyruvate in the form of a ketal. Such ketals have mainly been identified attached to neutral sugars or, in a few cases (so far), to a uronic acid. Two patterns of structural types have been recognized. A group of closely related strains and species may synthesize a polysaccharide in which the carbohydrate structure is constant but the acyl, and particularly the ketal, groups show considerable variation. This is seen in the production of the polysaccharide, colanic acid, by *Escherichia coli*, *Enterobacter cloaceae* and *Salmonella* spp. (Grant, Sutherland & Wilkinson, 1969). A similar pattern emerges for *Xanthomonas* spp. in which a wide range of isolates, mainly now designated as *X. campestris*, produce a polymer of the type indicated in Fig. 1. The type of ketal does not appear to vary, unlike colanic acid, but variations in the content of acetate and pyruvate occur. More commonly, each strain of a particular bacterial species may

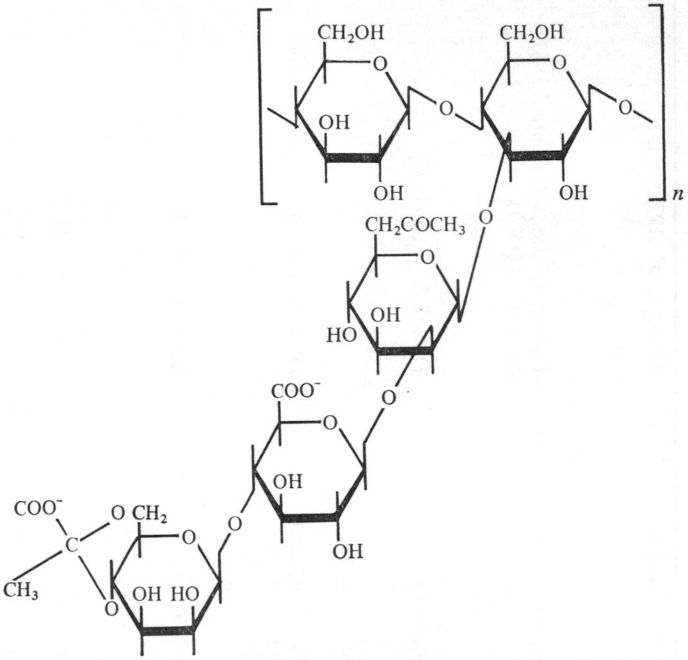

Fig. 1. Structure of xanthan gum.

produce a distinct polysaccharide chemotype. Thus, a wide range of components and structures are found among different isolates as can be exemplified by *Enterobacter* (*Klebsiella*) *aerogenes*, in which intensive studies have led to determination of many of the carbohydrate structures (Table 1). It should be remembered, however, that most of these studies have used a single strain of each serotype; there is thus a possibility that other strains might show slight differences, although this was not observed in the studies on several serotype 2 strains by Gahan, Sandford & Conrad (1967). It is clear that it is not possible to predict the composition or the structure of polysaccharide to be expected from most microbial isolates.

One further possibility to be considered in any assessment of exopolysaccharide-producing micro-organisms is the production of the same polysaccharide by different bacterial species; in effect a widening of the scope for production of a specific polymer from closely related species such as *Escherichia coli* and *Enterobacter aerogenes*. A recently discovered example of this was the observation of a common polysaccharide produced by *E. coli* type 42 and *E. aerogenes* type 63 (Niemann *et al.*, 1978). Another example, involving less closely related bacterial species, is the formation of bacterial alginate by certain *Pseudomonas aeruginosa*

Table 1. *Repeating unit structures of* Klebsiella *exopolysaccharides*

Type	Components	Structure	Reference
1	Fucose, glucose, glucuronic acid, pyruvate	-3Glcβ1-4GlcAβ1-4Fucβ1- 2||3 Pyr	Erbing et al. (1976)
2	Mannose, glucose, glucuronic acid (various acyl groups)	-4Glcα1-3Glcα1-4Manβ1- 1|3 Glcα	Gahan, Sandford & Conrad (1967) Sutherland (1967)
5	Mannose, glucose, glucuronic acid, pyruvate, acetate	-4GlcAβ1-4Glcβ1-3Man1- 4||6 Ac Pyr	Dutton & Yang (1973)
7	Mannose, glucose, galactose, glucuronic acid, pyruvate	-3GlcAβ1-2Manα1-3Glcβ1- 4|3 4||6 1|Gal Pyr Gal	Dutton, Stephan & Churms (1974)
8	Glucose, galactose, glucuronic acid (acetate, pyruvate)	-3Galβ1-3Glcα1-3Glcβ1- | GlcA	Sutherland (1970)
11	Galactose, glucose, glucuronic acid, pyruvate	-3Glcβ1-3GlcAβ1-3Galα1- 4|1 Galα 4||6 Pyr	Thurow et al. (1975)

Type	Components	Structure	Reference
16	Fucose, glucose, glucuronic acid, galactose	-3Glcβ1-4GlcAα1-4Fucα1- $\qquad\qquad\qquad$ |4 $\qquad\qquad\qquad$ 1| $\qquad\qquad\qquad$ Galβ	Chakraborty et al. (1977)
20	Mannose, glucose, galactose, glucuronic acid	-2Manα1-3Galα1- $\qquad\qquad$ |3 $\qquad\qquad$ 1| $\qquad\qquad$ Galα $\qquad\qquad$ |3 $\qquad\qquad$ 1| $\qquad\qquad$ GlcAβ	Bebault et al. (1973)
21	Mannose, galactose, glucuronic acid, pyruvate	-3GlcAα1-3Manα1-2Manα1-3Galβ1- $\qquad\qquad$ |4 $\qquad\qquad$ 1| $\qquad\qquad$ Galα $\qquad\qquad$ 4||6 $\qquad\qquad\ \,$ Pyr	Choy & Dutton (1973)
24	Mannose, glucose, glucuronic acid	-4GlcAα1-3Manα1-2Manα1-3Glcβ1- $\qquad\qquad$ |2 $\qquad\qquad$ 1| $\qquad\qquad$ Manβ	Choy, Dutton & Zanlunga (1973)
28	Mannose, glucose, galactose, glucuronic acid	-2Galα1-3Manα1-2Manα1-2Glcβ1- $\qquad\qquad$ |2 $\qquad\qquad$ 1| $\qquad\qquad$ GlcAβ $\qquad\qquad$ |3 $\qquad\qquad$ 1| $\qquad\qquad$ Glcβ	Curvall et al. (1975a)

Table 1.—*contd.*

Type	Components	Structure	Reference
32	Rhamnose, galactose, pyruvate	−4Rha1−4Rha1−3Gal1−3Rha1−4Rha1−3Gal1− 2‖3 Pyr	G. G. S. Dutton (unpublished)
36	Rhamnose, glucose, galactose, glucuronic acid, pyruvate	−3Galα1−3Rhaα1−3Rhaα1−2Rhaβ1− 2 1‖ GlcAβ 4 1‖ Glcβ 4‖6 Pyr	Dutton & Mackie (1977)
37	Glucose, galactose, glucuronic acid, lactate	−3Galβ1−4Glcβ1− 4 1‖ Glcα 6 1‖ Lac-4-O . . . . GlcAβ	Lindberg, Lindquist & Lönngren (1977)
47	Rhamnose, galactose, glucuronic acid	−3Galβ1−4Rhaα1− 3 1‖ GlcAβ 4 1‖ Rhaα	Björndahl et al. (1973)

Type	Components	Structure	Reference
52	Rhamnose, galactose, glucuronic acid	-4Rhaα1-2Galα1- 　　　　　　｜3 　　　　　GlcAβ1 　　　　　｜4 　　　　　Rhaα1 　　　　　｜2 　　　　　Galβ1 　　　　　｜3 　　　　　Galβ1	Björndahl et al. (1973)
54	Fucose, glucose, glucuronic acid (various acyl, no pyruvate)	-6Glcβ1-4GlcAα1-3Fuc1- 　　｜4 　　Glcβ1	Conrad et al. (1966) Sutherland (1967, 1970)
56	Rhamnose, glucose, galactose, pyruvate	-3Glcβ1-3Galβ1-3Galα1-3Galβ1- 　　　　　　　　　　｜2 　　　　　　　　　Rhaα1 4‖6 Pyr	Choy & Dutton (1973)
59	Mannose, glucose, galactose, glucuronic acid, acetate	-3Glcβ1-3Galβ1-2Manα1-3Manα1- 　｜4　　　　　⋮　　　　⋮ 　GlcAβ　O-6-Ac　O-6-Ac	Lindberg et al. (1975)
62	Mannose, glucose, galactose, glucuronic acid	-4Glcβ1-2GlcAα1-2Manβ1-3Galα1- 　　　　　　　　　　｜3 　　　　　　　　　1｜ 　　　　　　　　　Manα	Dutton & Yang (1977)
81	Rhamnose, galactose, glucuronic acid	-2Rhaα1-3Rhaα1-4GlcAβ1-2Rhaα1-3Rhaα1-3Galβ1-	Curvall et al. (1975b)

strains (Linker & Jones, 1966; Evans & Linker, 1973) and by *Azoto-bacter vinelandii* (Carlson & Matthews, 1966). Alginates also differ from most extracellular heteropolysaccharides in that their structure is not uniform, whereas, as can be seen from Table 1, most of these polymers are constructed from repeating units ranging in size from disaccharides to hexasaccharides or larger oligosaccharides.

INDUSTRIAL APPLICATIONS OF POLYSACCHARIDES

The industrial value of polysaccharides lies in their capacity for altering the rheological properties of aqueous solutions, either through gelling or through the alteration of their flow characteristics. The behaviour of the polysaccharides in solution may be Newtonian, pseudoplastic or plastic; many polysaccharides exhibit thixotropy, i.e. the solutions are characterized by high viscosity at low stress and decreased viscosity when increased stress is applied. Such characteristics have potential application in a number of industrial requirements. The various possible results of interaction between polysaccharide molecules can also be utilized. These aspects are considered in detail in the review by Rees (1975) and in further papers by Morris *et al.* (1977) and by Morris (1977).

Currently the industrial polysaccharide produced in largest amount, from any source, is starch. It is also the cheapest; US annual production is of the order of 1.4 Mt at a cost of £100–350 t^{-1}. Production and usage of other polysaccharides and gums is on a much smaller scale – alginates and xanthan are currently produced worldwide in quantities of about 8 kt a^{-1} and 5.5 kt a^{-1} respectively. Their cost is also much higher than that of starch, approximately £4400 t^{-1} and £6100 t^{-1} respectively, depending on grade etc.

The advantage of producing polysaccharides from micro-organisms rests in their assured production and quality, unaffected by marine pollution, tides, weather or, as in the case of gum acacia, by war, famine and drought. Polymer production can be controlled within precise limits and the scale of production can be geared to the market; location of production can also be arranged to utilize convenient or cheap substrates. One disadvantage is the high cost of installation and start-up of fermentation equipment together with large solvent requirements and the associated need for considerable amounts of energy. These have recently been considered elsewhere (Lawson & Sutherland, 1978), and it appears that most of the disadvantages can be overcome given sufficient application, money and ingenuity.

NON-FOOD USES

Enhanced oil recovery

The greatest single potential market for exopolysaccharide produced commercially is that provided by the oil industry if they adopt such polymers for tertiary oil recovery. A word of warning sounded earlier (*Oil and Gas Journal*, 1976) should not, however, be forgotten – 'Enhanced recovery progress may be tougher, more costly and slower than expected.' The polymers are required to improve water-flooding

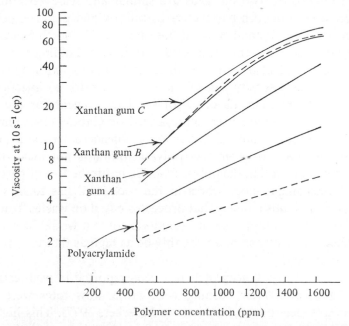

Fig. 2. Comparison of *Xanthamonas campestris* polysaccharide (xanthan) and polyacrylamide viscosities under reservoir conditions. Symbols: ——, 1% NaCl;– – – –, reservoir brine (1.64% NaCl, 0,14% CaCl$_2$; 0.18% MgCl$_2$.6H$_2$O). (From Sandvik & Maerker, 1977).

techniques in which the aqueous solution of polymer gives an increased efficiency of contact with, and displacement of, oil. Alternatively, they will be utilized in micellar–polymer systems. The role of the polymer is to reduce the flow capacity of the solution in the rock system, either by increasing the viscosity of the solution or decreasing the permeability of the system (Sandvik & Maerker, 1977). *X. campestris* polysaccharide has been compared with other polymers, such as polyacrylamide, at concentrations of 200–1500 ppm in salinities normally encountered in oil reservoirs (Fig. 2); the bacterial polysaccharide produced higher viscosity and lower sensitivity to saline than did the synthetic polymers.

Current drawbacks to its use include price. It has been suggested that polymer requirements might rise to 4.5 kt a^{-1}, but the exopolysaccharide would have to compete with established polymers, such as petroleum sulphonate, for which manufacturing plants with considerable capacity already exist (*Oil and Gas Journal*, 1976).

Until satisfactory field experiments have been carried out, the potential market for exopolysaccharides in this area cannot be predicted. There are numerous physical features which have to be quantified for any field and some of these characteristics are indicated by Sandvik & Maerker (1977). No two oil fields are similar and wide variations can be expected even in such parameters as salinity and pH. Temperature, too, can be quite different in locations such as the North Sea and the Persian Gulf; storage temperature of solutions in holding tanks prior to flooding would also differ in different locations, with consequent variations in the viscosity etc. of the solutions prior to injection into the field. The requirements of a polysaccharide agent for use in enhanced oil recovery have been summarized by Gabriel (1979) and it is clear that, although xanthan is currently considered to be a possible agent for this purpose, even this polysaccharide is not ideal. It must also be remembered that the polysaccharide has to be free of particulate matter, which might block pores in the rock, and has to be of high quality; it thus seems unlikely that processes based on wastes from other industries with high utilizable carbohydrate content would be suitable – polysaccharide quality should probably be as high as that used for food additive purposes.

Plugging of the rock near wells has been reported in well tests using xanthan and, even after flushing with brine, flow rates were below those initially determined (Sandvik & Maerker, 1977). This highlights the problems to be expected using polysaccharide preparations containing small amounts of particulate material, e.g. bacterial debris. In the samples used, the content of such particulate matter was estimated to be about 11%. Failure of polysaccharide to hydrate adequately may also have contributed to the problem of plugging. As has been emphasized, laboratory tests can only provide a poor approximation to the actual problem – the oil well.

Unfortunately, the concentration of polysaccharide in solution does not remain constant during the recovery process. Polymer molecules are removed from solution by chemical adsorption on to the permeable rock structure and by physical trapping in the rock. Both mechanisms effectively increase the quantity of polysaccharide required for the recovery process. While experiments have been carried out on the effect of these processes on polyacrylamides (Sandvik & Maerker, 1977), further data is required for microbial polysaccharides. Differences may also be expected for polysaccharides with differing charge:mass ratios.

Drilling muds

In the oil industry, drilling muds are aqueous suspensions containing clays and other colloidal materials. They perform two roles – lubrication of the drill and counterbalancing the upward pressure of oil. The muds must be flexible in their viscosity characteristics while at the same time lubricating the drilling bit. Any polysaccharide component of the drilling mud should exhibit high viscosity, pseudoplasticity and a viscosity spectrum which is relatively insensitive to temperature. Formulations including xanthan have been patented (Table 5) and some of that currently produced is incorporated into muds in use in various drilling locations. Glucan from *Sclerotium* is also marketed for this purpose (p. 127). Although less polysaccharide is used in drilling muds than would be needed for enhanced oil recovery, it still represents a novel use for this polysaccharide and a market which is not likely to diminish in the foreseeable future. The market size for microbial polysaccharide was recently quoted as 1830 t a^{-1} (Wells, 1977) although consumer resistance to price may lead to use of cheaper alternatives with poorer specifications.

Other non-food uses for microbial polysaccharides are likely to come from increased replacement of the traditional plant exudate gums or of modified polysaccharides such as carboxymethylcellulose. As some of these polymers are relatively cheap, the microbial polysaccharides, to compete successfully, should possess better characteristics in respect of stability at high or low pH values, gelling ability at lower temperature or in lower concentrations, or better synergism with other polysaccharides. The reactivity of the polysaccharide with salts or with dyestuffs and other highly charged ingredients is also extremely important. The properties of polysaccharides which have resulted in industrial applications are summarized in Table 2.

Most polysaccharides are used either as stabilizers, leading to improved suspension or dispersion of particulate material in aqueous mixtures, or as thickeners. Other major roles for the polymers are as film-forming agents or to assist water retention. These four roles probably account for 75–80 % of polysaccharide usage. The distribution of use by industry is much more uniform, as can be seen in Table 3.

Although perhaps less well understood, a further use of polysaccharides is as a lubricant. The polysaccharide solution, along with other

Table 2. *Some properties and applications of polysaccharides*

Use of colloidal properties

Stabilizer in foods such as ice cream, salad dressings and fruit drinks.
Stabilizer in emulsion paints.
Suspending agent to maintain uniform suspensions of particles such as solids in
    ceramic glazes, mica in wallpaper printing.

Gel formation

In foods – milk desserts, confectionery, jellies and animal foods.
In pharmaceutical and cosmetic preparations
In dental impressions.

Surface film formation

Anti-stick properties, releasing casts from moulds.
Binding pharmaceutical tablets.
Hydrophilic barrier creams.
Priming porous surfaces.
Temporary binding of sintered products.
Textile and paper sizing.
Dipping Christmas trees.

Thickening agent

Viscosifier in foods – sauces and syrups.
Root and seed dipping to retain moisture.
Thickening of cosmetic creams, lotions, shampoos and pharmaceutical products.
Thickening of latex, adhesives, textile printing pastes.

Fibre formation

Alginate yarns as temporary threads in assembling knitwear, etc.

solutes, flows as a plug at various flow rates. This property may be observed at high or low polysaccharide concentrations. It can be utilized in a number of ways. One of these is to assist in the extrusion of viscous pastes (squeezing toothpaste out of a tube); another potential

Table 3. *Industrial uses of polysaccharides*[a]

Industry	Amount used (% of total)
Detergent–laundry	16
Textile	14
Adhesive	12
Paper	10
Paint	9
Food	8
Pharmaceutical–cosmetic	7

[a] This represents all types of polysaccharides – plant exudate gums, modified polymers, microbial polysaccharides, etc.
After Wells (1977).

use is for transportation of slurries used for moulding and extrusion processes.

FOOD USES

Polysaccharides are incorporated into foods to function, among other purposes, as suspending agents, as thickeners and as inducers of gel formation. Another role, found in the many foods which are frozen and thawed, is that of controlling ice-crystal formation. Any such polysaccharides must conform to the food additive regulations of the country in which the food will be consumed. The annual market for polysaccharides, other than starch and derivatives of cellulose, is over 36 kt in the US alone, although the widely accepted use of 'convenience' foodstuffs in that country probably exaggerates the potential use in other areas of the world. Polysaccharides are also widely used for incorporation into pet foods either as viscosifying agents or as binders for semi-moist products. In these roles they are not subject to food additive regulations nor are there such strict requirements for good flow characteristics and for good 'mouth feel'.

As most prepared foods undergo heating at some stage in processing, it is important that the physical characteristics of the polysaccharides are not liable to deteriorate with heat treatment. The polysaccharide should also be unaffected by the normal ingredients and pH of the food into which it will be incorporated. It may also have to be compatible with other polysaccharides already used in the particular foodstuff, such as locust bean gum (galactomannan from the seeds of *Ceratonia siliqua*) and gum guar.

Although most polysaccharides are used in instant foods, salad dressings, sauces, whips, toppings, processed cheeses and dairy products, new uses are continually being developed. Incorporation into bread has been claimed to improve its texture and other properties. New blends of instant soups and frozen desserts are continually being developed. In this respect, it is probably easier to incorporate new polysaccharides into new food formulations than to use them to replace existing plant exudate or other gums, as all polysaccharides tend to be multifunctional. They are not only emulsifying agents but also increase the viscosity of the preparations into which they are incorporated. Thus incorporation into an existing formulation may also entail reduction of the content of other polysaccharides or thickening agents already present.

POLYSACCHARIDES MEETING INDUSTRIAL REQUIREMENTS

Bacterial homopolysaccharides

Curdlan A survey of micro-organisms capable of using petrochemicals led to the discovery by Harada and his colleagues (see Harada, 1977) of a strain of *Alcaligenes faecalis* var. *myxogenes* which was able to grow in media containing 10% ethylene glycol. This culture was unique in its ability to produce two polysaccharides, both having significant physical properties. One of them is a glucan containing succinyl and pyruvyl groups (Harada, personal communication); the other is a non-acylated β1–3-linked glucan which has been named *curdlan*. A mutant producing solely curdlan was isolated and shown to be stable. Using this mutant of *A. faecalis*, the polysaccharide was obtained in 50% yield from cultures in defined media.

Curdlan swells in water and forms an elastic, resilient gel on heating; gelling is not reversible. It has been proposed for a number of applications, primarily in the food industry, as a gelling agent, thickener or stabilizer. It was thought to be particularly useful in those foods which are heated with water during production or cooking. As it is not degraded by the animal body, it has been suggested as an ingredient for low calorie foods, particularly for jellies and similar desserts. Its safety as a food additive has been tested in Japan but it has not yet received approval for use by either the US Food and Drugs Administration or the corresponding authorities within the EEC. Non-food applications, utilizing the polysaccharide as a film or fibre as well as its function in binding and gelling agents, have also been suggested (Harada, 1977).

Dextran Although dextrans have been produced commercially from bacteria for longer than any other exopolysaccharides, their share of the polysaccharide market remains relatively small and shows little sign of increasing. This is in part because they have found few applications in the food industry and their authorization as food additives has been allowed to lapse. As a result, a proposal for their removal on the grounds of disuse has been made by the Food and Drugs Administration. Dextrans are α-linked glucans produced by several bacterial species including *Klebsiella* spp., *Acetobacter* spp., streptococci and *Leuconostoc* spp. Most applications have employed products from *Leuconostoc mesenteroides*, particularly the strain NRRL B512(F). The polysaccharide from this strain contains 95% α1–6-linked glucose residues, the remaining 5% being branches attached at the carbon-3 positions (Jeanes, 1974). It thus differs from other dextrans containing higher proportions of α1–3- and α1–4-linkages. The dextrans differ considerably in their physical properties because of the varying amounts of different linkages, as was indicated by Jeanes et al. (1954) (Table 4). The molecular weight range for dextran from a single bacterial strain can also vary widely (e.g. from 1×10^7 to 3×10^8; Arond & Frank, 1954) although that from strain B512 is primarily in the range $3–9 \times 10^7$ (Jeanes, 1974). For most purposes, carefully controlled, mild acid hydrolysis is used to degrade the high molecular weight product, which is then fractionated to give material of the desired size. It may also be possible to use hydrolysis with 'dextranases' as an alternative procedure to obtain lower molecular weight polyglucan (Lawson & Sutherland, 1978).

Table 4. *Properties of dextrans*

Group	Linkage class	Intrinsic viscosity at 25 °C (ml g^{-1})	Nature of polymer	Appearance of 1–2% aqueous solution
1	A,B	1.2–0.6	Very cohesive, tough gum or flocculant	Very turbid
2	A,C	0.5–0.2	Fine or flocculant precipitate	Opalescent solution
3a	A,B	0.9–0.5	Fine or flocculant precipitate or dense gum	Very turbid
3b	C	1.4–0.5	Flocculant precipitate or dense gum	Very turbid
4a	A,B	1.3–1.0	Soft gum	Slightly opalescent
4b	A,B	2.0–1.6	Cohesive, stringy gum	Slightly opalescent
4c	A,C	1.4–0.4	Stringy or fluid gums	Clear or slightly turbid
5a	A,B	1.2–0.6	Short or stringy gums	Turbid or slightly opalescent
5b	A,B	1.0–0.9	Flocculant precipitate or short gum	Slightly to very turbid

Data of Jeanes et al. (1954).

Dextrans have found uses in two major areas (i) as blood expanders; and (ii) as the basis for a wide range of adsorbents for use in the biochemical and pharmaceutical industries as well as in the research laboratories. In addition to their use for plasma substitutes, other applications have been devised, such as the recent suggestion that they be employed to form a hydrophilic layer at the wound surface of extensive burns, where their capacity to adsorb exuded fluids is particularly valuable (Paavolainen & Sindell, 1976). The concept of increasing the circulation life of enzymes in the body fluids by coupling them to form a glycoprotein conjugate has also recently been suggested by Sherwood et al. (1977) and Humphreys & Marshall (1978); the enzymes to be used include carboxypeptidase G and arginase.

Dextrans for use in biological adsorbents are prepared in the same way as other fractions. They may be modified by the introduction of cross-linking to yield a three-dimensional network of polysaccharide chains. As the polysaccharide still contains a high proportion of free hydroxyl groups, it is strongly hydrophilic and can be produced in a bead form capable of swelling in aqueous solutions. Alternatively, alkylation can be introduced to produce lipophilic properties (Pharmacia, 1966). Other derivatives of dextran have been prepared to enable the coupling of molecules for affinity chromatography and other purposes. Derivatives such as carboxymethyl and diethylaminoethyl dextrans form extremely useful ion-exchange adsorbents for the purification of enzymes and other proteins. Dextran derivatives may also be used to form the basic support for insolubilized enzymes.

The production of dextrans is essentially the use of a cell-free enzyme system to form an exopolysaccharide. The substrate does not enter the microbial cell, consequently the complex regulatory processes associated with such uptake mechanisms are not involved (see, for example, Sutherland, 1979). The production of the polysaccharide is rapid and the process is thus not prone to contamination problems, although careful control of the pH is required. Aeration is also unnecessary. Low molecular weight dextran is normally added to the culture fluid prior to inoculation in order to provide receptor molecules for polysaccharide formation. The product is fractionated with organic solvents prior to hydrolysis to yield the desired average molecular weight.

Bacterial heteropolysaccharides

Alginate Current commercial production of alginates uses seaweeds (*Phaeophyceae*) as the source, the copolymer of D-mannuronic acid and L-guluronic acid being found in varying amounts in different parts of the alga. The yield of alginates also depends on the season of the year, higher alginate contents having been found in winter, when conditions for harvesting marine algae are least favourable. The proportions of D-mannuronic acid and L-guluronic acid vary widely from one sample to another, with resultant differences in the properties of the alginates. The structure of the polysaccharide is constant, only 1–4-linkages joining the sugar residues. Recent developments have indicated that it is possible to produce *bacterial* alginates, using *Azotobacter vinelandii*, in which the ratio of the uronic acids can be controlled to some extent (Deavin *et al.*, 1977). An alternative source of bacterial alginate is *Pseudomonas aeruginosa* (Linker & Jones, 1966); because of the pathogenic properties of this bacterial species it has not been adopted for possible commercial use, although high yields of polysaccharide have been obtained from some strains under partially defined culture conditions. Alginates synthesized by *A. vinelandii* and *P. aeruginosa* differ from the traditional algal product in that they contain *O*-acetyl groups which appear to be associated with some of the D-mannuronic acid residues (Davidson, Sutherland & Lawson 1977). Bacterial alginates with appropriate specifications should be capable of replacing algal alginate in many of its applications. Production of the polysaccharide by microbial fermentation would have the considerable advantage of freedom from pollution and assured yields of known composition.

Alginates possess very characteristic properties. Calcium and other divalent cations cause precipitation of alginates from aqueous solution; the free acid form of the polymer is also insoluble. The polyguluronic acid portions of the molecule selectively bind calcium ions, and the mechanical strength of calcium alginate gels is mainly derived from the junctions formed by the polyguluronide blocks (Smidsrød, 1974). For many purposes, only alginates have the desired properties and no other polysaccharides have been found which could replace them. Yarns of calcium alginate can be prepared and can be used as temporary bonding agents in textile finishing, being readily removable in the final stages of manufacture. Calcium alginate can also be incorporated into wound dressings to provide a highly hydrophilic layer at the wound surface (Ciba, 1969). Most of the alginate currently produced is used in less specialized applications, although its unique properties make it particularly suitable for some processes. The textile and paper industries use alginates along with other materials as 'sizes' to alter the surface properties of cloths and paper. This is important prior to printing to enable satisfactory deposition and adherence of dyes and inks.

The largest single industrial use for alginate is in the food industry, which currently takes about 50% of the alginate produced. The major role of the polysaccharide is as a stabilizer for ice-creams, instant desserts, frozen custards, creams and cake mixes. Alginate has also found applications in the manufacture of beer and fruit drinks. In the former it enhances the foam or 'head' while in the latter it assists in the suspension of fruit pulp, thus making the product more appealing to the consumer than it would be if all the solids formed a sludge at the bottom of the container. For both these applications propyleneglycol alginate is used to ensure that the polysaccharide is not precipitated by the acidity of the beverages. Alginates are also used for a number of pharmaceutical preparations, primarily because of their value in forming stable emulsions.

More recently, new uses have been found for alginates. One such is in the coating of tree roots prior to planting to ensure a hydrophilic coating for the roots during transport from the nursery to the planting site. A similar treatment has been developed for Christmas trees prior to sale to extend the period before pine needles fall off and detract from the appearance of the tree – a problem associated with the high temperatures of centrally heated modern homes. Obviously the purity of the polymer is of considerably less importance than in pharmaceutical or food preparations but, as the requirement is basically for a hydrophilic layer, other polysaccharides might well replace alginates for this purpose. More detailed lists of the applications for alginates can be found in an article by McNeely & Pettitt (1972) and a publication from the Ciba company (1969).

Azotobacter polysaccharide Another exopolysaccharide which has recently been proposed as having possible industrial applications, is that from *Azotobacter indicum* (Kang & McNeely, 1977). This polymer contains glucose, rhamnose and a uronic acid in the approximate ratio 6.6:1.5:1, and has a relatively high acetyl content (8–10%). Although the identity of the uronic acid was not reported, an earlier study on an *A. indicum* polysaccharide revealed that the uronic acid present was L-guluronic acid (Haug & Larsen, 1970); this uronic acid was also identified in hydrolysates of polymer from another strain through the use of chromatography on DEAE-paper (Sutherland, unpublished results).

In solution, the *Azotobacter indicum* polysaccharide of Kang & McNeely (1977) exhibited pseudoplasticity and compatibility with various salts, except at high pH. The properties of the solutions resulted in the suggestion that they might be of value in the formulation of oil drilling muds.

Erwinia polysaccharide A soil isolate later identified as an *Erwinia* species, but whose identification with an authentic species has not yet been reported has been developed as a polysaccharide source (Kang *et al.*, 1977). The polymer is only produced in low yield when the bacteria are grown on glucose as carbon substrate but can be obtained optimally using lactose or an enzymic hydrolysate of starch. No structural details have been published for this polymer but it was reported to contain glucose, galactose, glucuronic acid and fucose in the molar ratio 6:4:3.2; acetate is also present. The chemotype is thus the same as that of *Enterobacter aerogenes* type 16 (Chakraborty *et al.*, 1977), of colanic acid and of *Sphaerotilus natans* (Gaudy & Wolfe, 1962), although these polymers differ in their acyl substituents. The viscosity of a 1.5% polysaccharide solution was claimed to be twice that of *Xanthomonas campestris* exopolysaccharide but the conditions of shear, temperature, etc. were not stated (Kang *et al.*, 1977). Unlike the *X. campestris* material, the *Erwinia* polymer (named Zanflo) was affected by pH and temperature changes but had good compatibility with most cations and also possessed good freeze–thaw stability. It has been suggested that, because of its compatibility with basic dyes, Zanflo might find applications in paint and other industries; few production details for this polysaccharide have been made available.

Xanthomonas polysaccharide   The most interesting polysaccharide to be produced from microbial sources in recent years has undoubtedly been the exopolysaccharide of *Xanthomonas campestris* and related strains of *Xanthomonas* formerly designated as species. This was first studied by the research group at the Northern Regional Research Laboratory at Peoria in a series of papers relating to production in batch and, latterly, continuous culture together with information about its physical properties (Jeanes, Pittsley & Senti, 1961; Rogovin, Anderson & Cadmus, 1961; Rogovin, Albrecht & Johns, 1965; Silman & Rogovin, 1970, 1972). The actual structure of the polysaccharide remained unclear until the definitive studies by Jansson, Kenne & Lindberg (1975) and by Melton *et al.* (1977). These indicated that the polymer was a substituted cellulose molecule with trisaccharide side-chains on alternate glucose residues (Fig. 1). The distribution of the pyruvate ketals and *O*-acetyl groups remains less certain, although the position of the ketal attached to the terminal-linked mannose of the side-chains is not in doubt. Sandford *et al.* (1977) obtained preparations of *X. campestris* polysaccharide which they designated high-pyruvate (4.4%) or low-pyruvate (2.5%) and remarked on the similarities of these values to the theoretically expected content, if every alternate or every fourth glucose, respectively, carried a ketal. There does appear to be considerable variation in the pyruvate content of exopolysaccharide from *X. campestris* grown under different conditions and in the polymer from different strains of *X. campestris*. These differences may reflect the availability of the presumed precursor, phosphoenolpyruvate (PEP), and the varying values obtained may derive from a mixture of pyruvylated and non-pyruvylated strands (Sutherland, 1979, and unpublished results). The importance of the pyruvate groups in relation to solution viscosity has been indicated in the studies of Cadmus *et al.* (1976).

X. campestris polysaccharide is distinguished from other exopolysac-charides so far examined not only in its chemical structure but also in its distinctive rheological properties in aqueous solution, as indicated by Morris (1977). The polysaccharide solutions yield stable emulsions with large shear-dependence of viscosity. Under the conditions normally employed in the food and related industries, there is little variation in viscosity with temperature, while, unlike many exopolysaccharides, there is also high salt-tolerance. Morris (1977) suggested that the occurrence of considerable intramolecular structuring led to the thixo-trophic behaviour and the emulsion stability, and that sufficient salt was present in many applications to maintain an ordered conformation over the temperature range normally encountered. The concept of a stiff chain of polysaccharide with an ordered secondary structure was also proposed by Holzworth (1976).

Although much of the work to date has been carried out with poly-saccharide from one particular *X. campestris* strain, NRRL B1469, there is mounting evidence to suggest that polymer from other strains is comparable in its physical and chemical properties (Lesley & Hochster, 1959; Morris *et al.*, 1977; Sutherland, unpublished results). The significance of adequate analytical procedures in testing such products is discussed later.

Industrial production of *X. campestris* polysaccharide has been by batch fermentation. Initial studies used fermenters of up to 900 l capacity with a culture medium containing corn syrup, distiller's solubles and salts (Rogovin *et al.*, 1961), with product recovery by organic solvent precipitation preceded by centrifugation to remove some of the bacterial cells. Although polysaccharide yields were highest when the medium contained more than 1 % glucose, greatest conversion of substrate to polymer occurred using 1 % glucose as carbon-substrate. The production of polysaccharide occurred principally during the first 72 h of culture, but further polymer production and consequently higher culture viscosities were obtained by prolonging incubation.

Xanthan (*X. campestris* polysaccharide) has been widely accepted for both food and non-food uses. The value of xanthan to the food industry is due to its unusual solution properties. It is currently approved by the US Food and Drug Administration, the only microbial exopolysac-charide presently on the list of food additives 'Generally Regarded As Safe' (GRAS-listed). The polysaccharide also has received draft approval from the EEC food regulations, category II (R/1233/73, Agri. 404). The major food use of xanthan is as a stabilizer and it is included as such in French dressing, fruit-flavoured beverages, processed cheese and other dairy products. It is also used in frozen and tinned foods as well as instant desserts, toppings and whips, the inclusion in such formulations frequently being in conjunction with locust bean gum to yield a stable gel. Non-food uses are numerous, a comprehensive list can be found in the survey by Jeanes (1974). This range is illustrated in Table 5.

Table 5. *Patented applications for* Xanthomonas campestris *polysaccharide and derivatives*

Application	Patent holder	Patent number
Oil drilling muds	Exxon	US 3251768 (1966)
Stabilizer – emulsion paints	Tenneco Chemicals	US 3438915 (1969)
Stabilizer – water-based paints	Kelco	US 3481889 (1969)
	Heyden Newport Chemicals	French 1395294 (1965)
Suspending agent for laundry starch	Henkel	US 3692552 (1972)
Carrier for agrochemicals	Kelco	US 3717452 (1973)
Agricultural and herbicidal sprays	Kelco	{ Canada 806643 (1969) US 3659026 (1972)
Metal pickling baths	Diamond Shamrock	US 3594151 (1971)
Gelled detergents	Chemed	US 3655579 (1972)
Gelled explosives	Kelco	US 3326733 (1967)
Waterproof dynamite	Ashland Oil and Refining	US 3383307 (1968)
Clay coatings for paper finishing	Kelco	US 3279934 (1966)
Flocculant for water clarification	Ashland Oil and Refining	US 3342732 (1967)
Derivatives		
Alkylene glycol ester for foods, cosmetics	Kelco	US 3256271 (1966)
Carboxyalkyl ethers for paper, textiles	Kelco	US 3236831 (1966)
Dialkylaminoalkoxy ethers for sizes	Kelco	US 3244695 (1966)
Hydroxyalkyl ethers for cosmetics	Kelco	US 3349077 (1967)
Sulphate for glue thickening	Kelco	US 3446796 (1969)

Fungal Polysaccharides

Many species of *Sclerotium* form exopolysaccharides in relatively simple culture media. The polysaccharide from one strain was identified as a β-linked glucan by Johnson *et al.* (1963) (Fig. 3) and was subsequently developed industrially under the name of 'Polytran' by the Pillsbury Co. This polymer forms a viscous gel in water at a concentration of 1.5% and exhibits pseudoplasticity over a broad range of

$$-3_D Glc\, \beta 1 - 3_D Glc\beta 1 - 3_D Glc\beta 1 -$$
$$|6$$
$$1|$$
$$_D Glc\beta$$

Fig. 3. Proposed structure of *Sclerotium* glucan. (From Johnson *et al.*, 1963.)

temperature and pH values. The polymer resembles the exopolysaccharide of *X. campestris* in being essentially viscostatic, i.e. the viscosity in aqueous solution is remarkably constant over the temperature range 15–90 °C; after exposure to 135 °C for a short time, solution viscosity at lower temperature was recovered. Sclerotium glucan solutions tolerate high salt concentrations and are also compatible with other polysaccharides such as starch, alginate, carrageenan and carboxymethylcellulose. Polytran has been proposed for various uses, some being common to those suggested for xanthan gum. These include addition to printing inks, latex paints, seed coatings, ceramic glazes and drilling muds. The addition of Polytran to glues used in composite boards is claimed to improve penetration control. It has also been recommended as an ideal suspending agent for agricultural herbicides and pesticides.

Table 6. *Microbial polysaccharides of commercial importance*

Name	State of development Present	Future	Trade name	Producer
Dextran	Production	Static or small expansion	Various	Pharmacia Dextran Products Polydex
Xanthan gum	Production	Expanding	Ketrol Kelzan Rhodigel 23	Kelco Rhone Poulenc/ General Mills
Pullulan	Development	Commercialization announced	Pullulan	Hayashibara Corp.
Erwinia exopolysaccharide	Production		Zanflo	Kelco
Scleroglucan	Development		Polytran F. S.	Pillsbury
Microbial alginate	Development		–	Tate & Lyle Ltd
Baker's yeast glycan	Development		BYR	Anheuser-Busch Inc.
Curdlan	Development		–	Takeda Chemical Ind.

Pullulan   The yeast-like fungus *Aureobasidium pullulans* yields an unusual homopolysaccharide which has been termed pullulan. This is composed of maltotriose units linked by $\beta 1$–6-bonds. Highest yields of polysaccharide were obtained from cultures in media containing semihydrolysates of starch, resulting in the conversion of 60–70% of the substrate to polymer. The product is claimed to be in the molecular weight range 10^4–10^6 (Yuen, 1974) depending on the strain and the cultural conditions used. Pullulan has been proposed as a biodegradable material for food coating and packaging; in particular, antioxidant properties have been claimed for films prepared from it. Derivatives could also be prepared by esterification and ethoxylation. Pullulan has also been considered as a suitable ingredient for low-calorie foods in place of starch (Yuen, 1974). The polysaccharide is readily degraded by a number of enzymes and its use would have to be confined to conditions where it was not exposed to enzymic hydrolysis and consequent loss of physical properties.

Other polysaccharides

Most of the individual polysaccharides considered in this section have at least reached the development stage (Table 6) but there are no doubt others which might be worth consideration. They would, however, have to possess properties either as good as those polysaccharides already developed or, in some particular respect, better. If the products are intended for food usage they would require thorough testing prior to acceptance, although there is at least the precedent of X. campestris polysaccharide.

Many of the polysaccharides which have been characterized in respect of their chemical structures are of no commercial value because of their poor salt compatibility or lability to extremes of pH. In Table 7 are listed several polymers which have been reported as having high

Table 7. *Polysaccharides which may be of potential commercial interest*

Organism	Chemotype	Properties of interest	Reference
Rhinocladiella elatior Y6272	N-Acetyl-D-Glucosamine, N-Acetyl-D-Glucosaminuronic Acid (2:1)	Viscosity	Sandford *et al.* (1973; 1975) Watson *et al.* (1976)
Cryptococcus laurentii var. *flavescens*	D-Mannose, D-Xylose, D-Glucuronic acid, Acetate	Plastic rheology, Thixotrophy	Jeanes *et al.* (1964)
Pullularia pullulans	D-Glucose	Film formation	Yuen (1974)
Arthrobacter viscosus	D-Glucose, D-Galactose, D-Mannuronic acid (1:1:1), Acetate	Viscosity	Cadmus *et al.* (1963); Jeanes *et al.* (1965)
Chromobacterium violaceum	D-Glucose, Amino sugar, Uronic acid	Gelatinous solution	Corpe (1960)
Pseudomonas sp. NCIB 11264	D-Glucose, D-Galactose, D-Mannose, (Rhamnose), Acetate, Pyruvate	Viscosity	Williams & Wimpenny (1977)
Pseudomonas sp. SY	D-Mannose, D-Glucose, D-Galactose, Uronic acid, Acetate	High viscosity	Sutherland & Williamson (unpublished)

viscosity or other properties which *might* lead to their further evaluation for industrial applications. There are no doubt many others which could be placed in this category but which have not so far been characterized chemically or physically.

INDUSTRIAL PRODUCTION

The current state of the art – batch culture

As xanthan is currently the major microbial exopolysaccharide produced industrially, it will be used to exemplify current methods. It is produced commercially in a conventional batch process developed from the work of Rogovin *et al.* (1965) on a very large scale using fermenters in the 50–200 m³ range. The carbohydrate substrate is either commerical-grade glucose or degraded starch, while the nitrogen source is an extract of either yeast or soy protein. The fermentation broth is buffered using K_2HPO_4; this salt also provides a source of potassium and phosphate for growth of the bacteria; $MgCl_2$ is also added.

The fermentation conditions are carefully controlled with respect to temperature, pH and length of fermentation. Mixing is especially important to ensure that the optimum rate of oxygen transfer is achieved; this in turn requires specifically designed impellers for the fermentation vessel. The time required for optimal production of polysaccharide and conversion of substrate to polymer is about 80 h; after this time, the culture broth is about pH 6.0 and its viscosity is about 30 000 cp. The polysaccharide is recovered by precipitation with methanol or with propan-2-ol. There is thus an integral requirement for a solvent recovery plant alongside the fermentation plant. Under the conditions described, the yield of polysaccharide, based on the amount of glucose substrate supplied, is between 75% and 80%.

The final product may be either drum- or spray-dried and milled to yield a fine, tan powder. Like all such highly water-soluble polymers, the product is very hygroscopic. The analysis of material produced in pilot plants showed typically: N, 0.4%; P, 0.1%; ash, 12%.

FUTURE AND POTENTIAL DEVELOPMENT

Various aspects of development can be considered. The exopolysaccharides themselves may be altered to improve their commercial acceptability. This, however, has to be set against the cost, especially in terms of energy requirement for any additional process stages. Alternatively, the techniques used to produce the polysaccharides may be altered, with the aim of both improving the product and lowering the cost through improved yields or reduced energy input. Both improved technology and 'genetic' improvement of strains appear to us to be worthy of consideration by the producer of industrial polysaccharides.

(1) *Improved methods of production*

Continuous culture The conventional method of exopolysaccharide production from micro-organisms has been to use batch cultures with relatively long incubation times to ensure maximum production of polysaccharide and maximum utilization of substrate. It would clearly be advantageous to use the continuous-culture mode in terms of production from a given size of fermenter but, until recently, few studies have been made on the application of continuous culture to microbial polysaccharide formation. Recently several groups have commenced such investigations, particularly in the UK where there is less resistance to the concept of continuous culture.

Production of alginic acid As already mentioned, alginate, a compound of considerable industrial importance because of its applications in the food and other industries, is normally isolated from brown algae. It is also produced by two bacterial species, and the employment of continuous culture of *Azotobacter vinelandii* to yield bacterial alginate has been studied in the UK by Tate and Lyle Ltd (Deavin *et al.*, 1977). The bacteria were grown with sucrose as the carbon-source and phosphate as the limiting substrate at a range of specific respiration rates. Polysaccharide concentration was essentially constant, decreasing only at very low respiration rates. The production of polysaccharide as a function of dilution rate was also examined. It was noted that polysaccharide *concentration* was highest at low dilution rates but the *specific rate of production* of the alginate was constant over the range of dilution rates studied.

The effect of differing growth-limiting nutrients was also studied. Molybdate and phosphate gave the most favourable specific rates of polysaccharide synthesis; it was also noted that even under sucrose-limited conditions, polysaccharide was still produced. This observation was most unexpected, as it was thought that when the bacteria were grown under carbon limitation all the available carbon would be used for growth. This result may indicate that polysaccharide production has some specific role in the physiology of *A. vinelandii*, although it seems unlikely that, as was suggested by Neijssel & Tempest (1976) for *E. aerogenes*, it acts as an ATP sink.

As the markets for alginates require a range of products with different physical properties (gelling qualities and viscosity in aqueous solution) there would be little advantage in turning to continuous culture of bacterial alginate unless product quality matched that from the traditional algal sources. It is clear from the studies of Deavin *et al.* (1977) that alginate from *A. vinelandii* grown in continuous culture possesses a comparable range of physical characteristics to those found in currently marketed alginates.

Some experiments with *P. aeruginosa* in continuous culture have also been performed (Mian, Jarman & Righelato, 1978) using nitrogen-limited conditions at dilution rates of 0.05–0.1 h^{-1}. Polysaccharide production at each dilution rate indicated a conversion rate of 56–64% polymer from substrate. The specific rate of polysaccharide production increased with increasing dilution rate. The product had an apparent mannuronic acid:guluronic acid ratio of 4:1 and an acetyl content of 5.2%.

Xanthan production   As far as can be ascertained, all the xanthan currently marketed is produced by batch fermentation, the growth conditions being controlled with respect to pH, oxygen tension etc. In their initial studies with single-stage continuous fermentation Silman & Rogovin (1972) showed that, using a complex medium containing excess glucose, xanthan was produced at $D = 0.026$ h^{-1} at a rate of 0.36 g kg^{-1} h^{-1}. The yield of polysaccharide, based on the glucose consumed, was 68% and the production rate for the xanthan was a function of the pH and the dilution rate. Further experiments showed that in longer runs the selection of variants producing less polysaccharide occurred between 6.5 and 8.7 culture generation times. These results with single-stage fermenters led these workers to suggest the use of multistage fermentation systems but these met with little success.

It should be recognized that all the growth experiments reported by the Peoria group had used complex culture media. Indeed, it had been stated that the use of complex media was mandatory for extracellular polysaccharide production. According to *Bergey's Manual*, complex media are required for the growth of *Xanthomonas* strains, however Starr (1946) has reported that some strains could grow in chemically defined media. Recently, Evans, Yeo & Ellwood (1979) reported that all the strains of *Xanthomonas* which they tested could grow in a simple salts medium with glucose as the sole carbon-source. This result allowed these authors to study the effect of different limitations and other environmental parameters in continuous culture on the production of exopolysaccharide by a number of strains of *Xanthomonas*. The results of these experiments will be considered in more detail.

Xanthomonas juglandis strain XJ107 grew well in the simple salts medium with glucose as the sole carbon-source. The bacteria produced two types of polysaccharide when grown in batch culture, the predominant one being composed of glucose and rhamnose, the other closely resembling that of *X. campestris* and yielding the same products when subjected to partial acid hydrolysis. When the bacteria were grown under carbon limitation, the cultures were not markedly viscous but there was formation of an exopolysaccharide composed of glucose and rhamnose. Exopolysaccharide was also produced under potassium and magnesium limitation, but analysis showed that the material was composed primarily of glucose, with some rhamnose and also some 'xanthan'-like polysaccharide. Nitrogen (NH_4^+), phosphate and sulphate limitations produced cultures which were all highly viscous. The exopolysaccharides were all mixtures of 'xanthan' together with the glucose–rhamnose polymer; the predominant component of all the mixtures was the 'xanthan'-like polysaccharide, the yield being 95% of the total extracellular polysaccharide when the bacteria were grown under sulphate limitation. No culture degeneration was detected in the cultures in sulphate-limited media, polysaccharide production remaining at the same level even when the culture had been continued for 1000 h. It thus seems that sulphate-limited growth of this *X. juglandis* strain precluded the selection of variants unable to produce exopolysaccharides or producing less such material. It has been suggested, although as yet there is no concrete evidence, that the genes coding for exopolysaccharide production are carried by a plasmid. If this is true, sulphate limitation provides conditions under which the plasmid is stabilized. Obviously if this result is applicable to all strains of Xanthomonas, it should be possible to produce xanthan on a large scale using continuous culture.

X. campestris is the 'species' commerically used for the production of 'xanthan'. It also grew well in a simple salts medium with glucose as the carbon-source. In continuous culture under different limitations, only the xanthan type of exopolysaccharide was produced. Again, sulphate limitation gave good yields of extracellular polymer and no culture deterioration was observed. It was also interesting to note that when the bacteria were grown under carbon limitation, xanthan was still produced, albeit in low yield. There is thus a parallel to the production of alginate by *A. vinelandii* in carbon-limited continuous culture. *Pseudomonas* NCIB 11264 was studied by Williams & Wimpenny

(1978) using single-stage continuous culture with glucose as the carbon-source. The optimum conditions of temperature and pH were found to be 30 °C and pH 7.0. Polysaccharide formation was maximal under nitrogen limitation and was dependent on the dilution rate. Phosphate limitation did not enhance the production of polysaccharide and, unlike *X. campestris* and *A. vinelandii*, no polysaccharide was formed when the cultures were carbon limited. The cultures could be run for up to 500 h without the development of mutant strains and they did not deteriorate in terms of polysaccharide production. The authors concluded that the structure of the polysaccharide was constant under all the conditions used. However, although this is probably true of the carbohydrate portion of the polymer, examination of the data suggests that the pyruvate content of the polysaccharide could vary quite markedly with different growth conditions.

The production of polysaccharide by strains currently rejected Many strains of micro-organisms are currently rejected in initial screening tests because although they produce polysaccharides on solid media they fail to do so in adequate yields. Some of the polymers are possibly of potential value, but the failure to convert sufficient substrate into polysaccharide renders the strains unacceptable. A further study of such strains might lead to the discovery of conditions under which higher yields of polysaccharide are obtained; alternatively mutagenesis might lead to altered control mechanisms giving the same result.

A more perplexing problem is the knowledge that many strains produce polysaccharide on solid media but fail to do so in liquid cultures. There is no clear indication as to why certain strains behave in this manner. They might be oxygen limited in liquid media; alternatively in liquid media, a possible acceptor for initiation of polysaccharide synthesis may be removed or surface proteins involved in polymer synthesis or nutrient uptake may be lost. Such possibilites might be worth further examination; alternatively, the cells might be tested as effective immobilized enzymes when attached to a suitable solid matrix.

(2) *Genetic improvement*

Mutation to improve polymer properties

Numerous environments have been screened by many laboratories for fresh sources of exopolysaccharide-producing micro-organisms but, despite considerable effort, few promising polymers have been found.

It should also be remembered that the time taken to develop the *Xanthomonas campestris* polysaccharide originally reported by the Peoria group (Jeanes *et al.*, 1961) through preliminary evaluation and pilot plant studies was of the order of ten to twelve years, further time being needed for scaling-up to an industrial process. Is this the only possible approach?

We now know much more about the physical properties of polysaccharides and the relationship between physical properties and chemical structure. For this information we are indebted to Atkins, Arnott, Rees and their colleagues (e.g. Atkins, Gardner & Isaac, 1977; Moorhouse *et al.*, 1977; Morris *et al.*, 1977). Polysaccharides may take the form of *ribbons* such as that found in the β1–4-linked glucans; of *helices* of the type represented by β1–3-linked glucans; or of *crumpled* structures found in β1–2-linked glucans (Rees, 1975). It should thus be possible to select a structure whose composition conforms to the desired physical type; lists of polymers of known structures can be found in the scientific literature and are exemplified by the *Enterobacter (Klebsiella) aerogenes* series listed in Table 1. Having selected a suitable microbial strain, this can be developed by mutagenesis and mutant isolation. As there are no selective pressures that can be exerted in the selection of such mutants, some ingenuity may be needed to identify the desired mutants. In our own laboratory, this started as serendipity but was developed into a deliberate procedure. Screening of *E. coli* mutants led to the discovery of some which, while still forming the exopolysaccharide colanic acid (a polymer with a hexasaccharide repeating unit (Lawson *et al.*, 1969; Sutherland, 1969), yielded either much more polymer than the wild-type or polymer with altered physical properties, presumably due to increased polymer chain length (Fig. 4, Table 8). A procedure which might assist in identifying colonies with altered physical properties was applied to two *Enterobacter aerogenes* serotypes. As a result a series of mutants was obtained, each producing a polysaccharide with enhanced viscosity relative to its antecedent (Fig. 5). Preliminary results suggest that the chemical composition of all the polymers in each 'family' is essentially the same; differences in physical properties do not, therefore, solely depend on the acetate and pyruvate content of the polymer – as was said to be the case for *X. campestris* polysaccharides (Cadmus *et al.*, 1976). While it is not implied that all polysaccharides could be changed from wild-type strains to mutant types with desirable physical characteristics, such an approach certainly seems worth considering.

Meanwhile, the search in other ecological niches continues and may

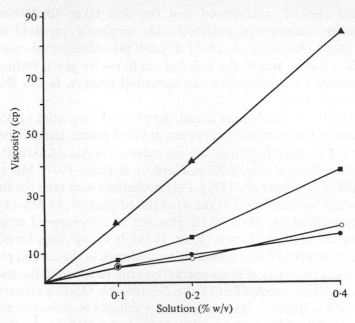

Fig. 4. Viscosity of colanic acid preparations (shear 70 s⁻¹, Ferranti viscometer, 20°C). Symbols: ▲, *E. coli* S614; ■, *E. coli*, S61; ●, *E. coli* S53; ○, *E. cloacae* 5920.

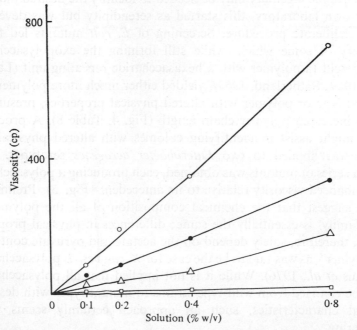

Fig. 5. Viscosity at low shear (10s⁻¹, 20 °C) for *E. aerogenes* type 8 mutants. Symbols: ○, 8SV3: ●, 8SV2; △, 8SV; □, 8S.

well yield further polymers of interest. A thorough assessment of the chemical composition and structure of any polymer is also required, as many of the earlier studies failed through lack of suitable techniques to give accurate representations of the polymer structures. This is seen in the original concept of a repeating unit of 16 monosaccharides for the *X. campestris* polymer (Sloneker, Orentas & Jeanes, 1963) compared with the pentasaccharide structure which is now recognized as correct (Jansson *et al.*, 1975; Melton *et al.*, 1976). Another example is seen in the recent revision of the composition of the succinoglucan from *A. faecalis* var. *myxogenes* to contain pyruvate as well as the succinate originally recognized (Harada, 1977).

Harvesting characteristics

Most of the microbial exopolysaccharides proposed for use are obtained initially as culture fluids with high viscosity (of the order of 30 000 cp at 25 °C and a shear rate of 10 s⁻¹). Polysaccharide may be prepared either by direct recovery from the fermentation fluids or after destruction or removal of the micro-organisms. Recovery of the microbial cells might have the advantage of providing a source of vitamins, amino acids, growth factors or even single-cell protein (SCP) albeit on a small scale. The recovery of the cells from the large volumes of highly viscous culture fluids requires a considerable expenditure of energy if centrifugation is employed. A better procedure might be to transfer the culture to a holding tank where autolysis could be induced or where enzymes degrading cell walls could be used; alternatively such procedures could be performed *in situ* in the fermenter. The use of conditional mutants, lysing under certain conditions, can be envisaged. Another approach would be the use of cells with altered surface properties. Not enough is known about the final stages of polysaccharide synthesis to determine whether altered surface characteristics in microbial cells affect the yield and properties of extracellular polysaccharides. Mutants of the type described by Norval & Sutherland (1969) would have the advantage that polysaccharide production is normal at the optimal temperature for growth of the bacteria but transfer to lower temperature leads to decreased lipopolysaccharide production and the formation of bacteria which autoagglutinate. Even in relatively viscous cultures the cells could be allowed to sediment in a holding tank at reduced temperature without affecting the polysaccharide product. Other types of mutation affecting the cell surface could be studied and it might be possible to obtain sedimentation through alteration of the ionic environment and so simplify the harvesting procedure still further.

Polysaccharides are currently recovered through precipitation with organic solvents. Some polymers are much more easily recovered than others. Those polymers which have a high content of acetyl or other lipophilic groups are more difficult to precipitate and require greater volumes of the organic solvent. The presence of the acetyl groups may not necessarily improve the characteristics of the polymer; if they do not confer any advantageous property, their removal by mild alkaline treament *prior* to product recovery could reduce the amount of solvent needed. An alternative, which would render the introduction of an additional stage in the harvesting process unnecessary, would be the isolation of mutants producing non-acetylated polysaccharide. This is possible without affecting the other components of the polysaccharide, as was demonstrated by Garegg *et al.* (1971). It is probably not necessary for polysaccharides containing approximately one acetyl group per 4–6 monosaccharides, but could be useful when the acetate:monosaccharide ratio is higher.

Elimination of polysaccharases

In general, exopolysaccharides other than levans and dextrans are not degraded by the micro-organisms responsible for their synthesis (Sutherland, 1977*b*). Although a few exceptions are known, they are mainly confined to *Streptococcus* species which can both synthesize and degrade hyaluronic acid – not a polysaccharide of potential commercial significance (Faber & Rosendal, 1954). However, alginate-producing strains of *Azotobacter vinelandii* are capable of producing alginases (Haug & Larsen, 1971). Normally culture conditions can be controlled to ensure that polysaccharase synthesis and activity is suppressed during culture and harvesting of the product. Such an approach is possible in *A. vinelandii* cultures which show a distinct lag between polysaccharide production and release of the alginase – an enzyme specific to polymannuronic acid. It would obviously be better to isolate mutants incapable of enzyme synthesis; this would obviate the need for such strict control over the conditions of polysaccharide production and harvesting.

At present, it seems unlikely that many of the polysaccharide-producing micro-organisms considered in Tables 6 and 7 are also able to degrade the polymers which they synthesize, but this problem might be encountered in new isolates.

Control mechanisms

At each stage in polysaccharide synthesis, control mechanisms can operate and can affect production of polymer (Sutherland, 1977a, 1979). The initial level of control is seen in the various mechanisms of substrate uptake. Few bacterial species have been extensively investigated but in two species which have, *Escherichia coli* (Herbert & Kornberg, 1976; Henderson, Giddens & Jones-Mortimer, 1977) and *Pseudomonas aeruginosa* (Whiting, Midgley & Dawes, 1976), it was clear that more than one uptake mechanism existed, even for a substrate such as glucose. Differences were also noted between strains, indicating the necessity of testing each species and strain intended for industrial use and showing that extrapolation of results from other micro-organisms was probably invalid. A knowledge of uptake mechanisms – whether permeases or group translocation mechanisms are involved (Postma & Roseman, 1976) – and, in particular, of the control mechanisms is particularly important if it is intended to use complex substrate mixtures and if attempts are to be made to use the micro-organisms as effective insolubilized enzyme systems attached to solid matrices. This last application would ideally necessitate the free access of the substrate to the site of polymer synthesis – the inner membrane of the microbial cell if it is a Gram-negative bacterium or the cytoplasmic membrane for Gram-positive species.

During or after uptake, the substrate is phosphorylated prior to conversion to sugar nucleotides. Control is also exerted at this stage through the sugar nucleotide pyrophosphorylases. Associated with this, in certain microbial species, is the regulation of storage polymers, such as glycogen. Although normally absent from microbial cells in the logarithmic phase of growth, glycogen can accumulate in relatively large amounts in non-proliferating bacteria (Holme, 1957). The accumulation of glycogen (or any other storage polymer composed of carbon) in an exopolysaccharide-producing strain represents wasted substrate and is therefore to be avoided. This could be accomplished by mutagenesis and selection of mutants lacking either ADP-glucose pyrophosphorylase or glycogen synthetase, otherwise any use of culture conditions in which glucose 1-phosphate would accumulate could lead to 'switch-on' of glycogen synthesis.

Regulation of the enzymes involved in sugar nucleotide production depends on the actual monosaccharides involved. Various systems have been examined and it has been shown that, as well as regulation of the *formation* of sugar nucleotides – the primary glycosyl donors in polysaccharide synthesis, there are various possible mechanisms for the removal and hydrolysis of these precursors (Ward & Glaser, 1969; Beacham *et al.*, 1973). It is possible that some of the enzymes for sugar nucleotide synthesis are regulated by plasmids, as was shown for colanic acid in the *Enterobacteriaceae* in the extensive studies by Markovitz and his colleagues (1977). Introduction of multiple copies of the plasmids could be employed to increase the levels of some of the enzymes involved and this approach could lead to increased polysaccharide synthesis. Unfortunately, few genetic studies have been made on other polysaccharide-producing micro-organisms but recent work on the genetic transfer of alginate-synthesizing *P. aeruginosa* species is of interest in this context (Fyfe & Govan, 1978).

The monosaccharides are transferred from the sugar nucleotides to an isoprenoid lipid intermediate by mechanisms analogous to those detected in peptidoglycan and lipopolysaccharide synthesis (Troy, Frerman & Heath, 1971). The amount of lipid intermediate in the microbial cell is limited and probably affords another means of controlling the rate and amount of polysaccharide production, especially as the lipid is also involved in the synthesis of the wall polymers. Evidence of such a control mechanism is difficult to obtain (Sutherland, 1977b) but a similar role for the longer-chain dolichol compounds in the glycosylation of glycoproteins was recently suggested by Hemming (1976). After the formation of oligosaccharide subunits of the extracellular polymer, attached to the isoprenoid lipid, they may either be excreted directly or transferred to some acceptor molecule at or near

Table 8. *Molecular weights of some exopolysaccharides*

Strain		Molecular weight	Reference
Acetobacter xylinum		5.67×10^5	Brown (1962)
Xanthomonas campestris		2×10^6	Dintzis, Babcock & Tobin (1970)
Streptococcus pneumoniae	Type 3	2.67×10^5	Koenig & Perrings (1955)
Escherichia coli K87		2.8×10^5	Tarcsay, Jann & Jann (1971)
Klebsiella pneumoniae	Type 1	2.94×10^6	Wolf *et al.* (1978)
Enterobacter aerogenes	Type 4	2.1×10^5	Churms, Merrifield & Stephan (1978)
	Type 5	1.29×10^6	Wolf *et al.* (1978)
	Type 6	2.5×10^6	
	Type 7	1.2×10^5	Churms *et al.* (1978)
	Type 8	1.13×10^6	Wolf *et al.* (1978)
	Type 9	1.2×10^6	Churms *et al.* (1978)
	Type 11	2×10^6	Wolf *et al.* (1978)
	Type 21	4.0×10^5	Churms *et al.* (1978)
	Type 27	9.4×10^5	
	Type 32	1.2×10^6	
	Type 54	1.2×10^6	
	Type 56	1.7×10^5	Wolf *et al.* (1978)
	Type 57	2.27×10^6	
	Type 64	1.7×10^6	Churms *et al.* (1978)
Pullularia pullulans		1.7×10^5	Taguchi *et al.* (1973)

the cell surface. This aspect of exopolysaccharide production has so far resisted investigation but it appears that under defined conditions and for any given strain, polymer of uniform molecular weight is produced (Table 8). The different processes which are controlled in exopolysaccharide synthesis are summarized in Fig. 6.

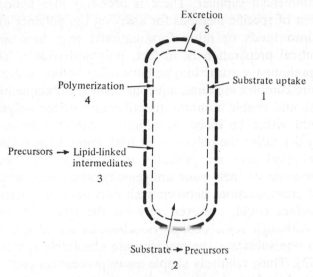

Fig. 6. Aspects of control affecting exopolysaccharide synthesis.

DEFINITION OF POLYSACCHARIDES OF INDUSTRIAL IMPORTANCE

Currently, methods for the definition of polysaccharides used industrially are totally inadequate, and this could well be an area for significant advances. Ideally, one requires specific methods for each proposed polysaccharide, particularly if it is intended for food and other high-grade usage. A typical definition is that for xanthan gum in the FAO list (1975), in which the product is indicated to contain D-glucose and D-mannose along with D-glucuronic acid and pyruvate, while limits are set for the quantities of heavy metals present.

Ideally, three parameters should be defined: chemical composition, including levels of ions etc. permitted; biological characteristics and physical characteristics.

Chemical composition Obviously the characterization of component sugars is required, and methods for carbohydrate analysis are readily available; non-carbohydrate constituents should also be determined and quantified. At the same time nitrogen should be determined as a measure of cell contamination (assuming that aminosugars are absent). Alternatively, ultraviolet absorption can provide an indication of the amount of contaminating nucleic acid. These tests can readily be performed either on the polysaccharide during production or on purchase from a commercial supplier. There is probably also a need for the development of specific methods for assaying the polymer *after* incorporation into foods or other formulations: (e.g. how much of a pharmaceutical preparation is, in fact, polysaccharide?). This would require a preliminary extraction, perhaps using boiling water or, in the case of more complex mixtures, aqueous phenol. Subsequently, specific methods should enable accurate quantification of the polysaccharide. These might either be based on affinity techniques or on enzymes specifically degrading the polymers, or both approaches might be used.

Immunological tests are probably of little value for exopolysaccharides because of their poor antigenicity when pure and the large number of cross-reactions between such polymers. Attempts to define polysaccharides could, however, rely on the specificity of polysaccharases. Although some endopolysaccharases are able to hydrolyse more than one substrate, the majority are absolutely specific (Sutherland, 1977b). Thus, relatively simple assay procedures could be developed for each polysaccharide based on their degradation by specific enzymes and quantitation of the fragments produced. Alginates present a rather more complex problem as the enzymes known to degrade them are either specific for polymannuronic acid (Davidson, Lawson & Sutherland, 1977) or for polyguluronic acid (Kashiwabara, Suzuki & Nisizawa, 1969; Davidson, Sutherland & Lawson, 1976; Boyd & Turvey, 1977). Consequently assays based on enzymes such as these would have to take account of the varying composition of alginates.

Biological characteristics Two main areas require to be checked – the safety of the material for those using, consuming and processing the product, and the presence of contaminating material which might not be determined by other tests. The user or processer should not be exposed to material which might, over a period of time, lead to sensitization, and there should obviously be no carcinogenic properties. These aspects can be tested using laboratory animals and other tests. If the micro-organisms from which the polysaccharide is obtained are Gram-negative bacteria, it may also be desirable to ensure that there is a minimum level of lipopolysaccharide (endotoxin) present. If there are monosaccharides in common between exopolysaccharide and cell wall, this level can be relatively difficult to determine, especially as the amount derived from the wall would in all probability be very much less than that from the extracellular polymer. The very sensitive Limulus assay tests for endotoxin (Yin *et al.*, 1972) could be used to provide a very sensitive and specific test.

Physical parameters The customer is probably most interested in the viscosity characteristics of the polysaccharides, especially at different pH and temperatures; alternatively gelling and synergistic properties may be of importance. As each laboratory tends to use its own favourite equipment, the conditions under which these tests are performed should be adequately defined. It will also be important to define the effects of various ions, both divalent and monovalent, acids and bases. Recovery of properties after heating, as well as retention of properties on prolonged exposure to high or low pH, may be important for particular applications. Although the viscosity in defined conditions may be adequate information for some purposes, the molecular weight may also have to be determined. This is less easy than for other macromolecules but a variety of procedures including ultracentrifugation and gel permeation chromatography have been successfully applied to obtain the values shown in Table 8.

CONCLUSIONS

Exopolysaccharides appear to be one of the potential growth areas of the fermentation industries. Although the market is currently limited to a small number of products, it should be possible, either through further screening or through a policy of deliberate development of polysaccharides with known characteristics, to produce new polymers with properties complementary to existing polymers such as xanthan.

More basic knowledge is required to enable the optimization of existing processes for polysaccharide production and to develop new techniques. It may still be some years in the future before one can anticipate the conversion of substrate to polymer by insolubilized enzyme systems, as can currently be achieved using dextran-synthesizing systems. Improved analytical techniques are needed which are specific for the particular polysaccharide in use and which can complement the existing methods for gross chemical analysis.

Acknowledgements
The authors are grateful to a number of workers, in particular

Professor S. Stirm and Dr S. C. Churms, for providing information prior to publication. Some of the work in the senior author's laboratory was supported by SRC Grant GRA 30049.

REFERENCES

ALBRECHT, W. J., SOHNS, V. E. & ROGOVIN, S. P. (1963). Pilot plant process for the isolation of a microbial polysaccharide with a quaternary ammonium compound. *Biotechnology and Bioengineering*, 5, 91–9.

AROND, L. H. & FRANK, H. P. (1954). Molecular weight, molecular weight distribution and molecular size of a native dextran. *Journal of Physical Chemistry*, 58, 953–7.

ATKINS, E. D. T., GARDNER, K. H. & ISAAC, D. H. (1977). X-ray diffraction by bacterial capsular polysaccharides: trial conformations for *Klebsiella* polyuronides K5, K57 and K8. In *Cellulose Chemistry and Technology*, ed. A. C. Arthur, pp. 56–72. Washington: American Chemical Society.

BAYER, M. & THUROW, H. (1977). Polysaccharide capsule of *Escherichia coli*: microscope study of its size, structure and site of synthesis. *Journal of Bacteriology*, 130, 911–36.

BEACHAM, I. R., KAHANA, R., LEVY, L. & YAGIL, E. (1973). Mutants of *Escherichia coli* K12 'cryptic' or deficient in 5'-nucleotidase and 3'-nucleotidase activity. *Journal of Bacteriology*, 116, 957–64.

BEBAULT, G. M., CHOY, Y. M., DUTTON, G. S. G., FURNELL, N., STEPHAN, A. M. & YANG, M. T. (1973). Proton resonance spectroscopy of *Klebsiella* capsular polysaccharides. *Journal of Bacteriology*, 113, 1345–7.

BJÖRNDAL, H., LINDBERG, B., LÖNNGREN, J., MESZAROS, M., THOMPSON, J. L. & NIMMICH, W. (1973). Structural studies of the capsular polysaccharide of *Klebsiella* type 52. *Carbohydrate Research*, 31, 93–100.

BOYD, J. & TURVEY, J. R. (1977). Isolation of a poly-α-L-guluronate lyase from *Klebsiella aerogenes*. *Carbohydrate Research*, 57, 163–71.

BROOKER, B. E. (1976). Surface coat transformation and capsule formation by *Leuconostoc mesenteriodes* NCDO 523 in the presence of sucrose. *Archives of Microbiology*, 111, 99–104.

BROOKER, B. E. (1979). In *Microbial Polysaccharides and Polysaccharidases*, eds. R. W. C. Berkeley, D. C. Ellwood & W. A. Hamilton (in press). London, New York: Academic Press.

BROWN, A. M. (1962). The mechanism of cellulose biosynthesis by *Acetobacter acetigenum*. *Journal of Polymer Science*, 59, 155–63.

CADMUS, M. C., LAGODA, A. A. & ANDERSON, R. F. (1962). Production of a new polysaccharide with *Cryptococcus laurentii* var. *flavescens*. *Applied Microbiology*, 10, 153–6.

CADMUS, M. C., GASDORF, H., LAGODA, A. A., ANDERSON, R. F. & JACKSON, R. W. (1963). New Bacterial Polysaccharide from *Arthrobacter*. *Applied Microbiology*, 11, 488–92.

CADMUS, M. C., ROGOVIN, S. P., BURTON, K. A., PITTSLEY, J. E., KNUTSON, C. A. & JEANES, A. (1976). Colonial variation in *Xanthomonas campestris* NRRL B-1459 and characterization of the polysaccharide from a variant strain. *Canadian Journal of Microbiology*, 22, 942–8.

CARLSON, D. M. & MATTHEWS, L. W. (1966). Polyuronic acids produced by *Pseudomonas aeruginosa*. *Biochemistry*, 5, 2817–22.

CHAKRABORTY, A. K., FRIEBOLIN, H., NIEMANN, H. & STIRM, S. (1977). Primary structure of the *Klebsiella* serotype 16 capsular polysaccharide. *Carbohydrate Research*, **59**, 525–30.

CHOY, Y. M. & DUTTON, G. G. S. (1973). The structure of the capsular polysaccharide from *Klebsiella* K type 21. *Canadian Journal of Chemistry*, **51**, 198–207.

CHOY, Y. M., DUTTON, G. G. S. & ZANLUNGA, M. (1973). The structure of the capsular polysaccharide of *Klebsiella* K type 24. *Canadian Journal of Chemistry*, **51**, 1819–25.

CHURMS, S. C., MERRIFIELD, E. H. & STEPHAN, A. M. (1978). *Carbohydrate Research* (in press).

Ciba Review (1969). *Alginates*, pp. 3–46. Manchester: Ciba.

CONRAD, H. E., BAMBURG, J. R., EPLEY, J. D. & KINDT, T. J. (1966). The structure of the *Aerobacter aerogenes* A3(SL) polysaccharide. II. Sequence analysis and hydrolysis studies. *Biochemistry*, **5**, 2808–17.

CORPE, W. A. (1960). The extracellular polysaccharide of gelatinous strains of *Chromobacterium violaceum*. *Canadian Journal of Microbiology*, **6**, 153–63.

CURVALL, M., LINDBERG, B., LÖNNGREN, J. & NIMMICH, W. (1975*b*). Structural studies of the capsular polysaccharide of *Klebsiella* type 28. *Carbohydrate Research*, **42**, 95–105.

DAVIDSON, I. W., SUTHERLAND, I. W. & LAWSON, C. J. (1976). Purification and properties of an alginate lyase from a marine bacterium. *Biochemical Journal*, **159**, 707–13.

DAVIDSON, I. W., SUTHERLAND, I. W. & LAWSON, C. J. (1977). Localization of *O*-acetyl groups of bacterial alginate. *Journal of General Microbiology*, **98**, 603–6.

DAVIDSON, I. W., LAWSON, C. J. & SUTHERLAND, I. W. (1977). An alginate lyase from *Azotobacter vinelandii* phage. *Journal of General Microbiology*, **98**, 223–9.

DEAVIN, L., JARMAN, T. R., LAWSON, C. J., RIGHELATO, R. C. & SLOCOMBE, S. (1977). The production of alginic acid by *Azotobacter vinelandii* in batch and continuous culture. In *Extracellular Microbial Polysaccharides*, eds. P. A. Sandford & A. Laskin, pp. 14–26. Washington: American Chemical Society.

DINTZIS, F. R., BABCOCK, G. E. & TOBIN, R. (1970). Studies on dilute solutions and dispersions of the polysaccharide from *Xanthomonas campestris* NRRL B-1459. *Carbohydrate Research*, **13**, 257–67.

DUGUID, J. P. (1951). The demonstration of bacterial capsules and slime. *Journal of Pathology and Bacteriology*, **63**, 673–85.

DUTTON, G. G. S. & MACKIE, K. L. (1977). Structural investigations of *Klebsiella* serotype K36 polysaccharide. *Carbohydrate Research*, **55**, 49–63.

DUTTON, G. G. S. & YANG, M. T. (1973). The structure of the capsular polysaccharide of *Klebsiella* K type 5. *Canadian Journal of Chemistry*, **51**, 1826–32.

DUTTON, G. G. S. & YANG, M. T. (1977). Structural investigation of *Klebsiella* serotype 62 polysaccharide. *Carbohydrate Research*, **59**, 179–92.

DUTTON, G. G. S., STEPHAN, A. M. & CHURMS, S. C. (1974). Structural investigations of *Klesiella* serotype 7 polysaccharide. *Carbohydrate Research*, **38**, 225–37.

ERBING, C., KENNE, L., LINDBERG, B., LÖNNGREN, J. & SUTHERLAND, I. W. (1976). Structural studies of the capsular polysaccharide from *Klebsiella* type 1. *Carbohydrate Research*, **50**, 115–20.

EVANS, C., YEO, R. G. & ELLWOOD, D. C. (1979). In *Microbial Polysaccharides and Polysaccharidases*, eds. R. W. C. Berkeley, D. C. Ellwood & W. A. Hamilton (in press). London, New York: Academic Press.

EVANS, L. R. & LINKER, A. (1973). Production and characterization of the slime polysaccharide of *Pseudomonas aeruginosa*. *Journal of Bacteriology*, **116**, 915–24.

FAO (1975). *Specifications for Identity and Purity of some Food Additives*. Nutrition Meetings Report Series No. 54B. Rome: UN Food and Agriculture Oranization.

FABER, V. & ROSENDAL, K. (1954). Streptococcal hyaluronidase. *Acta Pathologica et Microbiologica Scandinavica*, **35**, 159–64.

FYFE, J. A. M. & GOVAN, J. R. W. (1978). A genetic approach to the study of mucoid *Pseudomonas aeruginosa*. *Proceedings of the Society for General Microbiology*, **5**, 54.

GABRIEL, A. (1979). In *Microbial Polysaccharides and Polysaccharidases*, eds. R. W. C. Berkeley, D. C. Ellwood & W. A. Hamilton (in press). London, New York: Academic Press.

GAHAN, L. C., SANDFORD, P. A. & CONRAD, H. E. (1967). The structure of the serotype 2 capsular polysaccharide of *Aerobacter aerogenes*. *Biochemistry*, **6**, 2755–66.

GAREGG, P. J., LINDBERG, B., ONN, T. & HOLME, T. (1971). Structural studies on the M-antigen from two mucoid mutants of *Salmonella typhimurium*. *Acta Chemica Scandinavica*, **25**, 1185–94.

GAUDY, E. & WOLFE, R. S. (1962). Composition of an extracellular polysaccharide produced by *Sphaerotilus natans*. *Applied Microbiology*, **10**, 200–5.

GRANT, W. D., SUTHERLAND, I. W. & WILKINSON, J. F. (1969). Exopolysaccharide colanic acid and its occurrence in the *Enterobacteriaceae*. *Journal of Bacteriology*, **100**, 1187–93.

HARADA, T. (1977). Production, properties and application of curdlan. In *Extracellular Microbial Polysaccharides*, eds. P. A. Sandford & A. Laskin, pp. 265–83. Washington: American Chemical Society.

HAUG, A. & LARSEN, B. (1970). An extracellular polysaccharide from *Beijerinckia indica* containing L-guluronic acid residues. *Acta Chemica Scandinavica*, **24**, 1855–6.

HAUG, A. & LARSEN, B. (1971). Biosynthesis of Alginate. Part II. Polymannuronic acid C-5-epimerase from *Azotobacter vinelandii*. *Carbohydrate Research*, **17**, 297–308.

HEMMING, F. W. (1976). Dolichol phosphate, a coenzyme in the glycosylation of animal membrane-bound glycoproteins. *Biochemical Society Transactions*, **5**, 1223–331.

HENDERSON, P. J. F., GIDDENS, R. A. & JONES-MORTIMER, M. C. (1977). Transport of galactose, glucose and their molecular analogues by *Escherichia coli* K12. *Biochemical Journal*, **162**, 309–20.

HERBERT, D. & KORNBERG, H. L. (1976). Glucose transport as rate-limiting step in the growth of *Escherichia coli* on glucose. *Biochemical Journal*, **156**, 477–80.

HOLME, T. (1957). Continuous culture studies on glycogen synthesis in *Escherichia coli* B. *Acta Chemica Scandinavica*, **11**, 763–75.

HOLZWORTH, G. (1976). Conformation of the extracellular polysaccharide of *Xanthomonas campestris*. *Biochemistry*, **15**, 4333–9.

HUMPHREYS, J. D. & MARSHALL, J. J. (1978). Synthetic carbohydrate–protein conjugates and their potential applications in biochemical technology and medicine. Presented at *IXth International Symposium on Carbohydrate Chemistry, London*. London: Chemical Society.

JANSSON, P. -E., KENNE, L. & LINDBERG, B. (1975). Structure of the extracellular polysaccharide from *Xanthomonas campestris*. *Carbohydrate Research*, **45**, 275–82.

JEANES, A. (1974). Applications of extracellular microbial polysaccharide-polyelectrolytes: review of literature including patents. *Journal of Polymer Science* (*Symposium No. 45*), 209–27.

JEANES, A., HAYNES, W. C., WILHAM, C. A., RANKIN, J. C., MELVIN, E. H., AUSTIN, M., CLUSKEY, J. E., FISHER, B. E., TUCHIYA, H. M. & RIST, C. E. (1954). Characterization and classification of dextrans from ninety-six strains of bacteria. *Journal of American Chemical Society*, **76**, 5041–52.

JEANES, A., PITTSLEY, J. E. & SENTI, F. R. (1961). Polysaccharide B1459: A new hydrocolloid polyelectrolyte produced from glucose by bacterial fermentation. *Journal of Applied Polymer Science*, **17**, 519–26.

JEANES, A., PITTSLEY, J. E. & WATSON, P. R. (1964). Extracellular polysaccharide produced from glucose by *Cryptococcus laurentii* var. *flavescens* NRRL Y-1401: chemical and physical characterization. *Journal of Applied Polymer Science*, **8**, 2775–87.

JEANES, A. KNUTSON, C. A., PITTSLEY, J. E. & WATSON, P. R. (1965). Extracellular polysaccharide produced from glucose by *Arthrobacter viscosus* NRRL B1973: chemical and physical characterization. *Journal of Applied Polymer Science*, **9**, 627–38.

JOHNSON, J. KIRKWOOD, S., MISAKI, A., NELSON, T. E., SCARLETTI, J. V. & SMITH, F. (1963). Structure of a new glucan. *Chemistry and Industry*, **41**, 820–2.

KANG, K. S. & MCNEELY, W. H. (1977). In *Extracellular Microbial Polysaccharides*, eds. P. A. Sandford & A. Laskin, pp. 220–30. Washington: American Chemical Society.

KANG, K. S., VEEDER, G. T. & RICHEY, D. D. (1977). In *Extracellular Microbial Polysaccharides*, eds. P. A. Sandford & A. Laskin, pp. 211–19. Washington: American Chemical Society.

KASHIWABARA, Y., SUZUKI, H. & NISIZAWA, K. (1969). Alginate lyases of Pseudomonads. *Journal of Biochemistry*, **66**, 503–12.

KOENIG, V. L. & PERRINGS, J. D. (1955). Sedimentation and viscosity studies on the capsular and somatic polysaccharides of *Pneumococcus* type III. *Journal of Biophysical and Biochemical Cytology*, **1**, 93–8.

LAWSON, C. J. & SUTHERLAND, I. W. (1978). Polysaccharides. In *Economic Microbiology*, vol. 2, ed. A. H. Rose, pp. 337–92. London. New York: Academic Press.

LAWSON, C. J., MCCLEARY, C. W., NAKADA, H. I., REES, D. A., SUTHERLAND, I. W. & WILKINSON, J. F. (1969). Structural analysis of colanic acid from *Escherichia coli* by using methylation and base-catalysed fragmentation. *Biochemical Journal*, **115**, 947–58.

LESLEY, S. M. & HOCHSTER, R. M. (1959). The extracellular polysaccharide of *Xanthomonas phaseoli*. *Canadian Journal of Biochemistry*, **37**, 513–29.

LINDBERG, B., LÖNNGREN, J., RUDEN, U. & NIMMICH, W. (1975). Structural studies of the capsular polysaccharide of *Klebsiella* type 59. *Carbohydrate Research*, **42**, 83–93.

LINDBERG, B., LINDQUIST, B. & LÖNNGREN, J. (1977). Structural studies of the capsular polysaccharide of *Klebsiella* type 37. *Carbohydrate Research*, **58**, 443–51.

LINKER, A. & JONES, R. S. (1966). A new polysaccharide resembling alginic acid isloated from Pseudomonads. *Journal of Biological Chemistry*, **241**, 3845–51.

MCNEELY, W. H. & PETTITT, D. J. (1973). In *Industrial Gums*, 2nd Edition, ed. R. L. Whistler, pp. 49–81. London, New York: Academic Press.

MARKOVITZ, A. (1977). Genetics and regulation of bacterial capsular polysaccharide biosynthesis and radiation sensitivity. In *Surface Carbohydrates of the Pro-karyotic Cell*, ed. I. W. Sutherland, pp. 415–62. London, New York: Academic Press.

MELTON, L. D., MINDT, L., REES, D. A. & SANDERSON, G. R. (1976). Covalent structure of the extracellular polysaccharide from *Xanthomonas campestris:* evidence from partial hydrolysis studies. *Carbohydrate Research,* **46,** 245–57.

MIAN, F. A., JARMAN, T. R. & RIGHELATO, R. C. (1978). Biosynthesis of exopolysaccharide by *Pseudomonas aeruginosa. Journal of Bacteriology,* **134,** 418–22.

MOORHOUSE, R., WALKINSHAW, M. D., WINTER, W. T. & ARNOTT, S. (1977). In *Cellulose Chemistry and Technology,* ed. A. C. Arthur, pp. 133–52. Washington: American Chemical Society.

MORRIS, E. R. (1977). In *Extracellular Microbial Polysaccharides,* eds. P. A. Sandford & A. Laskin, pp. 81–9. Washington: American Chemical Society.

MORRIS, E. R., REES, D. A., WALKINSHAW, M. D. & DARKE, A. (1977). Order–disorder transition for a bacterial polysaccharide in solution. *Journal of Molecular Biology,* **110,** 1–16.

NEIJSSEL, O. M. & TEMPEST, D. W. (1976). Bioenergetic aspects of aerobic growth of *Klebsiella aerogenes* NCTC418 in carbon-limited and carbon-sufficient chemostat culture. *Archives of Microbiology,* **107,** 215–21.

NIEMANN, H., CHAKRABORTY, A. K., FRIEBOLIN, H. & STIRM, S. (1978). Primary structure of the *Escherichia coli* serotype K42 capsular polysaccharide and its serological identity with the *Klebsiella* K63 polysaccharide. *Journal of Bacteriology,* **133,** 390–1.

NORVAL, M. & SUTHERLAND, I. W. (1969). A group of *Klebsiella* mutants showing temperature-dependent polysaccharide synthesis. *Journal of General Microbiology,* **57,** 369–77.

OIL AND GAS JOURNAL (1967). Enhanced-recovery illusions fade, real prospects remain. *Oil and Gas Journal,* **74,** 106–25.

PAAVOLAINEN, P. & SINDELL, B. (1976). The effect of dextranomer on hand burns. *Annales Chirurgiae et Gynaecologiae Fenniae,* **65,** 313–17.

PHARMACIA (1966). *Sephadex-gel filtration in theory and practice.* Uppsala: Pharmacia.

POSTMA, P. W. & ROSEMAN, S. (1976). The bacterial phosphoenolpyruvate: sugar phosphotransferase system. *Biochimica et Biophysica Acta,* **457,** 213–57.

REES, D. A. (1975). In *Biochemistry of Carbohydrates,* ed. W. J. Whelan, pp. 1–42. London: Butterworths.

ROGOVIN, S. P., ANDERSON, R. F. & CADMUS, M. C. (1961). Production of polysaccharide with *Xanthomonas campestris. Journal of Biochemical and Microbiological Technology and Engineering,* **3,** 51–63.

ROGOVIN, P., ALBRECHT, W. & JOHNS, V. (1965). Production of industrial grade polysaccharide B-1459. *Biotechnology and Bioengineering,* **7,** 161–9.

SANDFORD, P. A., WATSON, P. R. & JEANES, A. R. (1973). An extracellular microbial polysaccharide composed of 2-acetamide-2-deoxy-D-glucose and 2–acetamide–2–deoxy–D–glucuronic acid: radiochemical and gas chromatographic analysis of the products of methanolysis. *Carbohydrate Research,* **29,** 153–64.

SANDFORD, P. A., BURTON, K. A., WATSON, P. R., CADMUS, M. C. & JEANES, A. (1975). Extracellular polysaccharide from the black yeast NRRL Y6272: improved methods for preparing a high-viscosity pigment-free product. *Applied Microbiology,* **29,** 769–75.

SANDFORD, P. A., PITTSLEY, J. E., KNUTSON, C. A., WATSON, P. R., CADMUS, M. C. & JEANES, A. (1977). Variation in *Xanthomonas campestris* NRRL B1459: characterization of xanthan products of varying pyruvic acid content. In *Extracellular Microbial Polysaccharides,* eds. P. A. Sandford & A. Laskin, pp. 192–210. Washington: American Chemical Society.

SANDVIK, E. I. & MAERKER, J. M. (1977). Application of xanthan gum for enhanced oil recovery. In *Extracellular Microbial Polysaccharides*, eds. P. A. Sandford & A. Laskin, pp. 242–64. Washington: American Chemical Society.

SHERWOOD, R. J., BAIRD, J. K., ATKINSON, A., WIBLIN, C. N., RUTTER, D. A. & ELLWOOD, D. C. (1977). Enhanced plasma persistence of a therapeutic enzyme by coupling to soluble dextran. *Biochemical Journal*, **164**, 461–4.

SILMAN, R. W. & ROGOVIN, P. (1970). Continuous fermentation to produce xanthan biopolymers: laboratory investigation. *Biotechnology and Bioengineering*, **12**, 75–83.

SILMAN, R. W. & ROGOVIN, P. (1972). Continuous fermentation to produce xanthan polymer: effect of dilution rate. *Biotechnology and Bioengineering*, **14**, 23–31.

SLONEKER, J. H., ORENTAS, D. G. & JEANES, A. (1963). Extracellular bacterial polysaccharide from *Xanthomonas campestris* NRRL B1459. Part III. Structure. *Candian Journal of Chemistry*, **42**, 1261–9.

SMIDSRØD, O. (1974). Molecular basis for some physical properties of alginates in the gel state. *Faraday Discussion of the Chemical Society*, **57**, 263–74.

STARR, M. P. (1946). The nutrition of phytopathogenic bacteria. *Journal of Bacteriology*, **51**, 131–43.

SUTHERLAND, I. W. (1967). Phage-induced fucosidases hydrolysing the exopolysaccharide of *Klebsiella aerogenes* type 54 [A3(SC)]. *Biochemical Journal*, **104**, 278–85.

SUTHERLAND, I. W. (1969). Structural studies on colanic acid, the common exopolysaccharide found in the Enterobacteriaceae, by partial acid hydrolysis. *Biochemical Journal*, **115**, 935–5.

SUTHERLAND, I. W. (1970). Structure of *Klebsiella aerogenes* type 8 polysaccharide. *Biochemistry*, **9**, 2180–5.

SUTHERLAND, I. W. (1972). The exopolysaccharides of *Klebsiella* serotype 2 strains as substrates for phage-induced polysaccharide depolymerases. *Journal of General Microbiology*, **70**, 331–8.

SUTHERLAND, I. W. (1977*a*). Microbial exopolysaccharide synthesis. In *Extracellular Microbial Polysaccharides*, eds. P. A. Sandford & A. Laskin, pp. 40–57. Washington: American Chemical Society.

SUTHERLAND, I. W. (1977*b*). *Surface Carbohydrates of the Prokaryotic Cell*. London, New York: Academic Press.

SUTHERLAND, I. W. (1979). In *Microbial Polysaccharides and Polysaccharidases*, eds. R. W. C. Berkeley, D. C. Ellwood & W. A. Hamilton. London, New York: Academic Press. (In press.)

TAGUCHI, R., KIKUCHI, Y., SAKANO, Y. & KOBAYASHI, T. (1973). Structural uniformity of pullulan produced by several strains of *Pullularia pullulans*. *Agricultural and Biological Chemistry*, **37**, 1583–8.

TARCSAY, L., JANN, B. & JANN, K. (1971). Immunochemistry of the K antigens of *Escherichia coli*. *European Journal of Biochemistry*, **23**, 505–14.

THUROW, H., CHOY, Y. M., FRANK, N., NIEMANN, H. & STIRM, S. (1975). The structure of *Klebsiella* serotype 11 capsular polysaccharide. *Carbohydrate Research*, **41**, 241–55.

TROY, F. A., FRERMAN, F. E. & HEATH, E. C. (1971). The biosynthesis of capsular polysaccharide in *Aerobacter aerogenes*. *Journal of Biological Chemistry*, **246**, 118–33.

WARD, J. B. & GLASER, L. (1969). Turnover of UDP-sugars in *E. coli* mutants with altered UDP-sugar hydrolase. *Archives of Biochemistry and Biophysics*, **134**, 612–22.

WATSON, P. R., SANDFORD, P. A., BURTON, K. A., CADMUS, M. C. & JEANES, A. (1976). An extracellular fungal polysaccharide composed of 2-acetamide-2-deoxy-D-glucuronic acid residues. *Carbohydrate Research*, **46**, 259–65.

150   IAN W. SUTHERLAND AND DEREK C. ELLWOOD

WELLS, J. (1977). In *Extracellular Microbial Polysaccharides*, eds. P. A. Sandford & A. Laskin, pp. 299–313. Washington: American Chemical Society.

WHITING, P. H., MIDGLEY, M. & DAWES, E. A. (1976). The regulation of transport of glucose, gluconate and 2-oxogluconate and of glucose catabolism in *Pseudomonas aeruginosa*. *Biochemical Journal*, **154**, 659–68.

WILLIAMS, A. G. & WIMPENNY, J. W. T. (1977). Exopolysaccharide production by *Pseudomonas* NCIB 11264 grown in batch culture. *Journal of General Microbiology*, **102**, 13–21.

WILLIAMS, A. G. & WIMPENNY, J. W. T. (1978). Exopolysaccharide production by *Pseudomonas* NCIB 11264 grown in continuous culture. *Journal of General Microbiology*, **104**, 47–57.

WOLF, C., ELSÄSSER-BEILE, V., STIRM, S., DUTTON, G. G. S. & BURCHARD, W. (1978). Conformational studies of bacterial polysaccharides. *Biopolymers*. (In press.)

YIN, E. T., GALANOS, C., KINSKY, S., BRADSHAW, R., WESSLER, S. & LÜDERITZ, O. (1972). Picogram sensitive assay for endotoxin: gelation of *Limulus polyphemus* blood cell lysate induced by purified lipopolysaccharides and lipid A from Gram-negative bacteria. *Biochimica et Biophysica Acta*, **261**, 284–9.

YUEN, S. (1974). Pullulan and its applications. *Process Biochemistry*, **9** (no. 1), 7–9.

MICROBIAL TECHNOLOGY: PRODUCTION OF VACCINES

A. JOHN BEALE* AND ROBERT J. C. HARRIS†

* Wellcome Research Laboratories, Beckenham, Kent BR3 3BS, UK

† Microbiological Research Establishment, Porton Down, Salisbury SP4 0JG, UK

INTRODUCTION

Vaccine production is an important application of microbiology, but the scale of manufacture is for the most part relatively small. At a time when vaccine technology is coming into its own, the pendulum seems to be swinging away again. The reasons are not hard to find. First, interest in 'interferons' is recovering and they are just beginning to be made on a large scale by modern technology. However, if the molecular biology of their production can be unambiguously identified there may be an opportunity of producing them by genetic manipulation without dependence on virus initiation in cell culture. Second, antibiotics may have reached the limit of their versatility and range of activity; and control of some infections, for example meningococci and pneumococci, is reverting to a vaccine approach. Third, anti-viral drugs are emerging, and with them a real prospect for control of some virus infections.

There are two aspects to developing a vaccine. First, the identification of the protective antigen or antigens; and second, the stimulation of the correct arm of the immune response (immunoglobulin or cell-mediated response).

The second set of problems is beyond the scope of this article; suffice it to say that living organisms are often needed to elicit this correct response, but with enough knowledge it should always be possible to achieve the result with the pure immunogen. The immune response can be modulated in a number of ways, for example by the use of an adjuvant. Unfortunately, the best of these is complete Freund adjuvant, which contains killed mycobacteria, but recent advances suggest that the important structure in mycobacteria is MDP (*N*-acetylmuramyl-L-alanyl-D-isoglutamine). Since MDP or its analogues (Petit & Lederer, 1978) can be injected by a variety of routes, and even given orally, there is reason to hope for application in vaccine technology. Chemical modification of antigens can also modulate the immune response, for example delayed hypersensitivity response at the expense of antibody production (Parish & Liew, 1972). The methods of producing high levels of circulating antibodies by exploiting the primary and secondary response were established in the twenties (Glenny, 1925). Local immunity is more difficult to achieve with a killed antigen, although progress has been made in stimulating such immunity in the gut (Porter, 1973; Pierce, 1976).

To revert, the isolation of immunogens on a large scale is a problem for microbial technology. Immunogens can be made by extraction from organisms grown conventionally, by genetic manipulation of foreign organisms, or by synthesis. At present the technology employed is often adapted from laboratory practice rather than fully developed as a branch of engineering.

DIFFERENCES BETWEEN TYPES OF VACCINE FROM THE PRODUCTION VIEWPOINT

The production of bacterial vaccines is very different from the production of viral vaccines because of the need to propagate viruses in cells. The two sorts of vaccine will therefore be considered separately. There are also considerable differences between the manufacture of living vaccines and killed vaccines. The identification and testing of a suitable attenuated seed strain is the vital consideration for living-vaccine production. The seed strain must be shown to be safe and effective and then to be identifiable in some way, so that its identity, and the identity of the vaccine derived from it, can be established by a series of 'finger-prints'. In addition one has to show freedom from all detectable extraneous agents. Detection is often initially available only as a research technique, but rapidly becomes a regulatory requirement. For live measles virus vaccine, for example, the chick embryo cell cultures are required to be derived from eggs from flocks known to be free from avian leukosis viruses as well as from other avian pathogens. This dates from around 1966, when the avian leukosis viruses became 'detectable' (Harris et al., 1966).

For successful killed-vaccine development it is necessary to know what antigen or antigens are required for clinical immunity. The details vary from case to case and information from one system does not necessarily apply to another. Thus, modified exotoxins of the bacteria are all that is required to prevent tetanus and diphtheria, but, so far at least, similar success has not been achieved with the toxins of *Vibrio cholerae* or *Escherichia coli*. Antigens must be produced in large quantities, purified and rendered inert, that is non-living and non-toxic, yet still retaining the capacity to produce immunity. In order to achieve this it is necessary to be able to measure the antigen specifically, preferably by chemical means, as can be done for example for bacterial polysaccharides, or immunologically, as for example for diphtheria toxin (Glenny & Okell, 1924) and D-antigen of polio virus (Beale, 1961); sometimes it is only possible to measure viable organisms or total mass. Large quantities of antigen are required and the strain selection is aimed at increasing antigen yield, but it is important to realize this may be unrelated to virulence. The major part of the problem is the processing of the material to yield antigen and the subsequent testing. Very large quantities of material need to be tested to ensure that it is free from residual viable organisms – a problem which first came to prominence with bacterial vaccines, but which became especially troublesome in the development of killed polio vaccine and led to the concept of consistency of manufacture as an indication of safety.

BACTERIAL VACCINES

Living vaccines

Attenuated living bacterial vaccines are not very frequently used, BCG is the most important example in human medicine. It is prepared from seed material, but over the years great differences in the seeds and in manufacturing methods have become obvious (IABS, 1977). The most important lesson, from the point of view of general vaccine manufacture, to emerge from experience with BCG was as a result of the Lübeck disaster (Wilson, 1967). This was caused by the use of virulent instead of attenuated organisms in the vaccine. The mistake was possible because the two organisms were handled in the same laboratory and incubator. This has led over the years to a whole paraphernalia of statutory precautions including inspection of establishment, proper qualification and training of staff, the keeping of records, and in certain cases the submission of manufacturing and safety testing protocols to national control authorities before release of vaccine lots for sale.

Other living vaccines are the modified *Salmonella* strains developed by Williams Smith (1956), for example *S. gallinarum*, *choleraesuis* and *dublin* which confer protection widely against salmonella serotypes because clinical immunity does not depend upon the antigens used for distinguishing Salmonella types. Other living attenuated vaccines exist against *Brucella abortus* (S-19 strain) and anthrax (Sterne spore vaccine). These are highly effective and cheap to prepare. However, in both cases effective killed vaccines can be prepared. These are the 45–20 strain vaccine for *B. abortus* and the pure toxin antigen from anthrax.

Killed vaccines

The classic examples of bacterial exotoxins as crucial protective antigens have already been mentioned. More recently the development of polysaccharide vaccines, first for some strains of meningococci and then for pneumococci, shows the power of the approach. Organisms can be grown in culture using strains and conditions that optimize the production of polysaccharides. These polysaccharides can then be purified and freeze-dried and dosage determined by weight of antigen of defined molecular weight. Successful polysaccharide vaccines have been made against type A and C meningococci (Gotschlich, Goldschneider & Artenstein, 1969), *Haemophilus influenzae* type b (Robbins *et al.*, 1973) and fourteen types of pneumococci (Austrian, 1977). There are a number of bacteria that produce similar or identical polysaccharides, for example *E. coli* K1 and *H. influenzae* type b and the more easily grown organism might be a better source of antigen as well as being capable of improvement by genetic manipulation.

Another example is the toxin of anthrax, at first only produced *in vivo* (Belton & Strange, 1954) until Smith and his colleagues devised means for its growth *in vitro* and use as a vaccine (Smith, 1964). Similarly, foot rot in sheep is considered to be caused by the combined action of *Fusobacterium necrophorum* and *Bacteroides nodosus*, yet a vaccine against *B. nodosus* alone is sufficient to prevent the disease. *B. nodosus* freshly isolated from lesions produces fimbriate colonies which are capable of reproducing the disease. On subculture there is a tendency for non-fimbriate avirulent colonies to appear (Short, Thorley & Walker, 1976; Thorley, 1976). Vaccines made from fimbriate cultures are more immunogenic than those made from non-fimbriate ones. Large-scale culture for vaccine production therefore requires the development of media and culture conditions which yield the immunogenic (fimbriate-colony) organisms.

The technology of production of bacteria requires either growth in large tanks for batch culture, exemplified by the Bilthoven unit (van Hemert, 1974), or the use of continuous culture techniques.

VIRAL VACCINES

The substrate

Whereas for the growth of bacteria, major effort is devoted to producing chemically defined media and conditions of growth to improve yields of antigen, for viruses a cell substrate is required. Virus vaccine technology was primitive for many years because of the need to use animals as the source of virus. For example, vaccinia was produced in the skin of calves and sheep for most of the time it was used for the smallpox eradication programme. Later the use of chick embryos was established as a source of viruses in large quantities, and later still the use of tissue culture cells was introduced.

The use of cells in culture was a major advance because it enabled large quantities of virus to be prepared with much less contamination by extraneous material from host tissues or other bacteria or viruses.

The problems have not been completely solved but there are two basic approaches. The first is to establish enclosed colonies or flocks which are monitored on a regular basis. This method has been used mainly for fowls, ducks and rabbits. A list of the tests used for a specific-pathogen-free flock of fowls for vaccine production is given in Table 1.

An alternative approach is to establish cell lines. These have the great advantage of bringing the same concepts and disciplines described for

Table 1. *Tests used to monitor fowls used in the production of a viral vaccine*

Virus	Strain	Test
Adenoviruses	Any avian strain producing a specific reaction	Agar gel precipitin
Avian encephalomyelitis	Van Roekel	Embryo susceptibility
Fowl pox	–	Clinical examination
Infectious bronchitis	Any strain producing a specific reaction, e.g. Beaudette or Massachusetts 41 and Connecticut and Iowa	Agar gel precipitin or complement fixation and serum neutralization
Infectious laryngotracheitis	Any strain producing a specific reaction	Serum neutralization
Influenza type A	Any avian strain producing a specific reaction	Agar gel precipitin
Newcastle disease	Any suitable lentogenic strain	Haemagglutination-inhibition
Infectious bursal disease	Any strain producing a specific reaction with the Cheville strain	Agar gel precipitin
Leukosis	RSV (RAV1 and RAV2)	Serum neutralization or Cofal
Reoviruses	Any avian strain producing a specific reaction	Agar gel precipitin

the seed lot system to the system of holding cell substrate. This entails storing aliquots of seed in a suitable manner, for example frozen in the presence of dimethyl sulphoxide at the temperature of liquid nitrogen, and only using the contents of an ampoule for a defined production programme.

The criteria for acceptance of such cell lines or strains for human vaccines have been defined by a special cell culture committee of the International Association of Biological Standardization in a series of reports (Proceedings, 1970, 1971; Garrett, Bishop & Reeson, 1978). The criteria are basically that the cells have the normal karyology for such cells and are not transformed, as judged by a number of parameters – mainly their inability to grow indefinitely in culture. Such cells have a defined, limited lifespan and are unable to grow in suspended cell culture. This makes the culture of living virus vaccine more difficult than it would be otherwise, but this is mitigated by the fact that usually such vaccines can be diluted several-fold to give an adequate vaccine dose. For living vaccine production, viruses are grown in cells attached to the surface of large bottles and, for fear that any change may alter the growth conditions and therefore the degree of attenuation, innovation has been limited, although there is significant variation in detailed practice between manufacturers.

When the seed lot system is used to initiate virus production for vaccine purposes then obviously detectable, extraneous, viable microbial agents must be absent. The virus in the final vaccine must not be more than ten tissue culture passages from virus used to prepare an immunogenic and safe vaccine. The safety is guaranteed by the quality of the seed and the care and consistency in the production procedure.

For the production of living veterinary vaccines the criteria for the cell substrate have been less generally agreed but, in the USA at least, a living vaccine virus can be propagated in a continuous cell line, provided the final product has been proved safe in the target species for the vaccine; also the cell should not give rise to tumours in the target species for the vaccine. The scale of production of living-virus vaccines again rarely requires the use of equipment larger than a series of roller bottles.

The cell substrate for killed vaccine for veterinary use is different. Here the choice can be made on technological grounds. Some big producers of killed foot and mouth disease vaccine employ BHK cells, which can be grown in suspension in large tanks. The major problem is the necessity to use large quantities of serum to support the growth of cell. This has three drawbacks: it is expensive; it is a potential source of bacterial, mycoplasmal and viral contamination; and it contaminates the final product, making purification both more difficult and more necessary. It is for this reason that some workers still advocate the use of anchorage-dependent cultures which can be maintained serum-free for foot and mouth disease, or have switched from suspended BHK cells to eggs for growth of killed Newcastle disease vaccine. The use of continuous cell lines for the production of human vaccines is not permitted at present, for fear that the cells may be able to transmit their oncogenic potential via the vaccine. It is possible that when the advantages of such cells, in terms of yield, purity of the final product and economy, are fully appreciated that better products will not be denied.

The choice of cell substrate having been made, the main areas of technology available for growth can be divided into two: (a) for cells in suspension, and (b) the anchorage-dependent cells. The growth of cells in suspension has been adopted for the growth of BHK cells used in the production of foot and mouth disease vaccine (Capstick et al., 1962). Large tanks of several thousand litres capacity have been used successfully.

The production of anchorage-dependent cells (for example the human diploid cell strains designated WI38 and MRC5 recommended for vaccine production by WHO and licensed by many national control authorities) is more difficult. There is as yet no standardized approach and many different techniques have been applied.

Glass and plastic bottles   This is the standard laboratory method and has been retained for some living attenuated vaccines. Bottles are limited in surface area, so that large numbers are needed. These occupy a lot of space and are labour intensive; also the hazards of contamination are severe.

Roller bottles   Improvements can be made by using roller bottles of different sizes and configuration.

Multiple plates   Such devices are made of titanium (Molin & Hèden, 1967), or glass (Weiss & Schleicher, 1968). Another similar approach to increase surface area is the use of plastic spirals inside a plastic or glass container of about one litre capacity (House, Shearer & Maroudas, 1972).

Glass beads   The 'Portacell' culture vessel described first by Gey (1933) uses the large surface area (in a small volume) of glass beads with a perfusion system. BHK cells have been used in this way for production of foot and mouth disease virus (Spier & Whiteside, 1976).

Sephadex beads This microcarrier system was initiated by van Wezel in 1967. Sephadex DEAE-A50 beads (100–150 μm diameter) are maintained in stirred suspension in a conventional suspension cell system with all the advantages of environmental control. One gram of Sephadex has a surface area of 6000 cm², enough for 30 Roux bottles. The culture medium ratio is 5 g beads:1 l medium. Human diploid cells (HDC) can be produced at *c.* 1×10^6 ml^{-1} (van Wezel, 1976). The Dutch produce inactivated poliomyelitis vaccine from monkey kidney cells grown in this way. The Sephadex beads can be reduced in surface charge, which has to be low otherwise cells tend to leave the bead.

Multitray system A new disposable, plastic unit provides 6000 cm² holding 1800 ml of medium in a very compact form.

Attenuated viral vaccines

The major advance in the production of these vaccines has been the ability to store the seed virus in small aliquots, so that consistent production is available. This was achieved first by glycerination of calf lymph for vaccinia virus and later by the advent of deep-freeze and freeze-drying techniques. These advances have been crucial both in enabling consistent seed virus to be maintained and also in increasing the stability of living vaccines distributed in the field. The development of freeze-drying for vaccinia virus and the development of stabilizer for viral vaccines, for example for yellow fever (Burfoot, Young & Finter, 1977), have been of great importance for their application.

The production of living vaccines against measles, mumps and rubella viruses also depends upon relatively small-scale growth for seed viruses. This can be accomplished in bottles, roller-culture apparatus, or multiple-plate systems. Production of the vaccines depends also on freeze-drying to yield a stable product.

Living attenuated poliomyelitis virus vaccines pose serious problems in manufacture. They are still usually propagated in monkey kidney cells, which contain a large variety of viruses (Hull, Minner & Mascoli, 1958). The polio virus cannot be freeze-dried but it can be stabilized by the addition of 1 M $MgCl_2$ buffered at pH 6.5 to 7.0.

Two approaches to the development of living viral vaccines have been attempted recently but have so far not been generally successful. The first is the use of recombinant techniques for developing living influenza vaccines (McCahon & Schild, 1972); the second is the use of mutagens to produce larger numbers of mutants with desirable characteristics, particularly 'ts' mutants.

There are many other living attenuated vaccines, especially for veterinary use. The major points are the same, the use of a proven seed virus, and either primary cells from a monitored enclosed colony of animals or continuously cultured cell strains or lines, and finally showing, by a variety of tests varying from plaque morphology to neurovirulence, that the vaccine batch and seed are the same.

Killed virus vaccines

The identification of protective antigens and their production in a substantially pure state has not been widely achieved for virus vaccines but advances in technology are now beginning to make this possible. The first vaccine where this approach has been successful is killed influenza vaccine. The major antigens responsible for protection are the haemagglutinin and neuraminidase, and sensitive and accurate methods for their measurement have been developed, especially for the major antigen, haemagglutinin (both for whole virus and subunit vaccine (Wood et al., 1977)). Virus can be grown in embryonated hens' eggs and purified by zonal centrifugation. After inactivation the virus can either be used whole or split into the two surface antigens, the so-called subunit vaccines. The subunit vaccines are less antigenic than whole-virus vaccines in individuals without prior experience of the related influenza viruses, but this can be overcome by repeated dosage or use of adjuvant.

Another vaccine where pure antigens can be employed is adenovirus. So far only small-scale studies have been carried out using crystalline viral antigens of adenovirus type 5 (Couch et al., 1973), but they convincingly demonstrate the potential of the approach.

Recently, killed poliomyelitis vaccine has been regaining credibility and indeed has continued use in Finland, Sweden, Holland and Canada (Institute of Medicine, 1977). Van Wezel (1967) has pioneered the use of cells on microcarriers for the large-scale production of killed poliomyelitis vaccine on monkey kidney cells. He has also developed purification techniques which ensure that the main component of the vaccine is polio virus particles containing the D-antigen.

The necessity to have the protective antigen present in the vaccine is well illustrated by vaccinia and measles. Killed vaccines for both these viruses have proved far inferior to the living ones, despite the fact that higher titres of neutralizing antibodies can be produced by the use of killed vaccines. Boulter *et al.* (1971) showed that a new antigen is produced during the release of vaccinia virus from infected cells. Vaccines based on the intracellular virus lack this antigen and are poorly protective, but when antibodies to the new antigen are stimulated there is protection. The extracellular virus is not only necessary for effective vaccine production, it is also the essential reagent for assessing meaningful antibody titres. The second example is measles, where a similar, though not identical, story has been uncovered. Measles virus has two surface antigens, the haemagglutinin and the haemolysin. Norrby (1975) found that natural measles and living measles vaccine produced antibodies to both antigens, whereas killed vaccine only produced antibodies to the haemagglutinin. These examples serve to show that the future direction will be towards defined, chemically identified antigens which may eventually be synthesizable.

EPILOGUE

The prospects for vaccine technology are bright. New knowledge of immunology is giving greater insight into the ways in which the immune system may be modulated to advantage. Moreover the possibility of improving the production and purity of immunogen is apparent. The goal of essentially pure antigens for many vaccines is in sight, and others should yield in future.

To counterbalance this are increasing fears about vaccine reactions, for example those thought to be associated with pertussis; the increase in regulations making improvements of old vaccine, and development of new vaccine, time-consuming and costly; the tendency to litigation, which appears to be spreading from the United States of America, also casts a shadow. It is no wonder that despite the exciting possibilities the number of organizations investing in vaccine research is diminishing.

REFERENCES

AUSTRIAN, R. (1977). Pneumococcal infection and pneumococcal vaccine. *New England Journal of Medicine*, **297**, 938–9.

BEALE, A. J. (1961). The D-antigen content in poliovaccine as a measure of potency. *Lancet*, ii, 1166–8.

BELTON, F. C. & STRANGE, R. E. (1954). Studies on protective antigen produced *in vitro* from *Bacillus anthracis*: medium and methods of production. *British Journal of Experimental Pathology*, **35**, 144–52.

BOULTER, E. A., ZWARTOUW, H. T., TITMUS, D. H. & MABER, H. B. (1971). The nature of the immune state produced by inactivated vaccinia virus in rabbits. *American Journal of Epidemiology*, **94**, 612–20.

BURFOOT, C., YOUNG, P. A. & FINTER, N. B. (1977). The thermal stability of a stabilized 17D yellow fever virus vaccine. *Journal of Biological Standardization*, **5**, 173–9.

CAPSTICK, P. B., TELLING, R. C., CHAPMAN, W. C. & STEWART, D. L. (1962). Growth of a cloned strain of hamster kidney cells in suspended cultures and their susceptibility to the virus of foot-and-mouth disease. *Nature, London*, **195**, 1163–4.

COUCH, R. B., KASEL, J. A., PEREIRA, H. G. & HAASE, A. T. (1973). Induction of immunity in man by crystalline adenovirus type 5 capsid antigens (37438). *Proceedings of the Society for Experimental Biology and Medicine*, **143**, 905–10.

GARRETT, A. J., BISHOP, D. & REESON, D. E. (1978). Variability in thymectomized irradiated mice that affects their responses to HeLa cells. *Journal of Biological Standardization*, **6**, 67–72.

GEY, G. O. (1933). An improved technic for massive tissue culture. *American Journal of Cancer*, **17**, 752–6.

GLENNY, A. T. & OKELL, C. C. (1924). Titration of diphtheria toxin and antitoxin by flocculation methods. *Journal of Pathology and Bacteriology*, **27**, 187–200.

GLENNY, A. T. (1925). The principles of immunity applied to protective inoculation against diphtheria. *Journal of Hygiene*, **24**, 301–20.

GOTSCHLICH, E. C., GOLDSCHNEIDER, I. & ARTENSTEIN, M. S. (1969). Human immunity to the meningococcus. IV. Immunogenicity of group A and group C meningococcal polysaccharide in human volunteers. *Journal of Experimental Medicine*, **129**, 1367–84.

HARRIS, R. J. C., DOUGHERTY, R. M., BIGGS, P. M., PAYNE, L. N., GOFFE, A. P., CHURCHILL, A. E. & MORTIMER, R. (1966). Contaminant viruses in two live virus vaccines produced in chick cells. *Journal of Hygiene*, **64**, 1–7.

HOUSE, W., SHEARER, M. & MAROUDAS, N. G. (1972). Method for bulk culture of animal cells on plastic film. *Experimental Cell Research*, **71**, 293–6.

HULL, R. N., MINNER, J. R. & MASCOLI, C. C. (1958). New viral agents recovered from tissue cultures of monkey kidney cells. III. Recovery of additional agents both from cultures of monkey kidney tissues and directly from tissues and excreta. *American Journal of Hygiene*, **68**, 31–40.

IABS (1977). An IABS enquiry into *in vitro* and *in vivo* studies of BCG vaccines. *Journal of Biological Standardization*, **5**, 79.

INSTITUTE OF MEDICINE (1977). *Evaluation of poliomyelitis vaccines*. Washington: Institute of Medicine, National Academy of Sciences.

McCAHON, D. & SCHILD, G. C. (1972). Segregation of antigenic and biological characteristics during influenza virus recombination. *The Journal of General Virology*, **15**, 73–7.

MOLIN, O. & HÈDEN, C. G. (1967). Progress in immunological standardization. Vol. 3 in *Proceedings of the 10th International Congress for Microbiological Standardization, Prague*. Basle: S. Karger.

NORRBY, E. (1975). Occurrence of antibodies against envelope components after immunization with formalin inactivated and live measles vaccine. *Journal of Biological Standardization*, **3**, 375–80.

PARISH, C. R. & LIEW, F. W. (1972). Immune response to chemically modified flagellin. III. Enhanced cell mediated immunity during high and low zone antibody tolerance to flagellin. *Journal of Experimental Medicine*, **135**, 298–311.

PETIT, J. F. & LEDERER, E. (1978). Immunostimulant properties of cell walls and their components. In *Relations between Structure and Function in the Prokaryotic Cell*, eds R. Y. Stanier, H. J. Rogers & J. B. Ward, pp. 177–200. *28th Symposium of the Society for General Microbiology*. Cambridge University Press.

PIERCE, N. F. (1976). Intestinal immunization with soluble bacterial antigens; the example of cholera toxoid. In *Acute Diarrhoea in Childhood*, Ciba Foundation Symposium 42, 129–43. Amsterdam: Elsevier, Excerpta Medica – North-Holland.

PORTER, P. (1973). Intestinal defence in the young pig – a review of the secretory antibody systems and their possible role in oral immunisation. *Veterinary Record*, **92**, 658–64.

Proceedings Minutes of 7th Meeting of the Cell Culture Committee, held at Geneva, Switzerland, 14 September 1970: pp. 54–60. International Association of Biological Standardization, 1970.

Proceedings Minutes 8th Meeting of the Cell Culture Committee, held at Chatham Bars, Cape Cod, Massachusetts, 4–5 October 1971, pp. 65–73. International Association of Biological Standardization, 1971.

ROBBINS, J. B., SCHNEERSON, R., ARGAMAN, M. & HANDZEL, Z. T. (1973). *Haemophilus influenzae* type b: disease and immunity in humans. *Annals of Internal Medicine*, **78**, 259–69.

SHORT, J. A., THORLEY, C. M. & WALKER, P.D. (1976). An electron microscope study of *Bacteroides nodosus*: ultrastructure of organisms from primary isolates and different colony types. *Journal of Applied Bacteriology*, **40**, 311–15.

SMITH, H. (1964). Microbial behaviour in natural and artificial environments. In *Microbial Behaviour 'in vivo' and 'in vitro'*, eds H. Smith & J. Taylor, pp. 1–29. *14th Symposium of the Society for General Microbiology*. Cambridge University Press.

SPIER, P. E. & WHITESIDE, J. P. (1976). The production of foot-and-mouth disease virus from BHK 21 C 13 cells grown on the surface of glass spheres. *Biotechnology and Bioengineering*, **18**, 649–57.

THORLEY, C. M. (1976). A simplified method for isolation of *Bacteroides nodosus* from ovine foot-rot and studies on its colony morphology and serology. *Journal of Applied Bacteriology*, **40**, 301–9.

VAN HEMERT, IR. P. (1974). Vaccine production as a unit process. In *Progress in Industrial Microbiology*, vol. 13, ed. O. J. Hockenhull. Harlow: Churchill-Livingstone.

VAN WEZEL, A. L. (1967). Growth of cell-strains and primary cells on microcarriers in homogeneous culture. *Nature, London*, **216**, 64–5.

VAN WEZEL, A. L. (1976). The large-scale cultivation of diploid cell strains in microcarrier culture. Improvements of microcarriers. In *Developments in Biological Standardization*, WHO/IABS Symposium, **37**, 143–7. Basle: Karger.

WEISS, R. E. & SCHLEICHER, J. B. (1968). A multisurface tissue propagator for mass scale growth of cell monolayers. *Biotechnology and Bioengineering*, **10**, 601–15.

WILLIAMS SMITH, H. (1956). The use of live vaccines in experimental *Salmonella gallinarum* infection in chickens with observations on their interference effect. *Journal of Hygiene*, **54**, 419–32.

WILSON, G. S. (1967). *The Hazards of Immunization*. London: Athlone Press.

WOOD, J. M., SCHILD, G. C., NEWMAN, R. W. & SEAGROATT, V. (1977). An improved single-radial-immunodiffusion technique for the assay of influenza haemagglutinin antigen: application for potency determinations of inactivated whole virus and subunit vaccines. *Journal of Biological Standardization*, **5**, 237–47.

CARBON CATABOLITE REGULATION OF SECONDARY METABOLISM

ARNOLD L. DEMAIN,* YVES M. KENNEL†
AND YAIR AHARONOWITZ‡

* *Department of Nutrition and Food Science,*
Massachusetts Institute of Technology,
Cambridge, Massachusetts 02139 USA

† *Rhone-Poulenc Industries, 94400 Vitry-sur-Seine, France*

‡ *Department of Microbiology, Tel-Aviv University, Tel-Aviv, Israel*

INTRODUCTION

Epps & Gale (1942) discovered that the synthesis of enzymes of primary metabolism (i.e., metabolism involved in growth) is influenced by glucose, and raised the question of the biological significance of this observation. Magasanik (1961) saw a survival advantage in this phenomenon and termed it 'catabolite repression', i.e. the inhibition of synthesis of inducible enzymes by intermediates produced in the rapid catabolism of glucose. A classic example involves the growth of a microorganism on a mixture of lactose and glucose. There is no advantage for it in making enzymes required for catabolism of lactose so long as ample supplies of glucose, a more favourable substrate, are present. 'More favourable' generally means able to support a higher growth rate. Cells provided with a control mechanism in which the better carbon source is used first have the advantage that the machinery for protein synthesis is used in an economical manner, and they will find the lactose ready to use when the glucose is exhausted.

The phenomenon of carbon catabolite repression in certain bacteria has been shown to act according to the scheme given in Fig. 1 (Pastan & Perlman, 1970). It involves the binding of a complex between cyclic adenosine 3′–5′-monophosphate (cAMP) and a cAMP receptor protein (CRP) to the promotor site of an operon; this binding stimulates the initiation of transcription by RNA polymerase. When glucose is present it inhibits adenylate cyclase, the enzyme that converts ATP to cAMP, thus decreasing the concentration of cAMP and inhibiting the transcription by RNA polymerase of operons subject to this control. These conclusions are derived from the work of many researchers

(Nakada & Magasanik, 1962; Makman & Sutherland, 1965; Ullman & Monod, 1968; de Crombrugghe *et al.*, 1970; Paigen & Williams, 1970; Pastan & Perlman, 1970; Aboud & Burger, 1971; Harwood & Peterkofsky, 1975; Náprstek *et al.*, 1975; Pastan & Adhya, 1976). Although glucose has been most extensively studied as the effector of carbon catabolite represssion, other rapidly used carbon sources are also active.

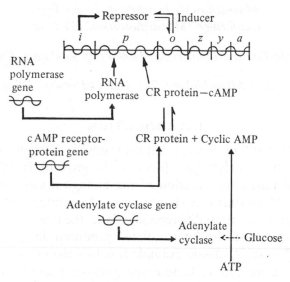

Fig. 1. Scheme depicting action of cyclic AMP on the *lac* operon. (After Pastan & Perlman, 1970.)

Although most discussions on carbon catabolite repression deal with enteric bacteria, the phenomenon also occurs in other bacteria, yeasts, and moulds. In yeasts, cAMP and CRP appear to be involved (van Wijk & Konijin, 1971; Sy & Richter, 1972; Schlanderer & Dellweg, 1974). Saruno, Tamura & Kato (1976) have recently claimed that carbon catabolite repression in the mould *Monascus purpureus* is reversed by cAMP, and that CRP is present in the mycelium. In contrast, in the genus *Bacillus* the phenomenon appears to be independent of cAMP, which has not been found in these organisms (Lopez & Thoms, 1977).

Another form of catabolite regulation exists, i.e. catabolite inhibition, in which metabolism of a favoured carbon source inhibits the activity of pre-existing enzymes (Evans, Handley & Happold, 1942; Romano & Kornberg, 1968; Morgan & Kornberg, 1969; Kapoor & Grover, 1970; Paigen & Williams, 1970).

CATABOLITE REGULATION OF
SECONDARY METABOLISM

It is known that the use of different carbon sources for production of a given metabolite results in different rates and extents of production. Thus an important practical aspect of any developmental programme for the industrial production of a new compound by fermentation is the assessment of the best carbon-source, or mixture of carbon-sources, from an economic point of view. However, the mechanisms by which a given carbon-source will affect positively or negatively the production of a secondary metabolite are not clearly understood.

At one time it was considered rather bizarre to think of secondary metabolism as being regulated by carbon-sources, or by any other type of effector. This was a natural consequence of the prevailing thought, which accepted the assumptions that secondary metabolites have no function in the survival of the producing organism; that production of secondary metabolites is a laboratory artefact and does not occur in nature; and that micro-organisms producing secondary metabolites, such as antibiotics, are resistant to the action of these potent molecules. Considerable evidence, however, has been put forth to support the opposite views, i.e. that such metabolites are produced in nature and aid in the survival of the producing organism (Brian, 1957; Krasil'nikov, 1958; Burkholder, 1959; Katz & Demain, 1977) and that producing organisms are sensitive to their own metabolites, at least in the early-exponential phase of growth (Demain, 1974b). Since many of these secondary metabolites are extremely potent antibiotics, it makes biological sense for cells to suppress the formation of these toxic compounds until rapid growth is nearing completion. Viewed from this angle, catabolite regulation of secondary metabolism, like that of primary metabolism, offers the cell a survival advantage. Thus, it is not surprising that glucose suppresses production of a large number of secondary metabolites (Table 1).

It should be noted that objections have been raised against the concept that catabolite regulation is involved in the negative effect of glucose on secondary metabolism. To be specific, Haavik (1974a) reported that the inhibitory effect of glucose on bacitracin formation was due to acid production. Further investigation (Haavik, 1974b), however, showed that the suppression of bacitracin formation was not due to low pH but to some other effect of the acetic and pyruvic acids produced. The acids were shown to suppress antibiotic synthesis only when in the undissociated state. Haavik (1974b) postulated that this was the result

Table 1. *Secondary metabolites whose production is reported to be suppressed by glucose*

Secondary metabolite	Organism	Reference
Gibberellic acid	*Fusarium moniliforme*	Darken, Jensen & Shu (1959)
Actinomycin	*Streptomyces antibioticus*	Gallo & Katz (1972)
	Streptomyces paravalus	Williams & Katz (1977)
Indolmycin	*Streptomyces griseus*	Hurley & Bialek (1974)
Siomycin	*Streptomyces sioyaensis*	Kimura (1967)
Chloramphenicol	*Streptomyces* sp.	Umezawa *et al.* (1948)
Penicillin	*Penicillium chrysogenum*	Johnson (1952)
Cephalosporin	*Cephalosporium acremonium*	Demain (1963)
Kanamycin	*Streptomyces kanamyceticus*	Satoh, Ogawa & Satomwa (1976)
Puromycin	*Streptomyces alboniger*	Pogell *et al.* (1976)
Neomycin	*Streptomyces fradiae*	Maxon & Chen (1966)
Ergot alkaloids	*Claviceps purpurea*	Taber (1967), Brar, Giam & Taber (1968)
Violacein	*Chromobacterium violaceum*	Friedheim (1932)
Prodigiosin	*Serratia marcescens*	Ramsey, Quadri & Williams (1973)
Mitomycin	*Streptomyces verticillatus*	Kirsch (1967)
Bacitracin	*Bacillus licheniformis*	Haavik (1974a, b)
Enniatin	*Fusarium sambucinum*	Andhya & Russell (1975)
Streptothricins (racemomycins)	*Streptomyces lavendulae,* *Streptomyces racemochromogenus,* *Streptomyces albidoflavus*	Sawada *et al.* (1976)
Benzodiazepine alkaloids	*Penicillium cyclopium*	Luckner, Nover & Böhm (1977)

of the undissociated acid's ability, once excreted, to pass back through the cell membrane, create a low internal pH upon dissociation and suppress the biosynthetic process. However, lowering the pH with a mixture of hydrochloric and phthalic acids did not interfere with bacitracin synthesis. We feel that Haavik did not eliminate a direct repression or inhibitory effect of pyruvic and/or acetic acid on bacitracin synthetase, a possibility that would support the concept of carbon catabolite regulation. To eliminate pH as a factor it would be necessary to conduct these experiments with pH control. It is important to note that, in certain cases, pH has in fact been eliminated as a factor in the glucose effect (Sankaran & Pogell, 1975; Williams & Katz, 1977).

Table 2 gives examples of processes in which carbon-sources other than glucose are best for secondary metabolism. It is suspected that this is due to catabolite regulation by glucose, although the suppressive effect of glucose has not yet been tested in these cases.

In situations in which carbon-sources other than glucose are favoured for growth, the favoured carbon-source suppresses secondary metabolism. For example, β-carotene production by *Mortierella ramanniana* var. *ramanniana* is best on fructose, a poor carbon-source, and poorest on galactose, the best carbon-source tested (Attwood, 1971). Novobiocin production by *Streptomyces niveus* in a defined medium containing citrate and glucose occurs only after the preferred carbon-source, citrate, is exhausted. After a diauxic growth lag, glucose is used, and the antibiotic is produced (Kominek, 1972).

Table 2. *Secondary metabolite processes[a] in which other carbon-sources are preferred over glucose*

Secondary metabolite	Micro-organism	Preferred carbon-source	Reference
Tetracycline	*Streptomyces aureofaciens*	Sucrose, starch	Van Dyck & DeSomer (1952)
Erythromycin	*Streptomyces erythreus*	Sucrose	Corum *et al.* (1954)
Kasugamycin	*Streptomyces kasugaensis*	Soy bean oil	Yagi *et al.* (1971)
Butirosin	*Bacillus circulans*	Glycerol	Howells *et al.* (1972)
Fortimicin	*Micromonospora* sp.	Starch, dextrin	Nara *et al.* (1977)

[a] Other than those mentioned in Table 1.

In most of the above cases, the mechanism by which the carbon-source controls secondary metabolism is unknown. However, catabolite repression has been shown to be important in the suppression of actinomycin biosynthesis by glucose (Gallo & Katz, 1972). In the classical defined medium containing glucose and galactose, *Streptomyces antibioticus* delays antibiotic synthesis until glucose is exhausted. If additional glucose is added prior to the initiation of actinomycin synthesis, antibiotic synthesis is severely suppressed. Rapid growth accompanies glucose assimilation. Formation of phenoxazinone synthase, an enzyme required for actinomycin synthesis, is repressed by glucose but not by galactose. Other rapidly used carbon-sources also repress phenoxazinone synthase. A similar phenomenon has been observed in puromycin fermentations. *O*-Demethylpuromycin methyltransferase, the final enzyme of puromycin biosynthesis, is repressed by glucose (Pogell *et al.*, 1976). Several carbon-sources cause catabolite repression of *N*-acetylkanamycin amidohydrolase in *S. kanamyceticus*. This enzyme, thought to be the final one of kanamycin biosynthesis, is repressed by glucose, fructose, lactose, mannose and maltose (Satoh *et al.*, 1976). The report that cAMP relieves glucose repression in

S. kanamyceticus is the only one indicating that the cyclic nucleotide has such an effect in *Streptomyces*.

Carbon catabolite repression is also important in the interconversion of members of antibiotic families. α-D-Mannosidase, which converts the undesirable mannosidostreptomycin to streptomycin in *Streptomyces griseus*, is repressed by glucose (Demain & Inamine, 1970). The conversion of cephalosporin C to the less active desacetylcephalosporin C in *Cephalosporium acremonium* is repressed by glucose, maltose and sucrose; glycerol and succinate are not repressive (Hinnen & Nüesch, 1976).

As mentioned earlier, glucose or other readily utilizable carbon sources may inhibit the action of pre-existing enzymes rather than repress their synthesis (Paigen & Williams, 1970). With respect to secondary metabolism, cells actively producing neomycin (Maxon & Chen, 1966; Majumdar & Majumdar, 1971), siomycin (Kimura, 1967) or benzylpenicillin (Demain, 1968) are suppressed in antibiotic formation by glucose addition. Unfortunately, due to the long duration of these experiments, and thus the possibility of new protein synthesis, it is difficult to decide whether the effect is indeed a result of immediate catabolite inhibition or of catabolite repression of an enzyme(s) being rapidly degraded and resynthesized.

CATABOLITE REGULATION OF β-LACTAM BIOSYNTHESIS

The antibiotics belonging to the β-lactam group have proved to be of great interest in chemotherapy, mainly because of their selective mode of action. Their specific inhibition of cell wall synthesis renders them non-toxic.

The biosynthesis of cephalosporins and penicillins has been studied over the years (see Demain, 1974a for review) and, although the enzymology is not yet completely known, the work makes it possible to assess the general pathways of β-lactam production by micro-organisms. A practical point of interest lies in controlling the rate of production and yield of these molecules, via a better understanding of the process in terms of its metabolic regulation.

The earliest recognition of a negative effect of glucose on a secondary process involved benzylpenicillin production (Johnson, 1952). Glucose was found to be excellent for the growth of *Penicillium chrysogenum* but poor for penicillin production; lactose showed the opposite pattern (Soltero & Johnson, 1953). A medium containing both glucose and lactose was devised (Jarvis & Johnson, 1947) in which growth occurred only on glucose until hexose was exhausted. At that point, the extensive mycelial mass developed on glucose began to produce antibiotic on lactose. It was subsequently found (Davey & Johnson, 1953) that penicillin production could be increased simply by feeding glucose intermittently so that its level in the medium never became high enough to interfere with antibiotic production. Continuous feeding of glucose or other sugars was soon established as a routine practice in industry. Resting mycelia with a fully developed penicillin-synthesizing system were later found (Demain, 1968) to be suppressed in benzylpenicillin synthesis by glucose but not by lactose. Although this suggested catabolite inhibition rather than repression, the experiments were somewhat long in duration, and no precautions were taken to prevent new protein synthesis.

Similar findings were obtained in the case of cephalosporin C formation by *Cephalosporium acremonium* (Demain, 1963). The best defined medium for formation of this antibiotic contained glucose and sucrose. During trophophase, glucose was rapidly depleted and growth occurred. During idiophase, the slowly assimilated sucrose was consumed, and both cephalosporin C and penicillin N were made.

We recently examined the carbon-source control of cephalosporin biosynthesis in the actinomycete, *Streptomyces clavuligerus* (Aharonowitz & Demain, 1976). This strain produces four β-lactam antibiotics (Nagarajan, 1972; Brown *et al.*, 1976; Howarth, Brown & King, 1976): penicillin N, clavulanic acid, and two cephalosporins, 7-(5-amino-5-carboxyvaleramido)-3-carbamoyloxymethyl-3-cephem-4-carboxylic acid and 7-(5-amino-5-carboxyvaleramido)-7-methoxy-3-carbamoyloxyme-thyl-3-cephem-4-carboxylic acid (the latter is also known as cephamycin C). We first developed a chemically defined medium for growth and antibiotic production and described the phosphate control of cephalosporin production in this strain (Aharonowitz & Demain, 1977). We next found the organism incapable of using glucose in the defined medium, although it could grow on starch, maltose, glycerol, succinate, α-ketoglutarate and fumarate. Growth on the carbohydrates was much more extensive than that on the organic acids.

Growth on glycerol and maltose led to a high level of biomass, but the specific production of cephalosporins decreased as carbohydrate concentration was increased. As can be seen in Fig 2, glycerol addition increased both the rate and extent of growth, indicating that carbohydrate was the growth-limiting factor. A glycerol concentration of 0.2% accelerated the growth rate. Higher levels had no effect on growth rate

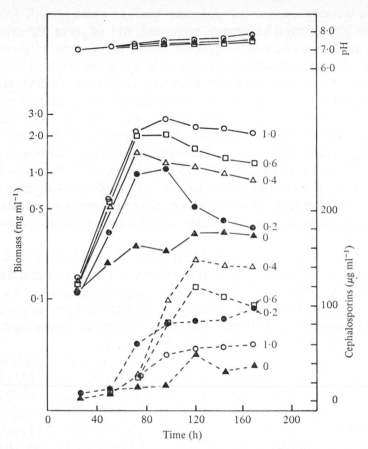

Fig. 2. Growth (solid lines) and cephalosporin production (broken lines) of *Streptomyces clavuligerus* as a function of time in the presence of different glycerol concentrations (0 to 1.0% v/v) and 0.4% L-asparagine.

but further increased the extent of growth. The culture reached the stationary growth phase after 70–80 h, and antibiotic production continued until 120 h. When the maximum values for biomass, volumetric cephalosporin titre and specific cephalosporin titre were plotted against glycerol concentration (Fig. 3), it could be seen that the extent of growth increased linearly with increasing concentration of glycerol. Whereas volumetric production of cephalosporins was optimal at 0.4% glycerol, specific cephalosporin production of the antibiotic was inversely proportional to glycerol concentration. These data show that growth on glycerol, the preferred carbon-source, inhibits or represses cephalosporin production. Therefore, to achieve the highest volumetric yield, it is not sufficient only to provide conditions for high specific production,

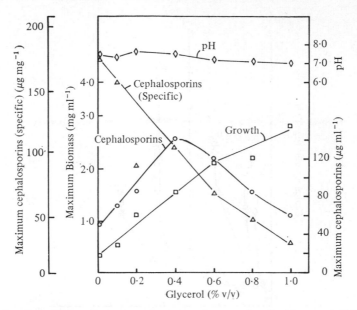

Fig. 3. The effect of glycerol concentration on maximum values for biomass, cephalosporin titre, specific cephalosporin titre and pH, in the presence of 0.4% L-asparagine. Maximum specific cephalosporin titre is the maximum cephalosporin titre per ml (in μg) divided by the maximum biomass per ml (in mg). The organism is *Streptomyces clavuligerus*.

a critical level of biomass is also needed. Glycerol at 0.4% provided that degree of biomass development.

Fig. 4 represents a differential plot of the data obtained for Fig. 3. It can be seen that increasing glycerol concentrations not only affected the volumetric titre but also the relationship between the growth and production phases. At low glycerol concentrations, production appeared to be associated with growth. However, at concentrations above 0.2%, we observed a distinct separation between growth and secondary metabolism. Similar results were obtained when maltose was used as the major carbon-source. When starch was used as carbon-source, production was more closely associated with growth than was the case with glycerol or maltose. We postulate that this is due to the slow hydrolysis of starch, which creates a situation of carbon limitation and thus releases the culture from carbon-source regulation. When *S. clavuligerus* was grown on organic acids such as α-ketoglutarate or succinate (which are relatively poor carbon-sources), antibiotic production resembled that on starch, i.e. specific production was high and associated with growth (Fig. 5).

Our results strongly suggest some type of carbon catabolite control of cephalosporin production in *S. clavuligerus*. Whether the regulation is

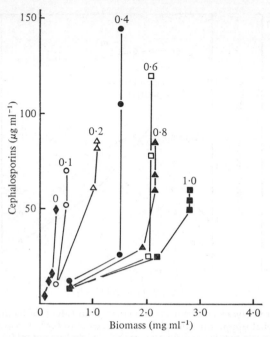

Fig. 4. The effect of glycerol concentration (0 to 1% v/v) on the differential rate of cepha-losporin production by *Streptomyces clavuligerus*. The volumetric titre of cephalosporin is plotted against the highest biomass reached up to that particular point, i.e. lysis is ignored in this plot.

due to carbon catabolite repression or inhibition is not yet known. It is clear, however, that the negative effect of preferred carbon-sources is not due to trivial effects such as acid production or oxygen limitation, since pH was relatively constant and virtually unaffected by carbo-hydrate utilization and the negative effect occurred at low biomass concentrations where oxygen could not have been limiting. Thus, we are indeed dealing with a true regulatory effect.

The data obtained with *S. clavuligerus* suggest that the definition of the term 'secondary metabolite' should not include the stage in the growth cycle in which it is produced. It is particularly irritating to see secondary metabolites defined as those compounds produced after growth has ceased or in the absence of growth (although we have no argument with investigators who state that secondary metabolites are *often* produced after growth in batch culture). Of course, those using the stage of the growth cycle as part of the definition have long ignored the production of antibiotics by steady-state (i.e., exponentially growing) cultures in the chemostat (Pirt & Righelato, 1967) and the production of primary metabolites such as glutamic acid, citric acid and vitamin B_{12}

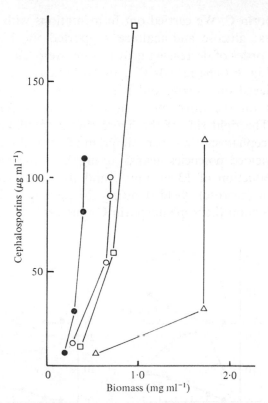

Fig. 5. The effect of different carbon sources on the differential rate of cephalosporin production by *Streptomyces clavuligerus*. The concentrations of glycerol, starch, succinate and α-ketoglutarate were 0.5%. The volumetric titre of cephalosporin is plotted against the highest biomass reached up to that particular point, i.e., lysis is ignored in this plot. Symbols: ●, α-ketoglutarate; ○, succinate; □, starch; △, glycerol.

during the stationary phase of their respective cultures. They have also ignored the findings that antibiotic production is often growth-associated in one medium but produced after growth in another medium (Pirt & Righelato, 1967; Ito, Aida & Uemura, 1969; Malik & Vining, 1970). The important point is that trophophase–idiophase dynamics are a function of nutrition or, more precisely, of nutritional regulation, and not a function of the type of molecule produced. In our opinion, the only valid definition of a secondary metabolite is that it is a molecule not required for growth of the organism that produces it.

As mentioned above, cephalosporin C production by *C. acremonium* in a glucose- and sucrose-containing defined medium showed fermentation dynamics suggesting carbon-source control (Demain, 1963; 1974a). Recently we had an opportunity for closer examination of *C. acremonium* and the production of its two β-lactam antibiotics, penicillin N and cephalosporin C. We carried out fermentations with five different carbon sources: glucose and maltose supported the highest growth rates; then, in order of decreasing growth rate, were fructose, galactose and sucrose. Fig. 6 (left) reveals that the volumetric cephalosporin C potencies achieved on glucose and maltose were about 100 μg ml⁻¹, whereas those on the three more slowly utilized sugars were over 200 μg ml⁻¹. The right side of the figure shows specific potencies, i.e. microgram of cephalosporin C per milligram of dry cell weight. Glucose and maltose yielded potencies near 5 μg mg⁻¹, fructose and galactose supported production of 12 μg mg⁻¹, and the sugar supporting the slowest growth (sucrose), yielded nearly 20 μg mg⁻¹. Data on total β-lactam production (i.e. cephalosporin C plus penicillin N) reveal the

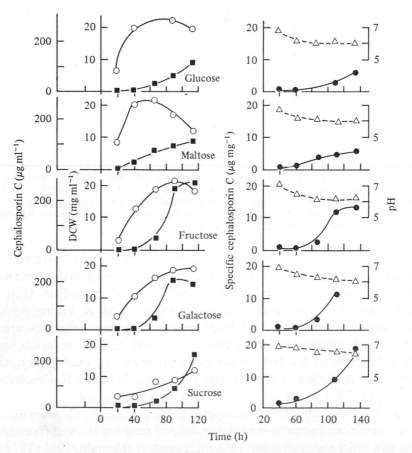

Fig. 6. Growth (○), cephalosporin C production (■), specific cephalosporin C production (●) and pH (△) in a *Cephalosporium acremonium* culture grown in media containing different sugars.

same general trend (Table 3). Specific antibiotic production on these carbon-sources decreased in the following order: sucrose, galactose, fructose, maltose, glucose. Thus, a roughly inverse correlation was found between growth rate and antibiotic formation.

Table 3. β-*Lactam antibiotic production by* Cephalosporium acremonium *on different carbon-sources*

Carbon-source	Maximum β-lactam antibiotic production (μg ml^{-1})	Maximum dry cell weight (mg ml^{-1})	Specific β-lactam antibiotic production (μg mg^{-1})
Glucose	830	22.5	36.9
Maltose	1130	21.8	51.9
Fructose	1250	21.5	58.1
Galactose	1650	19.1	86.4
Sucrose	1040	11.9	87.4

Experiments were next conducted with three different levels of glucose. As shown in Fig. 7, growth ceased in each case when glucose was exhausted from the medium. Of particular interest was the observation that β-lactam antibiotics were formed at a very slow rate during

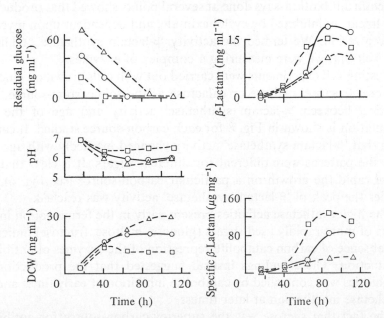

Fig. 7. Growth and total β-lactam antibiotic formation by *Cephalosporium acremonium* in media containing different concentrations of glucose.

growth and, at least in the case of the fermentations containing low and intermediate amounts of glucose, production accelerated only at the point of glucose depletion. When glucose levels were initially high, accelerated β-lactam production never occurred. The highest specific production was attained with lowest initial glucose concentration and vice versa All these data support a strong regulation of β-lactam antibiotic production by the carbon source used for growth.

Two questions were raised by the above fermentation experiments:

(1) Why was more antibiotic made per cell on sucrose than on glucose?

(2) Why was more antibiotic made per cell at the lowest glucose level than at the highest glucose level?

To answer these questions, we attempted to determine whether carbon catabolite control was being exerted at the level of repression or inhibition. Resting-cell experiments were used to obtain such information, because production of antibiotic under conditions where no protein synthesis can occur can be attributed only to the enzymes present at the time of the cell harvest. In this technique, washed mycelia were incubated in phosphate buffer plus cycloheximide for up to 3 h. Total β-lactam antibiotic assays done at several points showed that production was linear, uninhibited by cycloheximide, and dependent upon mycelial concentration. We termed the activity 'β-lactam synthetase', realizing full well that we were measuring a complex of enzymes.

Resting-cell experiments were carried out with cells grown on glucose, sucrose, fructose, maltose or galactose as the sole carbon-source. The relation between 'β-lactam synthetase' activity and age of the cell population is shown in Fig. 8 for each carbon-source studied. It can be seen that 'β-lactam synthetase' activity changed markedly with age and that the patterns were different for different sugars. It is clear that the more rapid the growth on a particular carbon-source (see Fig. 6), the earlier the peak of 'β-lactam synthetase' activity was reached.

The high synthetase activities present early in the fermentation in the cases of more rapidly used sugars (glucose, maltose, fructose) indicated the absence of carbon catabolite repression of the enzymes of antibiotic production. These findings instead suggested that the production of antibiotics was controlled by catabolite inhibition at early times and by synthetase inactivation at later times.

The fact that sucrose was the superior carbon-source for antibiotic production by *C. acremonium* (see Fig. 6, Table 3 and Demain, 1971)

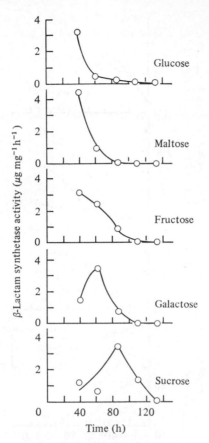

Fig. 8. 'β-Lactam synthetase' activity of *Cephalosporium acremonium* as a function of time in fermentations in media containing different sugars.

could be explained by the later appearance of the synthetase activity and its presence during the idiophase of the fermentation. This would be especially relevant if the early synthetase activity (e.g. activity at 40 h in the cases of glucose, maltose and fructose) was non-functional due to catabolite inhibition. For example, Fig. 6 indicates very poor production of β-lactam antibiotics at 40 h in fermentations with high glucose levels. If peak synthetase activity was present at this time, why were β-lactam antibiotics not rapidly made? To investigate this further, synthetase activities were determined in cells growing at three different concentrations of glucose.

At all three glucose concentrations, 'β-lactam synthetase' activity peaked at 40 h, and cell dry weights were roughly equivalent in all three cases (Fig. 9). Yet only the 'low-glucose cells' started their rapid

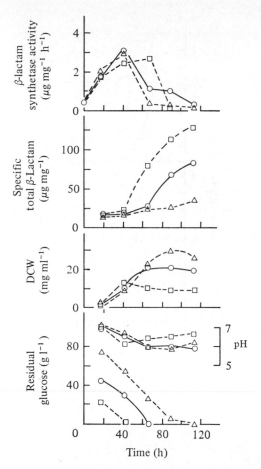

Fig. 9. Growth, total β-lactam antibiotic formation and 'β-lactam synthetase' activity of *Cephalosporium acremonium* in media containing different concentrations of glucose.

β-lactam production at this time. Since they were the only cells that had exhausted their sugar, catabolite inhibition by glucose was indicated as the mechanism of regulation. At 65 h, the cells grown at the intermediate glucose concentration exhausted their glucose supply and started rapid antibiotic production; at this time, there was still considerable 'β-lactam synthetase' left. By the time the cells grown with a high sugar concentration depleted their sugar (90 h), no 'β-lactam synthetase' activity remained; thus, these cells never increased their β-lactam production rate. On a specific basis, β-lactam antibiotic production was highest at the lowest glucose concentration and lowest with the highest sugar level.

One possible explanation for the burst of production when glucose was depleted is that metabolism of high levels of glucose leads to low pH values, which impair antibiotic production by 'β-lactam synthetase'. Observation of pH changes, however, showed that the decreases were minimal, and it is doubtful that the pH changes could account for the observed differences in antibiotic production. More likely is the possibility that antibiotic production was inhibited during growth, due to catabolite inhibition of 'β-lactam synthetase'. To examine this possibility, we compared the rate of antibiotic production by resting cells in the presence and absence of sugars but found no marked inhibition of 'β-lactam synthetase' by glucose or any other sugar (Table 4). However,

Table 4. '*β-Lactam synthetase*' activity in the presence of different sugars[a]

Addition to buffer	'β-Lactam synthetase' activity (μg per mg cell per h)
None	0.53
1% Sucrose	0.53
1% Glucose	0.55
1% Fructose	0.53
1% Maltose	0.65

[a] Cells were harvested after growth on defined medium, then washed and incubated in buffer with the sugars indicated.

the combination of glucose plus ammonium sulphate did inhibit production (Table 5). When complete medium was used, the rate was

Table 5. '*β-Lactam synthetase*' activity in the presence of different additives[a]

Incubation mixture	β-lactam synthetase activity (μg per mg cells per h)	Activity as % of control	pH
Buffer	1.24	100	7.11
Buffer + 6.3% glucose	1.07	86	7.00
Buffer + 0.75% $(NH_4)_2SO_4$	1.11	89	7.05
Buffer +0.75% $(NH_4)_2SO_4$ + 6.3% glucose	0.83	67	6.70
Buffer + 0.75% $(NH_4)_2SO_4$ + 6.3% glucose + 0.3% methionine	0.75	61	6.70
Growth medium	0.72	58	7.10

[a] Cells were harvested after growth on defined medium containing 6.3% glucose as carbon-source, washed and incubated as indicated.

only slightly lower than that with ammonium sulphate plus glucose. A slight pH change was observed when the buffer contained ammonium sulphate, but the inhibition observed could not be attributed to pH, since we still saw the inhibition in complete medium without significant pH change. We thus conclude that it is mainly the combination of ammonium sulphate and glucose that is responsible for the inhibition of 'β-lactam synthetase' in defined production medium.

SUMMARY

Catabolite regulation of primary metabolism is a mechanism that ensures the economical use of cell metabolic machinery, and it thus provides a survival advantage. Likewise, catabolite regulation of secondary metabolism has significance for survival if it retards antibiotic formation until exponential growth is near completion, i.e. it helps prevent suicide by the antibiotic-producing micro-organism. Production of many secondary metabolites is suppressed by glucose or other preferred carbon-sources.

The earliest report of a negative effect of glucose was published during the 1950s and involved the production of benzylpenicillin by *Penicillium chrysogenum*. During the next decade, a glucose effect was also detected in the production of cephalosporins by *Cephalosporium acremonium*. In neither case was the mechanism elucidated. We recently examined the effect of various sugars on the biosynthesis of these antibiotics by *Streptomyces clavuligerus* and *C. acremonium*. In the streptomycete, cephalosporin production is regulated by some type of carbon catabolite control. Preferred carbon sources such as glycerol and maltose decreased production of the antibiotics. Poorer carbon sources such as α-ketoglutarate and succinate led to high specific production of cephalosporins and shifted the dynamics of the fermentation to a greater degree of association with growth. In *C. acremonium*, we also found the carbon-source to regulate biosynthesis of the β-lactam antibiotics penicillin N and cephalosporin C. This was shown by fermentations using one of five sugars as the major carbon-source; an inverse relationship was found between rate of growth on a particular sugar and specific antibiotic titre. Fermentations using three different concentrations of glucose showed that antibiotic formation markedly accelerated only when glucose was exhausted from the medium. An inverse relationship was found between glucose concentration and specific antibiotic titre. A short-time, resting-cell system was used to assay 'β-lactam synthetase' activity. Surprisingly, the specific activity of this complex of enzymes reached its peak in the mycelia during rapid growth on each sugar, even though little or no β-lactam was being formed at that time in the fermentation. Thus, carbon catabolite repression is probably not the factor controlling β-lactam formation in this case; rather it appears to involve catabolite inhibition by a combination of carbon- and nitrogen-sources.

Although our data suggest that glucose causes catabolite inhibition of one or more enzymes of β-lactam formation, we realize that this suggestion must be taken with a great deal of caution, due to the complexity of the process involved. The production of an antibiotic involves a great number of reactions over and above those responsible for converting α-aminoadipate, cysteine and valine to a penicillin or a cephalosporin. Of great importance are transport systems, the energy status of the cells, the activity of the pathways of primary metabolism and the intracellular concentrations of the antibiotic precursors. Any repressive or inhibitory effect of the carbon-source on these metabolic parameters could reflect itself in a lower rate of antibiotic production, despite the presence of a fully developed antibiotic-synthesizing system. Furthermore, the use of resting cells producing antibiotic from endogenous reserves in buffer as a measure of 'β-lactam synthetase' should be accepted with caution. For example, decreases in the pool concentrations of the amino acid precursors could reduce the antibiotic-synthesizing activity without necessarily affecting the concentration of enzymes present in the resting cells. We thus present our conclusions with trepidation and hope that the efforts of our laboratory and our colleagues around the world will soon result in the elucidation of the enzymology of β-lactam biosynthesis. Only when this is accomplished will we be able to subject our hypotheses concerning carbon regulation to a true test.

Acknowledgements

The research on *S. clavuligerus* was supported by National Science Foundation Grant No. GB-43351 (BMS-18791). We thank Rhone-Poulenc Industries for support of Yves M. Kennel during his stay at Massachusetts Institute of Technology. The work on *C. acremonium* was funded by the Industrial Division of the Bristol–Myers Company.

REFERENCES

ABOUD, M. & BURGER, M. (1971). Cyclic 3′, 5′ adenosine monophosphate-phospho-diesterase and the release of catabolite repression of β-galactosidase by exogenous cyclic AMP in *Escherichia coli*. *Biochemical and Biophysical Research Communications*, **43**, 174–82.

AHARONOWITZ, Y. & DEMAIN, A. L. (1976). Metabolic regulation of cephamycin biosynthesis. *Interscience Conference on Antimicrobial Agents and Chemotherapy*, 16th Conference (Abstracts), p. 49. Washington: American Society for Microbiology.

AHARONOWITZ, Y. & DEMAIN, A. L. (1977). Influence of inorganic phosphate and organic buffers on cephalosporin production in *Streptomyces clavuligerus*. *Archives of Microbiology*, **115**, 169–73.

ANDHYA, T. K. & RUSSELL, D. W. (1975). Enniatin production by *Fusarium sambucinum*: primary, secondary, and unitary metabolism. *Journal of General Microbiology*, **86**, 327–31.

ATTWOOD, M. M. (1971). The production of β-carotene in *Mortierella ramanniana* var. *rammaniana* M29: the effect of changes in the environment upon growth and pigmentation. *Antonie van Leeuwenhoek, Journal of Microbiology and Serology*, **37**, 369–78.

BRAR, S. S., GIAM, C. S. & TABER, W. A. (1968). Patterns of in vitro ergot alkaloid productivity by *Claviceps paspali* and their association with different growth rates. *Mycologia*, **60**, 806–26.

BRIAN, P. W. (1957). The ecological significance of antibiotic production. In *Microbial Ecology*, eds R. E. O. Williams & C. C. Spicer, pp. 168–88. *7th Symposium of the Society for General Microbiology*. Cambridge University Press.

BROWN, A. G., BUTTERWORTH, D., COLE, M., HANSCOMB, G., HOOD, J. D., READING, C. & ROLINSON, G. N. (1976). Naturally occurring β-lactamase inhibitors with antibacterial activity. *Journal of Antibiotics*, **29**, 668–9.

BURKHOLDER, P. R. (1959). Antibiotics. *Science*, **129**, 1457–65.

CORUM, C. J., STARK, W. M., WILD, G. M. & BIRD, H. L., Jr. (1954). Biochemical changes in a chemically defined medium by submerged cultures of *Streptomyces erythreus*. *Applied Microbiology*, **2**, 326–9.

DARKEN, M. A., JENSEN, A. L. & SHU, P. (1959). Production of gibberellic acid by fermentation. *Applied Microbiology*, **7**, 301–3.

DAVEY, V. F. & JOHNSON, M. J. (1953). Penicillin production in corn steep media with continuous carbohydrate addition. *Applied Microbiology*, **1**, 208–11.

DE CROMBRUGGHE, B., VARMUS, H. E., PERLMAN, R. L. & PASTAN, I. (1970). Stimulation of *lac* mRNA synthesis by cyclic AMP in cell free extracts of *Escherichia coli*. *Biochemical and Biophysical Research Communications*, **38**, 894–901.

DEMAIN, A. L. (1963). Biosynthesis of cephalosporin C and its relation to penicillin formation. *Transactions of the New York Academy of Sciences*, Series II, **25**, 731–42.

DEMAIN, A. L. (1968). Regulatory mechanisms and the industrial production of microbial metabolites. *Lloydia*, **31**, 395–418.

DEMAIN, A. L. (1971). Overproduction of microbial metabolites and enzymes due to alteration of regulation. *Advances in Biochemical Engineering*, **1**, 113–42.

DEMAIN, A. L. (1974a). Biochemistry of penicillin and cephalosporin fermentations. *Lloydia*, **37**, 147–67.

DEMAIN, A. L. (1974b). How do antibiotic-producing micro-organisms avoid suicide? *Annals of the New York Academy of Sciences*, **235**, 601–12.

DEMAIN, A. L. & INAMINE, E. (1970). Biochemistry and regulation of streptomycin and mannosidostreptomycinase (α-D-mannosidase) formation. *Bacteriological Reviews*, **34**, 1–19.

EPPS, H. M. R. & GALE, E. F. (1942). The influence of the presence of glucose during growth on the enzymic activities of *Escherichia coli*: comparison of the effect with that produced by fermentation acids. *Biochemical Journal*, **36**, 619–23.

EVANS, W. C., HANDLEY, W. C. R. & HAPPOLD, F. C. (1942). The tryptophanase-tryptophan reaction. 5. Possible mechanisms for the inhibition of indole production by glucose in cultures of *B. coli*. *Biochemical Journal*, **36**, 311–18.

FRIEDHEIM, E. A. H. (1932). La fonction respiratoire du pigment du *Bacillus violaceus*. *Comptes rendus des séances de la Societé de biologie*, **110**, 352. Cited in DeMoss, R. D. (1967). Violacein. In *Antibiotics. II. Biosynthesis*, eds D. Gottlieb & P. D. Shaw, pp. 77–81. Berlin, New York: Springer-Verlag.

GALLO, M. & KATZ, E. (1972). Regulation of secondary metabolite biosynthesis: catabolite repression of phenoxazinone synthase and actinomycin formation by glucose. *Journal of Bacteriology*, **109**, 659–67.

HAAVIK, H. I. (1974a). Studies of the formation of bacitracin by *Bacillus licheniformis*: effect of glucose. *Journal of General Microbiology*, **81**, 383–90.

HAAVIK, H. I. (1974b). Studies on the formation of bacitracin by *Bacillus licheniformis*: role of catabolite repression and organic acids. *Journal of General Microbiology*, **84**, 321–6.

HARWOOD, J. P. & PETERKOFSKY, A. (1975). Glucose-sensitive adenylate cyclase in toluene-treated cells of *Escherichia coli* B. *Journal of Biological Chemistry*, **250**, 4656–62.

HINNEN, A. & NÜESCH, J. (1976). Enzymatic hydrolysis of cephalosporin C by an extracellular acetylhydrolase of *Cephalosporium acremonium*. *Antimicrobial Agents and Chemotherapy*, **9**, 824–30.

HOWARTH, T. T., BROWN, A. & KING, T. J. (1976). Clavulanic acid, a novel β-lactam isolated from *Streptomyces clavuligerus*: X-ray crystal structure analysis. *Journal of the Chemical Society: Chemical Communications*, 266–7.

HOWELLS, J. D., ANDERSON, L. E., COFFEY, G. L., SENOS, G. D., UNDERHILL, M. A., VOLGER, D. L. & EHRLICH, J. (1972). Butirosin, a new aminoglycosidic antibiotic complex: bacterial origin and some microbiological studies. *Antimicrobial Agents and Chemotherapy*, **2**, 79–83.

HURLEY, L. H. & BIALEK, D. (1974). Regulation of antibiotic production: catabolite inhibition and the dualistic effect of glucose on indolmycin production. *Journal of Antibiotics*, **27**, 49–56.

ITO, M., AIDA, K. & UEMURA, T. (1969). Studies on the bacterial formation of peptide antibiotic, colistin. 2. On the biosynthesis of 6-methyloctanoic and iso-octanoic acids. *Agricultural and Biological Chemistry*, **33**, 262–9.

JARVIS, F. G. & JOHNSON, M. J. (1947). The role of the constituents of synthetic media for penicillin production. *Journal of the American Chemical Society*, **69**, 3010–18.

JOHNSON, M. J. (1952). Recent advances in penicillin fermentation. *Bulletin of the World Health Organization*, **6**, 99–121.

KAPOOR, M. & GROVER, A. K. (1970). Catabolite-controlled regulation of glutamate dehydrogenases of *Neurospora crassa*. *Canadian Journal of Microbiology*, **16**, 33–40.

KATZ, E. & DEMAIN, A. L. (1977). The peptide antibiotics of *Bacillus*: chemistry, biogenesis and possible functions. *Bacteriological Reviews*, **41**, 339–74.

KIMURA, A. (1967). Biochemical studies on *Streptomyces sioyaensis*. 2. Mechanism of the inhibitory effect of glucose on siomycin formation. *Agricultural and Biological Chemistry*, **31**, 845–52.

KIRSCH, E. J. (1967). Mitomycins. In *Antibiotics. II. Biosynthesis*, eds D. Gottlieb & P. D. Shaw, pp. 66–72. Berlin, New York: Springer-Verlag.

KOMINEK, L. A. (1972). Biosynthesis of novobiocin by *Streptomyces niveus*. *Antimicrobial Agents and Chemotherapy*, **1**, 123–4.

KRASIL'NIKOV, N. A. (1958). *Soil Micro-organisms and Higher Plants*. Moscow: Academy of Science.

LOPEZ, J. M. & THOMS, B. (1977). Role of sugar uptake and metabolic intermediates on catabolite repression in *Bacillus subtilis*. *Journal of Bacteriology*, **129**, 217–24.

LUCKNER, M., NOVER, L. & BÖHM, H. (1977). *Secondary Metabolism and Cell Differentiation*, p. 53. Berlin, New York: Springer-Verlag.

MAGASANIK, B. (1961). Catabolite repression. *Cold Spring Harbor Symposium on Quantitative Biology*, **26**, 249–56.

MAJUMDAR, M. K. & MAJUMDAR, S. K. (1971). Synthesis of neomycin by washed mycelium of *Streptomyces fradiae* and some physiological considerations. *Folia Microbiologica*, **16**, 285–92.

MAKMAN, R. S. & SUTHERLAND, E. W. (1965). Adenosine 3′, 5′-phosphate in *Escherichia coli*. *Journal of Biological Chemistry*, **240**, 1309–14.

MALIK, V. S. & VINING, L. C. (1970). Metabolism of chloramphenicol by the producing organism. *Canadian Journal of Microbiology*, **16**, 173–9.

MAXON, W. D. & CHEN, J. W. (1966). Kinetics of fermentation product formation. *Journal of Fermentation Technology*, **44**, 255–63.

MORGAN, M. J. & KORNBERG, H. L. (1969). Regulation of sugar accumulation by *Escherichia coli*. *FEBS Letters*, **3**, 53–6.

NAGARAJAN, R. (1972). β-Lactam antibiotics from *Streptomyces*. In: *Cephalosporins and Penicillins, Chemistry and Biology*, ed. E. H. Flynn, pp. 636–61. London, New York: Academic Press.

NAKADA, D. & MAGASANIK, B. (1962). Catabolite repression and the induction of β-galactosidase. *Biochimica et Biophysica Acta*, **61**, 835 7.

NÁPRSTEK, J., JANAČEK, J., SPIŽEK, J. & DOBROVÁ, Z. (1975). Cyclic 3′5′ adenosine monophosphate and catabolite repression in *Escherichia coli*. *Biochemical and Biophysical Research Communications*, **64**, 845–50.

NARA, T., YAMAMOTO, M., KAWAMOTO, I., TAKAYAMA, K., OKACHI, R., TAKASAWA, S., SATO, T. & SATO, S. (1977). Fortimycins A and B, new aminoglycoside antibiotics. 1. Producing organism, fermentation and biological properties of fortimycins. *Journal of Antibiotics*, **30**, 533–40.

PAIGEN, K. & WILLIAMS, B. (1970). Catabolite repression and other control mechanisms in carbohydrate utilization. *Advances in Microbial Physiology*, **4**, 251–324.

PASTAN, I. & ADHYA, S. (1976). Cyclic adenosine 5′ monophosphate in *Escherichia coli*. *Bacteriological Reviews*, **40**, 527–51.

PASTAN, I. & PERLMAN, R. (1970). Cyclic adenosine monophosphate in bacteria. *Science*, **169**, 339–44.

PIRT, S. J. & RIGHELATO, R. C. (1967). Effect of growth rate on the synthesis of penicillin by *Penicillium chrysogenum* in batch and chemostat cultures. *Applied Microbiology*, **15**, 1284–90.

POGELL, B. M., SANKARAN, L., REDSHAW, P. A. & McCANN, P. A. (1976). Regulation of antibiotic biosynthesis and differentiation in streptomycetes. In: *Microbiology-1976*, ed. D. Schlessinger, pp. 543–7. Washington: American Society for Microbiology.

RAMSEY, H. H., QUADRI, S. M. H. & WILLIAMS, R. P. (1973). Inhibition by glucose of prodigiosin biosynthesis in *Serratia marcescens*. (Abstracts) *Annual Meeting of the American Society for Microbiology*, p. 180, P239.

ROMANO, A. H. & KORNBERG, H. L. (1968). Regulation of sugar utilization by *Aspergillus nidulans*. *Biochimica et Biophysica Acta*, **158**, 491–3.

SANKARAN, L. & POGELL, B. M. (1975). Biosynthesis of puromycin in *Streptomyces alboniger*: regulation and properties of *O*-dimethylpuromycin *O*-methyltransferase. *Antimicrobial Agents and Chemotherapy*, **8**, 721–32.

SARUNO, R., TAMURA, F. & KATO, F. (1976). Adenosine 3′,5′-cyclic monophosphate control of the enzyme of nuclease metabolism in *Monascus purpureus*. Abstracts, 5*th International Fermentation Symposium*, ed. H. Dellweg, p. 155. Berlin: Westkreuz.

SATOH, A., OGAWA, H. & SATOMWA, Y. (1976). Regulation of N-acetylkanamycin amidohydrolase in the idiophase in kanamycin fermentation. *Agricultural and Biological Chemistry*, **40**, 191–6.

SAWADA, Y., SAKAMOTO, H., KUBO, T. & TANIYAMA, H. (1976). Biosynthesis of streptothricin antibiotics. II. Catabolite inhibition of glucose on racemomycin-A production. *Chemical and Pharmaceutical Bulletin*, **24**, 2480–5.

SCHLANDERER, G. & DELLWEG, H. (1974). Cyclic AMP and catabolite repression in yeasts. In *Schizosaccharomyces pombe* glucose lowers both the intracellular adenosine 3′,5′-monophosphate levels and the activity of catabolite-sensitive enzymes. *European Journal of Biochemistry*, **49**, 305–16.

SOLTERO, F. V. & JOHNSON, M. J. (1953). The effect of the carbohydrate nutrition on penicillin production by *Penicillium chrysogenum* Q-176. *Applied Microbiology*, **1**, 52–7.

SY, J. & RICHTER, D. (1972). Content of cyclic 3′,5′-adenosine monophosphate and adenylylcyclase in yeast at various growth conditions. *Biochemistry*, **11**, 2788–91.

TABER, W. A. (1967). Fermentative production of hallucinogenic indole compounds. *Lloydia*, **30**, 39–66.

ULLMAN, A. & MONOD, J. (1968). Cyclic AMP as an antagonist of catabolite repression in *Escherichia coli*. *FEBS Letters*, **2**, 57–60.

UMEZAWA, H., TAKAKI, T., KANARI, H., OKAMI, Y. & FUKIYAMA, S. (1948). *Japanese Medical Journal*, **1**, 358. Cited in Smith, C. G. & Hinman, J. W. (1963). Chloramphenicol. *Progress in Industrial Microbiology*, **4**, 137–63.

VAN DYCK, P. & DESOMER, P. (1952). Production and extraction methods of aureomycin. *Antibiotics and Chemotherapy*, **2**, 184–98.

VAN WIJK, R. & KONIJIN, T. (1971). Cyclic 3′,5′-AMP in *Saccharomyces carlsbergenesis* under various conditions of catabolite repression. *FEBS Letters*, **13**, 184–6.

WILLIAMS, K. & KATZ, E. (1977). Development of a chemically defined medium for the synthesis of actinomycin D by *Streptomyces parvulus*. *Antimicrobial Agents and Chemotherapy*, **11**, 281–90.

YAGI, Y., KITAMURA, I., OKAMURA, K., OZAKI, A., HAMADA, M. & UMEZAWA, H. (1971). Production of kasugamycin by *Streptomyces kasugaensis*. 1. Oils and fatty acids as carbon sources. *Journal of Fermentation Technology*, **49**, 117–21.

THE FERMENTATION OF MILK BY LACTIC ACID BACTERIA

ROBERT C. LAWRENCE AND TERENCE D. THOMAS

New Zealand Dairy Research Institute, Palmerston North, New Zealand

INTRODUCTION

Micro-organisms have played a part from early times in the making of dairy products. The 'natural' souring of milk was found to preserve what was otherwise a highly perishable food and coincidentally resulted in a range of desirable flavours and textures. Milk contains 85–88% water, and the traditional characteristics of the many different varieties of cheese and other fermented milk products depend mainly upon how much water is removed and the way in which different cultures are used to carry this out. The major requirement of a culture is the fermentation of the lactose in milk to lactic acid. The rate of acid production has become more critical with the amalgamation of many small plants into large manufacturing units and the introduction of mechanized equipment, since the product is now usually made to a strict time-schedule. In addition, consumption of cheese and other fermented milk products rises steadily each year. In order to handle the greater milk throughputs, the traditional methods of manufacture have changed dramatically in the last twenty years.

It is not, therefore, as paradoxical as it might seem to introduce a paper on milk fermentation by emphasizing the single most important factor which retards fermentations, that is, lysis of the culture bacteria by bacteriophage. The relationship between the lactic streptococci and their phages influences most milk fermentations in a way that has almost no parallel in other microbial fermentations. In individual plants, up to 10^{18} bacterial cells and 10^{16} phage particles may be produced each day; the opportunity for mutation and genetic exchange to occur must be unique. When a newly isolated strain is introduced into a cheese plant it is usually only a short time before a phage that can attack it appears. The selection of suitable cultures is likely to remain difficult, therefore, until the source of phage has been determined and the range of phage types that may appear in a plant can be anticipated with greater certainty. One must also know how to differentiate the various closely related strains and how to formulate compatible mixtures of different strains.

Cultures were being used, however, long before phages were known to exist. It was common practice for cheese-makers to hold back each day some sour milk to 'start' the next day's cheese-making. With experience they learnt to recognize which sour milks to use by their smell and taste. Commercial firms began to supply such starter cultures about 80 years ago and it is important to recognize that these undefined mixtures generally worked fairly well, certainly until the large manufacturing plants came into existence in the 1960s. Over the years these cultures changed as a result of the interaction between the bacteria in the milk and their phages. Much of the world's cheese and other fermented milk products is still made with the descendants of these early cultures.

To most microbiologists the lactic acid bacteria and their phages must present a confusing picture, and indeed there are many areas where basic data are still lacking. Undoubtedly a major obstacle to progress in the logical use of lactic acid bacteria in the past has been the combination of secrecy and fragmentation of effort, which is understandable when many commercial interests are involved. The inhibition of free exchange of information has led to the illusion that many of the commercial problems that exist are much more complex than they really are.

REQUIREMENTS OF CULTURES IN PRODUCT MANUFACTURE

Metabolic products

The primary requirement of most cultures is the production of lactic acid. In many milk fermentations this is carried out by the mesophilic lactic streptococci, in particular *Streptococcus cremoris* and *Streptococcus lactis* (Table 1). These organisms are used because they possess very limited metabolic diversity and are homolactic during batch growth in milk, forming only small amounts of secondary products such as acetic acid and diacetyl (discussed in the section on flavour development). In addition, both *S. cremoris* and *S. lactis* are only weakly proteolytic and lipolytic. For some products, such as yoghurt and many Swiss and Italian cheeses, the temperatures reached during manufacture are relatively high and the thermophilic organisms, *Streptococcus thermophilus*, *Lactobacillus bulgaricus* or *Lactobacillus helveticus*, are then used (Table 1). The CO_2-producing species, *Streptococcus dia-*

cetylactis and *Leuconostoc* strains, are often added to cultures of *S. cremoris* and *S. lactis*, particularly for the manufacture of cheeses such as Gouda where some 'eye' formation is considered desirable.

Table 1. *Cultures used in the manufacture of some fermented milk products*

An indication is given of the relative rates of acid production and the likelihood of phage problems using modern processing methods.

Product	Relative rate of acid production[a]	Likelihood of phage problems[a]	Species used[b]	Important microbial products
(1) Cottage cheese	5+	2+	S. cremoris	Lactic acid,
Cheddar	4+	4+	S. lactis	diacetyl[c]
Gouda	2+	+	S. diacetylactis	acetaldehyde[c],
Cheshire	5+	3+	Leuconostoc spp.	CO_2[c]
(2) Camembert	4+	2+	As (1) plus Penicillium camemberti	Lactic acid plus protein breakdown
Blue cheese	4+	2+	As (1) plus Penicillium roqueforti	Lactic acid plus fat breakdown
(3) Yoghurt	5+	(+)	S. thermophilus L. bulgaricus	Lactic acid, acetaldehyde
Swiss cheese	+	(+)	S. thermophilus L. helveticus (or L. bulgaricus) Propionibacteria spp.	Lactic acid Propionic acid, CO_2, proline

[a] Dependent upon conditions of manufacture of a given product, especially the organisms used, inoculum size, temperature range, and method of salting.
[b] One or more strains may not be present.
[c] Not important in all products.

It is important that all the lactose in cheese is metabolized by the starter bacteria. The presence of residual lactose may permit the growth of contaminant organisms possessing greater biochemical diversity, with resultant production of off-flavours and changes in the characteristic flavour and texture of the product.

Inhibition of spoilage bacteria and pathogens

Many dairy products depend upon the microbiological activities of starter cultures to control the growth of spoilage bacteria and pathogens. Inhibition results from competition for nutrients, decrease in pH and redox potential, and in some cases the formation of specific inhibitory compounds. Acetate and H_2O_2 appear to be particularly important, even though the amounts produced are relatively small, but lactate and antibiotics have also been implicated (Speck, 1976; Babel, 1977). An increasingly common innovation in North America is the addition of a low inoculum of a lactic culture (less than 0.5%) to milk as it is received at the plant. This limits the growth of psychrotrophic bacteria during storage at low temperatures.

Certain strains of *S. lactis* produce the antibiotic nisin, and a similar inhibitory substance, diplococcin, is produced by some strains of *S. cremoris*. Nisin is a narrow-spectrum, polypeptide antibiotic active against Gram-positive organisms. Its sensitivity to some digestive enzymes, freedom from toxicity, and heat stability in acid solutions has led to its use as a food additive in many countries. Lactobacilli are also reported to produce various types of antibiotics. The production of inhibitory substances by some strains of lactic acid bacteria means that care is required when combining strains. The inhibitory interactions between the component strains must be taken into account, as well as their competitive growth abilities.

Nutritional aspects of fermented milk products

In recent years there has been much speculation about the possible therapeutic benefits that result from the interaction between lactic acid bacteria and man. Several species of lactobacilli can be isolated from the intestinal tract but only *Lactobacillus acidophilus* is present in sufficient numbers to be of importance (Speck, 1976). Neither *L. bulgaricus* nor *S. thermophilus*, the species used for the manufacture of yoghurt, can be detected in the intestinal tract. There is considerable evidence that replenishment of the intestinal lactobacilli is beneficial (see Speck, 1976) but relatively little has been done to find effective ways in which lactobacilli can be used as dietary components.

ECOLOGY AND CLASSIFICATION OF STRAINS

It is clear that the genus *Streptococcus* is closely related to the genera *Lactobacillus* and *Leuconostoc* (London, 1976) and a distant relationship has been demonstrated between this cluster and the genus *Propionibacterium*. Many of the lactic acid bacteria and related organisms have become highly specialized, as manifested by their complex nutritional requirements, lack of biochemical diversity, and restricted habitats. While most species of streptococci are potentially pathogenic, and almost all are parasitic, Hirsch (1952) has pointed out that the lactic streptococci are a noteworthy exception.

The natural habitat of the lactic streptococci is still uncertain. They have not been detected in milk drawn aseptically from the udder but are invariably found in normally produced raw milk. *S. lactis* is the predominant organism, *S. cremoris* strains being present in such low numbers that they are difficult to isolate. The inability to hydrolyse arginine has long been used to differentiate *S. cremoris* from *S. lactis*. This distinction may be more apparent than real, however, since the repeated subculturing of *S. lactis* strains isolated from raw milk led to the appearance of variants that could not utilize arginine (Heap, Limsowtin & Lawrence, 1978). Similarly, it is possible that *S. lactis* strains may be variants of *S. diacetylactis* that are unable to ferment citric acid, since citrate permease-negative strains of *S. diacetylactis* have been described (see Lawrence, Thomas & Terzaghi, 1976).

Considerable evidence is accumulating to suggest that *S. cremoris*, *S. lactis* and *S. diacetylactis* (and even perhaps *S. thermophilus*) form a phenotypically and genotypically continuous spectrum of variation. The three species are similar in DNA base-composition (Knittel, 1965) indicating that they are closely related. The lactic streptococci belong to the serological Group N, with the exception of *S. thermophilus*, for which no group antigen has yet been demonstrated. The Group N antigen appears to be a membrane-associated lipoteichoic acid. The teichoic acid moiety consists of glycerol phosphate units, approximately half of which are substituted with galactosyl residues, whereas the glycolipid moiety contains glucosyl residues (Wicken & Knox, 1975). The cell walls of Group N streptococci and *S. thermophilus* consist of peptidoglycans of similar composition and structure, and a poly-saccharide of unknown structure, containing rhamnose, glucose, glucosamine and galactose (Schleifer & Kandler, 1967).

A readily fermentable carbohydrate, especially lactose or glucose, is required for rapid growth of lactic acid bacteria since, unlike *Streptococcus faecalis*, they cannot grow on gluconeogenic substrates. The lactic acid bacteria useful in milk fermentations do not normally possess a cytochrome system or the enzymes of the tricarboxylic acid cycle and fermentation during batch growth on lactose under anaerobic conditions is homolactic. However, under other growth conditions these organisms may behave quite differently, indicative of a complement of enzymes not normally expressed.

SELECTION AND DIFFERENTIATION OF STRAINS

Starter-culture collections tend to reflect a selection of organisms which can produce lactic acid rapidly in milk at cheese-making temperatures. Organisms with a diversity of metabolic activities are undesirable and, even within the closely related lactic streptococci, *S. cremoris* is more useful than *S. lactis* since it is less likely to give bitter and fruity flavour defects in cheese (Lawrence & Pearce, 1972).

It is also important in most milk fermentations to select strains that are attacked by different groups of phages. This is not a simple matter, however, since selection on the basis of acid-producing activity and apparent resistance to certain phages is no guarantee that mutation of a given phage will not enable it to attack more than one of the strains isolated. The recent report (Limsowtin, Heap & Lawrence, 1978) of restriction/modification systems operating within variants of the same lactic streptococcal strain may partly explain why phage-typing strains of lactic streptococci can be misleading. Different isolates from commercial cultures often appear initially to be distinct strains in terms of their phage-sensitivity patterns but may be derived from the same ancestral cell and differ only in their restriction systems. Most of these 'resistant' isolates are attacked by phage when subcultured repeatedly in the presence of filtered wheys from cheese plants (Heap & Lawrence, 1976), suggesting that many strains are sensitive to overlapping groups of phages and that the sensitivity is masked in some cases in the first few generations only.

Production of mutant strains

The difficulty in isolating new starter strains from natural sources has prompted considerable research into the isolation of mutants with specially desirable characteristics. There have been numerous reports in recent years of the use of ultraviolet (UV) light, mutagenic chemicals, X-rays and γ-rays to produce mutants with higher rates of acid, diacetyl and proteinase production than the parent strain in *S. lactis* and *S. diacetylactis*, and enhanced nisin production in *S. lactis* strains (see Lawrence *et al.*, 1976).

The acquisition of one characteristic, however, can also lead to changes in other properties. Thus phage-resistant mutants can usually be isolated in the absence of a mutagen simply by growing the parent culture with phage, but few can be used in fermentations since they usually lack the acid-producing activity of the parent. This suggests that phage resistance and acid production are inversely related in most strains. Mutations that alter phage-receptor sites may interfere with transport and proteinase systems, which are also located at the cell surface. Similarly Sinha (1977) has found that selection of *S. cremoris* strains for resistance to acriflavin resulted in cross-resistance to a number of drugs and antibiotics. Furthermore the mutants showed resistance to lytic phages to which the parent strain was sensitive and, unlike the parent, the mutants grew well at higher temperatures (40 °C). The acriflavin-resistant mutation that resulted in the temperature-resistant phenotype is thought to involve structural modification of the cell membrane since the mutant also showed increased resistance to treatment with sodium dodecyl sulphate.

GROWTH AND METABOLISM OF LACTIC ACID BACTERIA IN MILK

The commercial use of lactic acid bacteria is based on their ability to grow in milk and ferment lactose to lactic acid, but only a limited understanding of their nutrition and metabolism exists at present. They require a sugar such as lactose for growth, and a wide range of amino acids, vitamins and other growth factors. Milk is a satisfactory medium except that amino acids and low molecular weight peptides are present at only low concentrations. Rapid growth and acid production in milk therefore requires efficient systems for the degradation of milk protein as well as for lactose fermentation.

Commercial milk fermentations are usually batch culture operations and are essentially anaerobic. At the optimum temperature for growth (*c.* 30 °C), lactic streptococci have mean generation times in milk of 60–70 min and grow to a maximum cell density of about 0.5 mg dry wt bacteria per ml. In complex broth media the same organisms have doubling times of 35–45 min. Growth of lactic streptococci in milk stops as a result of the acid developed (terminal pH about 4.5 corresponding to 0.5% (w/v) lactic acid) leaving about 4.5% (w/v) unfermented lactose in the culture. The longer the cells in these stationary-phase cultures are exposed to high acid concentrations the more prolonged is the lag on subculture. In contrast to the growth of mesophilic streptococci in milk, the commercially important lactobacilli have temperature optima near 40 °C. These organisms are also relatively 'acid tolerant' and the culture pH may fall to 3.5 corresponding to the formation of about 1.5% (w/v) lactic acid.

The cellular environment in liquid cultures is markedly different from that which usually exists during the manufacture of most fermented milk products, where bacterial growth takes place with the cells embedded in a gel. Neither those streptococci which grow in long chains, nor the lactobacilli are evenly distributed throughout the gel but tend to grow as discrete microcolonies. Under these conditions, nutrient availability and end-product dispersal must be different from the situation when cells are homogeneously suspended in liquid milk. It must also be remembered that most of the lactose originally present in milk is removed when the whey is expelled from the curd at the 'cooking' and draining stages of cheese-making. In addition to low pH, growth inhibition also arises in cheese-making from the use of elevated cooking temperatures and the addition of NaCl. These stresses have less effect on the rate of acid production than on the growth rate of lactic streptococci (Turner & Thomas, 1975). This uncoupling may be important in Cheddar cheese-making since current views on flavour development suggest that it is desirable to obtain the required acid concentration from the minimum density of bacterial cells (Lowrie, Lawrence & Peberdy, 1974). *L. bulgaricus* and *L. lactis* also show a similar uncoupling since molar growth yields on glucose at 40 °C were less than half of those of 37 °C (Dirar & Collins, 1972). The 'slip mechanisms' involved in these uncoupling processes in the lactic acid bacteria have not been investigated.

Nitrogen nutrition

Difficulty is usually experienced in growing lactic streptococci, especially *S. cremoris* strains, in chemically defined media and yet these strains grow fairly well in milk. Growth to high cell densities (0.5 mg dry wt per ml or about 10^9 colony-forming units per ml), and hence rapid coagulation of milk, requires the synthesis of about 0.25 mg bacterial protein per ml (i.e. 0.025% total amino acids). Milk, however, contains only 0.004–0.01% total free amino acids and small quantities of low molecular weight peptides. Some growth in milk may occur without proteolysis since proteinase-deficient variants grow at low cell densities at the same rate as the parent strain (see Lawrence *et al.*, 1976). However, these variants become growth-limited at 10–25% of the maximum cell densities reached by the parent culture, presumably due to depletion of the amino acids and peptides initially present, and milk coagulation is markedly delayed. When casein digest is added to milk, the growth of these variants is indistinguishable from that of the parent strain, indicating that proteolysis is essential for growth to high cell densities. Although the actual protein(s) involved has not yet been identified, casein appears to be important, since its removal from milk results in a whey medium which must be supplemented with protein hydrolysate to achieve growth comparable to that in the original milk. In addition, casein is hydrolysed by the cell wall proteinases of all strains studied except S. cremoris ML_1 (Exterkate, 1976). This strain grows normally in milk, suggesting an alternative nitrogen source.

Although the level of proteinase in lactic streptococci is low, the overall system for the supply of free amino acids from milk protein is efficient enough to permit logarithmic growth to high cell densities with mean generation times of 1 h at 30 °C. Nevertheless some step in this system limits the growth rate in milk, since addition of digested milk protein, and specifically small peptides, increases the growth rate and stimulates acid production. The utilization of milk protein for growth appears to involve proteinase(s) located in the cell wall (Thomas, Jarvis & Skipper, 1974; Exterkate, 1976), and peptidases located both in the plasma membrane (Exterkate, 1975) and the cytoplasm (Mou, Sullivan & Jago, 1975; Law, 1977), where they may be associated with ribosomes (Schmidt, Morris & McKay, 1977). It is not clear whether any of the surface-bound proteinase is released from the cell during growth in milk. Although at least part of this system is loosely bound (Mills & Thomas, 1978), the attachment is Ca^{2+}-dependent and milk probably contains sufficient Ca^{2+} to prevent release. Peptidases capable of releasing all the free amino acids essential for growth have been reported in lactic streptococci (Mou et al., 1975). Growth of S. thermophilus was stimulated by peptides and the intracellular peptidases in this species have been extensively studied (see Desmazeaud & Juge, 1976). A satisfactory appraisal of the role of these enzymes in supplying amino acids for growth of lactic acid bacteria in milk awaits determination of the actual number of enzymes involved, their enzymatic properties, especially the substrate specificities of individual enzymes, together with unambiguous evidence on their cellular location and physiological role. The organisms for such studies should obviously be grown in milk, and techniques have been described for this purpose (Thomas & Turner, 1977). The proteinase activity associated with lactic streptococci is so low compared with that in most bacteria that obtaining a sufficiently sensitive assay has been a major obstacle (Lawrence et al., 1976). Most studies would be facilitated by the development of assays that allowed detection of individual enzymes in crude cell fractions.

Stimulatory interactions occur between the different cell types that are usually present in a culture of a single lactic streptococcal strain. Some cells apprently lack cell-wall proteinase activity (Pearce, Skipper & Jarvis, 1974) and their active growth in milk is presumably dependent upon the supply of proteolysis products from the parent organism. These negative variants often become dominant, indicating that they must have a competitive advantage over the parent strain. Similar stimulatory interactions involving nitrogen nutrition may also occur when cultures containing several different strains of the same species are grown together in milk. For instance, mixed cultures of lactic streptococci which gave satisfactory rates of acid production in plants manufacturing lactic casein (coagulation time about 18 h at 22 °C from 0.1% inoculum) were dominated (60–80%) by *S. cremoris* organisms which, when isolated and cultured alone, coagulated milk very slowly (2–7 days at 22 °C from 0.1% inoculum (Thomas & Lowrie, 1975)). The various interactions found with the lactic acid bacteria are modified by changes in the environment and would therefore best be studied under controlled conditions using continuous culture techniques.

Greatest stability of certain mixed cultures seems to occur when each component of the mixture is dependent on another for an essential growth factor(s). Further work in this field is obviously required but one example is well documented. When growing in milk, the yoghurt bacteria *S. thermophilus* and *L. bulgaricus* produce acid at a slower rate on their own than when they grow in combination (Accolas *et al.*, 1977). It appears that *L. bulgaricus* produces from the milk protein amino acids and small peptides which stimulate the growth of *S. thermophilus*, while this organism produces formic acid which stimulates the growth of *L. bulgaricus* (see Shankar & Davies, 1977). The interdependence is sufficiently strong for a continuous procedure for making yoghurt to be developed (Driessen, Ubbels & Stadhouders, 1977).

Lactose metabolism

Lactic streptococci produce about 1 µmole lactate per mg dry wt bacteria per min in milk at optimum growth temperatures (Turner & Thomas, 1975); this corresponds to nearly 10% of their weight per min. Only about 3% of the fermented lactose is incorporated into cellular material. During homolactic fermentation, end-products other than lactic acid are also produced but in small amounts only, and this section is confined to the quantitatively significant pathways of lactose metabolism.

Present data suggest that the lactic streptococci ferment lactose as shown in Fig. 1. Although it has been assumed that the mechanism of lactose uptake is similar to that proposed for *Staphylococcus aureus*, there is now considerable doubt about this assumption. The transport of lactose by *Staph. aureus* involves a phosphoenolpyruvate (PEP)-

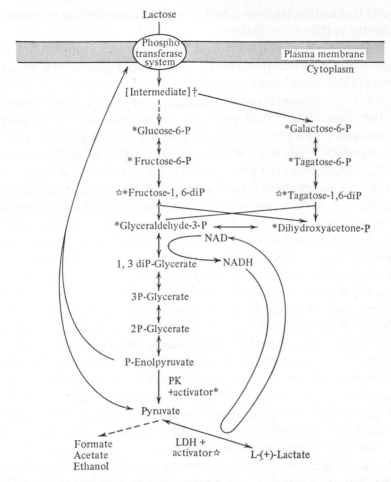

Fig. 1. Quantitatively significant pathways of lactose metabolism in lactic streptococci. Intermediates that activate pyruvate kinase (PK) are marked with an asterisk (*), those that activate lactate dehydrogenase (LDH) are marked with a star (★). †[Intermediate] is reported to be 6-phosphogalactosyl-(β1–4)-glucose in *Staphylococcus aureus* (Postma & Roseman, 1976) but may be lactose diphosphate in some lactic streptococci (see text).

dependent phosphotransferase system (PTS) and phosphorylation occurs during translocation (see Postma & Roseman, 1976). The lactose phosphate (6-phosphogalactosyl-(β1–4)-glucose) is hydrolysed intracellularly by phospho-β-D-galactosidase (P-β-gal) apparently to glucose and galactose-6-phosphate. Complementation studies of the multi-component system involved in lactose metabolism showed a similarity between *Staph. aureus* and *S. lactis* (McKay *et al.*, 1970). Furthermore, the presence of the enzymes of the D-tagatose-6-phosphate pathway (Fig. 1) in the Group N streptococci (Bissett & Anderson, 1974) suggests that lactose transport, and subsequent metabolism to pyruvate, are similar in the two organisms.

Recent studies at the New Zealand Dairy Research Institute, however, suggest that during translocation by *S. lactis* both glucose and galactose moieties of the disaccharide are phosphorylated simultaneously at the C_6 position (J. Thompson, unpublished). In non-growing cells of this organism, the intracellular 'PEP potential' (concentration of PEP plus 2-phospho-and 3-phosphoglycerate) can be maintained at 40–45 mM – a situation which facilitates the quantitative analysis of sugar transport and PEP utilization in the physiologically intact cell. Under these conditions hexoses accumulated (as phosphorylated derivatives) to a maximum intracellular concentration of 35–45 mM, ceasing when the endogenous PEP source was exhausted. In contrast, maximum accumulation of lactose (about 16 mM) was approximately half that of monosaccharides, suggesting that a lactose-diphosphate was formed. Enzymatic and autoradiographic analyses of intermediates revealed no trace of either lactose-diphosphate or lactose-phosphate (which has been postulated for *Staph. aureus*) within the cells, even at sampling times of less than 1 s, but high levels of both glucose-6-phosphate and galactose-6-phosphate were present. In these experiments the cells had been preincubated with iodoacetate to block glycolysis and no intracellular ATP could be detected. The data suggest that in *S. lactis* a transitory lactose-diphosphate intermediate is formed (during translocation) which is immediately cleaved to glucose-6-phosphate and galactose-6-phosphate. The two hexose phosphates may then be metabolized directly via the glycolytic and D-tagatose-6-phosphate pathways respectively.

Thus in the lactic streptococci both compounds derived from cleavage of the lactose derivative are broken down simultaneously. These organisms grow in batch culture at the same rate with homolactic fermentation in media containing either lactose or glucose (Thomas, 1976). However, growth on galactose is relatively slow and fermentation is heterolactic, indicating an important difference between galactose supplied as lactose or in the free form. When glucose plus galactose are supplied to *S. lactis* the glucose is used preferentially, even by galactose-adapted cells (Thompson, Turner & Thomas, 1978). The mechanism involved in this sequential utilization of sugars appears to involve fine control of enzyme activity by catabolite inhibition.

Lactose-defective variants arise spontaneously in cultures of lactic streptococci as a result of loss of some of the components making up the PEP:lactose-PTS and P-β-gal systems which may be coded on plasmid DNA (Anderson & McKay, 1977). These variants have impaired acid-producing ability in milk, and lactose fermentation is heterolactic (Demko, Blanton & Benoit, 1972). Other variant organisms, such as *S. lactis* 7962 (see Thompson & Thomas, 1977) do not possess a functional lactose-PTS or P-β-gal but take up lactose as the free sugar, and after hydrolysis with β-galactosidase, ferment galactose to glucose-6-phosphate via the Leloir pathway. *S. lactis* 7962 is of no use in milk fermentations because the growth rate on lactose is slow (mean generation time in broth containing lactose 98 min compared with 33 min on glucose) and only about 15% of the lactose is converted to lactic acid (Thomas, 1976). The PEP:lactose-PTS and P-β-gal appear to be prerequisites for rapid and homolactic lactose fermentation by lactic streptococci and, together with the parallel pathways for glucose-6-phosphate and galactose-6-phosphate metabolism (Fig. 1), may represent the most efficient process for lactose fermentation. There is no evidence for other pathways of lactose metabolism to pyruvate which have any quantitative significance in lactic streptococci, although these organisms do have the enzymes necessary for the conversion of glucose-6-phosphate to the ribose-5-phosphate required for nucleotide synthesis.

The nature of lactose uptake in the thermophilic lactic acid bacteria has not been investigated in any detail, but it seems likely that most strains do not possess a high-affinity lactose-PTS similar to that present in the lactic streptococci. Lactose-fermenting lactobacilli, with the exception of *L. casei*, possess β-galactosidase but little if any P-β-gal activity (Premi, Sandine & Elliker, 1972). In addition, lactose fermentation by some thermophilic lactic acid bacteria is unbalanced, so that the cells excrete free galactose into the medium and preferentially ferment the glucose moiety of the lactose molecule (O'Leary & Woychik, 1976). It has been suggested that part of the lactose system in *L. casei* is coded on plasmid DNA (Hofer, 1977).

In-vitro data (Thomas, 1976) suggest that the last two enzymes involved in the conversion of lactose to lactate have important regulatory properties. Lactic streptococci possess a single allosteric pyruvate kinase which is markedly activated by all eight intermediates of lactose metabolism from the hexose-6-phosphates to the triose phosphates (Fig. 1). The control of pyruvate kinase activity by these intermediates earlier in the pathway may regulate the intracellular PEP concentration and hence provide a coupling between lactose transport via the PEP-dependent PTS and the subsequent metabolism of this sugar. Strong support for a regulatory role of pyruvate kinase *in vivo* has come from experiments with physiologically intact cells (Thompson & Thomas,

1977; J. Thompson, unpublished). Cells metabolizing glucose contained high intracellular concentrations of pyruvate kinase activators, especially fructose-1,6-diphosphate, and relatively low concentrations of the pyruvate kinase substrate (PEP) and the immediate precursors (2-phospho- and 3-phosphoglycerate); see Table 2. Upon exhaustion of

Table 2. *Intracellular concentrations of glycolytic intermediates in* S. lactis ML_3 *when metabolizing glucose and after glucose exhaustion from the medium*

Glycolytic intermediate	Intracellular concentration (mM)	
	Glucose present in medium	Glucose exhausted from medium
Glucose-6-phosphate	1.6 ± 0.1	<0.1
Fructose-6-phosphate	0.3 ± 0.1	<0.1
Fructose-1, 6-diphosphate	15.8 ± 0.3	<0.1
Dihydroxyacetone phosphate ⎫ Glyceraldehyde-3-phosphate ⎬	6.7 ± 0.1	<0.1
3-Phosphoglycerate	6.2 ± 0.6	28.9 ± 0.3
2-Phosphoglycerate	1.2 ± 0.1	5.3 ± 0.9
Phosphoenolpyruvate	3.0 ± 0.3	11.3 ± 0.3
Pyruvate	2.3 ± 0.1	1.1 ± 0.1

Unpublished data of J. Thompson.

glucose the intracellular pyruvate kinase activators were completely depleted while the concentration of glycolytic intermediates immediately before pyruvate kinase increased. PEP utilization, and hence pyruvate kinase activity, was 600-fold slower in these cells without exogenous glucose than in the glycolysing cells, suggesting that depletion of pyruvate kinase activators *in vivo* switches off the enzyme, leaving the cell essentially primed for many hours with PEP – the energy donor for sugar uptake via the PTS. Slow utilization of PEP, 2-phospho- and 3-phosphoglycerate may provide a continuing supply of ATP for the maintenance of these cells which, apart from glycolytic intermediates, have no known endogenous energy sources (see Thompson & Thomas, 1977).

The terminal enzyme involved in lactate formation by lactic streptococci is an NAD-dependent lactate dehydrogenase. This enzyme is also activated *in vitro* by intermediates of lactose metabolism (Fig. 1) but in this case, however, specifically by the ketohexose diphosphates (Thomas, 1976). Both pyruvate kinase and lactate dehydrogenase from S. lactis were inhibited by phosphate, which had a profound effect on the properties of these enzymes *in vitro*. The fructose-1,6-diphosphate concentration required for half-maximal velocity of lactate dehydrogenase was increased 2000-fold (Crow & Pritchard, 1977) by the presence of inorganic phosphate at a concentration similar to that in the cytoplasm of *S. faecalis*.

Most lactic acid bacteria used in commercial milk fermentations are classically regarded as homolactic fermenters of lactose. However, although anaerobic growth in batch culture results in the conversion of more than 90% of the fermented lactose to lactate, one cannot assume that the same organisms growing in the very different environments associated with the various stages of cheese-making will always produce the same products. Recent investigations (Table 3) have shown

Table 3. *Residual glucose and fermentation products with*
S. lactis ML_3

S. lactis ML_3 grown in a chemostat at pH 6.5 with a $N_2:CO_2$ (95:5) atmosphere, the chemically defined medium contained 28 mM glucose.

Dilution rate (h^{-1})	Residual glucose (mM)	Lactic acid	Formic acid	Acetic acid	Ethanol	Carbon recovery (%)
		(mmole product per 100 mmole glucose fermented)				
0.76	5.11	187	ND	ND	ND	94
0.64	1.55	193	ND	ND	ND	97
0.56	0.03	185	16	13	9	103
0.48	ND	134	29	26	53	98
0.36	ND	38	134	73	88	95
0.25	ND	15	141	99	94	95
0.18	ND	7	125	111	103	96
0.11	ND	2	156	103	87	90

ND = not detectable.
Unpublished data of T. D. Thomas, D. C. Ellwood and V. M. C. Longyear.

that the fate of pyruvate may be dramatically altered by changing the growth environment and these 'homolactic' streptococci may produce almost no lactic acid. As the dilution rate was reduced in chemostat cultures, so that glucose or lactose became the limiting nutrient, the end-products changed to formic acid, acetic acid and ethanol. The products suggest that under conditions of limiting energy supply the cells were able to obtain ATP from pyruvate metabolism, by a mechanism involving pyruvate formate-lyase (Fig. 2). This phenotypic switch was observed with five out of six strains of *S. lactis* and *S.*

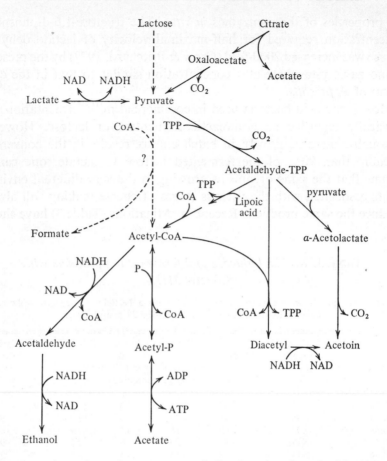

Fig. 2. Alternative pathways for pyruvate metabolism in various 'homolactic' acid bacteria used in starter cultures. Citrate is metabolized only by *Streptococcus diacetylactis* and *Leuconostoc cremoris*.

cremoris growing in a chemostat with either glucose or lactose limitation (T. D. Thomas, D. C. Ellwood & V. M. C. Longyear, unpublished). The end-products changed in response to growth limitation by carbohydrate since fermentation was homolactic at low dilution rates when the sugar was present in excess. It is not clear what regulates the pathways of pyruvate metabolism. However, transfer of heterolactic cells from the chemostat to a non-growing system with excess glucose was accompanied by an immediate switch back to homolactic metabolism, suggesting that a fine control of enzyme activity rather than control of enzyme synthesis is involved. Use of a rapid sampling procedure to extract glycolytic intermediates from cells growing in the chemostat showed that the intracellular fructose-1,6-diphosphate remained relatively high in heterolactic cells (Thomas *et al.*, unpublished), indicating that heterolactic metabolism was not due to the lack of an activator for LDH. Clearly lactic streptococci have the potential to carry out what may be undesirable fermentations, but whether they are important in the development of off-flavour in cheese is unknown. A similar switch in end-products of glucose fermentation was observed with *L. casei* when this 'homolactic' organism was grown in a chemostat at low dilution rates with glucose limitation (de Vries *et al.*, 1970).

Milk fermentations involving lactic streptococci are usually considered to be 'anaerobic', but these organisms have a limited ability to induce NADH oxidase, utilize oxygen and produce peroxides (Anders, Hogg & Jago, 1970). Aerobic metabolism may therefore lead to growth inhibition and diversion of pyruvate from lactate to other end-products, unless active steps are taken to preclude the entrainment of air during handling of milk prior to the fermentation.

FLAVOUR DEVELOPMENT IN FERMENTED MILK PRODUCTS

Production of volatile compounds

The formation of C_2 and C_4 compounds by lactic streptococci has been the subject of intensive research in recent years because of their probable contribution to flavour and aroma in dairy products. In the manufacture of cottage cheese, sour cream, cultured butter and buttermilk, *S. diacetylactis* has a dual role in producing both lactic acid and flavour compounds, particularly diacetyl and acetaldehyde, while *Leuconostoc* species are considered to be important for flavour development only (Table 1). In these products the over-production of acetaldehyde is undesirable and good flavour is dependent upon a correct balance between acetaldehyde and diacetyl. Acetaldehyde, however, is thought to be the most important flavour component of natural yoghurt and *c*. 10 μg ml^{-1} is present when *S. thermophilus* and *L. bulgaricus* are used in its manufacture (Bottazzi & Vescovo, 1969). Similarly, diacetyl concentrations are low in fermented products and seldom exceed 5 μg ml^{-1}. In contrast, acetoin, which does not contribute to flavour, is often found in relatively high concentrations (up to 500 μg ml^{-1}).

While glycolysis is the major pathway to pyruvate in lactic acid bacteria, some organisms also produce pyruvate as an intermediate in citrate metabolism. Milk contains about 1.7 mg ml^{-1} citrate, which can be utilized by *S. diacetylactis*, *Leuc. cremoris* and certain lactobacilli (see Cogan, 1975) but not by *S. lactis*, *S. cremoris* or the 'yoghurt bacteria', which cannot transport citrate. The most likely mechanisms by which pyruvate is metabolized are shown in Fig. 2 (see Speckman & Collins, 1973; Jonsson & Pettersson, 1977). The extent to which flavourful compounds such as diacetyl and acetaldehyde are formed appears to be regulated by the supply of acetyl-CoA available (Collins & Bruhn, 1970) and by the oxidation–reduction balance of the system. The presence of diacetyl reductase in some strains of *S. diacetylactis* may explain the loss of flavour observed on prolonged incubation of some fermented milks made with such cultures.

Acetate is normally produced from acetyl-CoA, since all lactic streptococci possess phosphotransacetylase and acetate kinase (Collins & Bruhn, 1970; Lees & Jago, 1976), but may also be formed by the cleavage of citrate (Fig. 2). In addition, lactic streptococci possess aldehyde and alcohol dehydrogenases for acetyl-CoA reduction (Fig. 2) and produce small amounts of acetaldehyde and ethanol from glucose (Lees & Jago, 1976). No pyruvate carboxylase has been detected, suggesting that acetaldehyde is not formed by direct decarboxylation of pyruvate. *L. bulgaricus* and *S. thermophilus* possess the enzyme for reduction of acetyl-CoA to acetaldehyde, but in most strains no alcohol dehydrogenase activity could be detected (Lees & Jago, 1976), which is consistent with the accumulation of acetaldehyde in yoghurt cultures. In contrast, *Leuc. cremoris*, which has high alcohol dehydrogenase activity, produces little acetaldehyde but considerable amounts of ethanol.

When the amount of lactose or glucose present is growth limiting, 'homofermentative' lactic streptococci and lactobacilli ferment these sugars almost exclusively to formate, acetate and ethanol, as already discussed. Heterolactic fermentation by lactic streptococci may also result from growth on galactose (Thomas, 1976), in the presence of oxygen (see Anders *et al.*, 1970) and at low temperatures (Bang, 1949), or from mutation leading to loss of either a functional lactose-PTS (Demko *et al.*, 1972) or lactate dehydrogenase (McKay & Baldwin, 1974*b*). Thus although the lactic acid bacteria useful in milk fermentations are homolactic when grown anaerobically with excess lactose or glucose, they have considerable latent potential for alternative metabolism of pyruvate (Fig. 2). Little is known about the factors which regulate these alternative pathways, although this information is required for an understanding of both the flavours and off-flavours that develop in many fermented milk products.

Cheese flavour

Although the metabolism of viable lactic streptococci leads to the production of aroma compounds in short-life dairy products, their importance in hard cheeses is more difficult to evaluate, since the typical flavours only develop after several months' storage. Law, Castanon & Sharpe (1976) have concluded that the intracellular starter enzymes released after the lysis of the streptococcal cells play no direct part in flavour formation but may produce such low molecular weight precursors of aroma compounds as cysteine and methionine. Reports that *S. diacetylactis* and *Leuconostoc* strains are necessary for the production of characteristic Gouda flavour (Stadhouders, 1974) and that leuconostocs are important in Cheddar cheese (see Collins, 1962) do not now appear justified. Recent findings demonstrate clearly that the flavours characteristic of Gouda (Kleter, 1976) and Cheddar (Lowrie *et al.*, 1974) develop when *S. cremoris* is the only organism present. There is increasing evidence that the major role of a culture in Cheddar cheese-making may be to provide a suitable environment of acidity, moisture and low redox potential, which allows the elaboration of characteristic Cheddar flavour (Lowrie *et al.*, 1974; Law *et al.*, 1976). So little is known about the properties required of cultures to achieve a desired flavour and texture that their initial selection can only be made at the present time on the basis of acid-producing activity. If use of a selected culture then results in a product with the required characteristics, its composition and activity is conserved, usually by freezing and storage at temperatures below $-35\ ^{\circ}\mathrm{C}$.

Clearly many different factors contribute to cheese flavour, and the mechanisms involved vary according to cheese type. The bacterial proteinases and peptidases, together with the rennet added to cheese milk, hydrolyse part of the casein causing textural changes while the nature and amount of the various compounds formed influence the cheese taste. Although the chemical changes which take place during cheese ripening are relatively well characterized (Desmazeaud & Gripon, 1977) the compounds that lead to a desirable flavour have not been clearly defined. Even if these compounds are present, their contribution to flavour may be obscured by other metabolic products. Thus excessive growth of starter during cheese-making and survival of large numbers of viable cells in cheese may lead to flavour defects such as bitterness. The growth of organisms other than the bacteria producing lactic acid is also sometimes important. For example in Swiss-type cheese, propionibacteria metabolize the lactate produced by lactic acid bacteria, forming propionate and acetate and the CO_2 required for the formation of 'eyes' – a commercial criterion of primary importance. These organisms also produce proline, which is thought to contribute to the flavour of this type of cheese. Cheese curd is therefore a complex environment and the many interactions taking place often preclude the unambiguous interpretation of data.

GENETIC RELATIONSHIPS IN THE LACTIC ACID BACTERIA

The extensive literature available concerning the metabolism of the lactic acid bacteria contrasts with the limited information available on their genetics. The relative ease with which cultures of lactic streptococci produce variants makes reproducible data difficult to obtain and is no encouragement to the investigator interested in basic genetic relationships. Furthermore, until recently no laboratory system of genetic exchange was available. It was a considerable advance, therefore, when a transducing system for *S. lactis* strain C_2 was reported (McKay, Cords & Baldwin, 1973). The initial choice of *S. lactis* C_2 was most fortuitous, since the lactose-fermenting ability (*lac*) was found to be transducible. However, transduction has only been observed to date with this one strain and further studies should therefore concentrate on the characterization of other transducing phages and on the use of additional genetic markers. Three other types of genetic interchange (i.e. transformation, conjugation and lysogenic conversion) are known in some serological groups of streptococci, but none has yet been demonstrated with strains of Group N streptococci.

Lysogeny

Lysogeny appears to be a common property of bacteria, but a rigorous demonstration of lysogeny in the lactic streptococci has yet to be made. Since lactic streptococci grow in chains rather than as single cells, it may prove exceedingly difficult to isolate cured cells and then re-lysogenize them. The production of phage particles in cultures irradiated with UV light, which is presumptive evidence for lysogeny, has, however, often been observed (Lawrence *et al.*, 1976; Huggins & Sandine, 1977), although indicator strains for the induced phage have been found in only a few cases. Only temperate phages which are readily induced by UV light have been identified to date, but undoubtedly non-inducible prophage and defective prophage also exist. Further studies are needed to determine the optimum conditions required for induction of prophages, the number of different prophages and their location within the cell.

Plasmids

Reports on the number and size of plasmids, or the presence of co-valently closed circular DNA, indicate strain and species variation within the lactic streptococci (see Lawrence *et al.*, 1976). There is now considerable evidence to suggest that *lac* and proteinase activity (*prt*) of *S. lactis* strains are mediated through two distinct plasmids (Efstathiou & McKay, 1976). Anderson & McKay (1977) have demonstrated that the genetic determinants for EII-lac, FIII-lac and P-β-gal are located on a single plasmid in *S. cremoris* B1, and it also seems likely that plasmids are involved in nisin production (*nis*) in some *S. lactis* strains (Fuchs, Zajdel & Dobrzanski, 1975). There is still confusion, however, as to whether one genetic determinant has a pleiotropic effect. Loss of a common control mechanism or a change in the cell membrane might explain the apparent simultaneous loss of *lac* and *nis* together (Fuchs *et al.*, 1975), of *lac* and *prt* together (McKay & Baldwin, 1974*a*) and the ability of one phage to transduce *lac*, *mal* and *prt* (McKay *et al.*, 1973; McKay & Baldwin, 1974*a*). Drug-resistant mutants, phage-resistant mutants and restriction/modification mutants have all been reported in the lactic streptococci and should be considered for demonstrating plasmid linkage in lactic streptococci.

Variability within starter cultures

It is important to recognize that both the acid-producing ability of a culture and its phage sensitivity may change on repeated subculturing. Limsowtin *et al.* (1978), showed that the extent to which a particular variant accumulated was dependent upon the strain and the subculturing medium. Different isolates from the same 'pure' cultures required incubation at 22 °C from 18 h to at least 10 days before they could coagulate milk. Since plasmids are considered to carry part if not all of the genes controlling *lac* and *prt* (Efstathiou & McKay, 1976), a simple explanation for differences between strains in the incidence of variants is that the strains vary in the ease with which they lose plasmids. If the genetic information for restriction/modification systems in lactic streptococci is also carried on plasmids, as occurs in *E. coli* (see Reanney, 1976), the diversity of strains that can be identified on the basis of sensitivity to phage becomes readily understandable.

It is obvious that the potential exists for a diversity of lactic strepto-coccal strains and that 'pure' cultures which have been frequently transferred may contain a mixture of variant cells. For this reason it seems unlikely that continuous milk fermentations involving lactic streptococci will find practical use in product manufacture. In contrast, a continuous method of making yoghurt has proved satisfactory (Driessen *et al.*, 1977) which suggests that at least some strains of *S. thermophilus* and *L. bulgaricus* are stable.

Genetic manipulation

Mutation and the various mechanisms for DNA transfer have undoubtedly resulted in a reservoir of naturally occurring strains of lactic streptococci which vary considerably in genotype. There has recently been much speculation about artificial manipulation of the genetic material of starter cultures but this will not be a simple matter. The first concern must be to provide a starter that is resistant to phage and still has a high rate of acid production in milk. The limited information available about these characteristics, and also those involved in the development of flavour and texture of fermented products, clearly makes it difficult to 'engineer' starter strains with any certainty at present.

In particular, it must be emphasized how little is known about the basic biology of phage resistance in the lactic acid bacteria. Mutations resulting in changes in phage-adsorption characteristics, lysogenic conversion and plasmid-induced resistance no doubt all play a part. Nevertheless there appear in theory to be a number of approaches by which an 'ideal' starter with respect to phage resistance can be obtained. Such a strain could be one that restricted all phages or alternatively carried only a defective prophage(s). One must first determine the nature of the restriction/modification systems present in the Group N streptococci, the number and specificity of the endonucleases and where they are located. The genes responsible for restriction have been shown in other genera to be located in plasmids as well as on the bacterial chromosome (see Reanney, 1976) so that it might be possible to add restriction systems as plasmid DNA. The development of defective lysogens would involve the introduction of defective prophage into a cured cell. A defective lysogen would presumably be insensitive to a number of related phages but one still needs to know how many defective prophages must be present to ensure resistance to all the phages that are likely to arise in a cheese plant.

LACTIC STREPTOCOCCAL PHAGES

Differentiation

It appears that only a small number of morphologically distinct phage types attack the Group N streptococci, probably no more than five

(Tsaneva, 1976; Jarvis, 1977). The phages can be differentiated by head size and shape into small isometric, large isometric and prolate. The most common type has a small isometric head and a short tail, and over one-third of these also have a collar. Host-controlled modification and mutation of these basic phage types have, however, undoubtedly led to a much wider spectrum of phages, and morphologically similar phages have been found to grow on a wide variety of lactic streptococcal hosts. Thermophilic starters seem to be much less sensitive to phage than mesophilic starters (Table 1). Those phages attacking strains of *S. thermophilus* and *L. helveticus* that have been characterized to date show marked specificity (Sozzi & Maret, 1975).

Of the common methods of phage identification used at the present time only serology is definitive enough to establish whether a newly isolated phage is different from or the same as phages already known. Improved electron microscope techniques and data on such characteristics as nucleic acid content, sedimentation velocity and adsorption should prove to be helpful in further differentiation of lactic streptococcal phages.

Source of phage in commercial plants

The dimensions of the phages induced by UV light from starter cultures (Lawrence *et al.*, 1978) and from 'wild' lactic streptococci isolated from raw milk (Heap *et al.*, 1978) are similar to those of the virulent phages present in cheese wheys (Jarvis, 1977). Only small-headed isometric phages have been identified so far in UV lysates (H. A. Heap, unpublished), suggesting that large isometric and prolate phages may be mutants of the small isometric phage. This is supported by the relatively low frequency with which virulent large isometric and prolate phages are found in cheese wheys, and is in line with the report that mutants whose phenotypic expression affects the head length of the T-even phages of *E. coli* have been isolated (Cummings & Bolin, 1976).

When the same starter strains are introduced into widely separated cheese-making plants, the phages that appear are very similar in morphology and host range. This suggests a common source for the phage, probably the cultures themselves or the raw milk, which invariably contains lactic streptococci. Phages have been detected in raw milk and some at least survive pasteurization. Phages for newly introduced strains are usually detected first in the largest cheese plants, as one would expect if a degree of chance were involved. The likelihood that induced phages will be present appears to depend upon the volume of milk being processed. This will determine both the number of lactic streptococci arriving at the plant in the raw milk and, more importantly, the number of starter cells propagated during manufacture. The first phages to be detected in a cheese plant sometimes have slow rates of replication which may then increase, presumably as a result of mutation. Not all phages increase in virulence to the same extent. *S. lactis* ML_8, for instance, has been used continuously in New Zealand cheese plants for over twenty years, but the rates of replication of the phages that attack strain ML_8 are no faster now than they were when this strain was first used.

It was of more than philosophical interest to know whether the lactic streptococci are lysogenic, since this may explain why it is difficult to achieve a 'phage-free' situation in commercial plants. The significance of lysogeny, as far as the initial phage infection of cultures in cheese plants is concerned, is difficult to evaluate. In other genera both temperature and hydrogen peroxide are known to induce phage but the effect of these factors on the lactic streptococci is not yet known. Nevertheless, strains that spontaneously release up to 10^4 phage per ml have been reported (Lawrence *et al.*, 1976; Huggins & Sandine, 1977). The assumption that phage comes from either the starter cultures themselves or from the lactic streptococci in raw milk may prove to be an oversimplification, since Reanney (1976) has pointed out that phage can transfer genes between taxonomically distant genera. Regardless of its origin, however, phage levels in a commercial plant can be greatly reduced by removing any strains which act as indicators for the phages present (Lawrence *et al.*, 1978).

MICROBIOLOGICAL ASPECTS OF PRODUCT MANUFACTURE

The manufacturing procedures used are too varied for discussion here and the reader is referred to Kosikowski (1977) for details. Basically the lower the pH and final water content, and for some products the higher the ratio of salt to water, the longer the product will keep. Thus Parmesan cheese with a moisture content of about 25% keeps almost indefinitely, whereas Camembert with a moisture content of about 52% must be eaten within a few weeks of manufacture. In any specific plant, the rate of acid production by the cultures is affected by the temperature reached during manufacture, the amount of salt added and by changes in the composition of the milk with the stage of lactation and the weather. Inhibitory substances in the milk, resulting either from the poor bacteriological quality of the raw milk or from the treatment of bovine mastitis with antibiotics, may stop acid production altogether. Some starter strains are so sensitive to antibiotics that the addition of penicillinase to the milk is standard practice in many yoghurt and cheese plants. The most important inhibitory factor, however, in the manufacture of many fermented milk products is phage attack.

The relative importance of the rate of acid production in milk fermentations varies, depending both on the product and its method of manufacture (Table 1). When there is a requirement for rapid acid production and a low terminal pH, and the cultures used are potentially vulnerable to phage attack, maximum precautions against phage infection must be taken. This has become increasingly necessary in plants where the daily number of manufacturing cycles has increased, and with the recent trend to a reduction in the total time taken to reach the desired acidity in the manufacture of most fermented milk products. Efficient hygiene procedures for cleaning equipment, good ventilation and the use of well designed culture propagation vessels and manufacturing equipment are all important. However, since it is likely that the phage detected in manufacturing plants has originated from lactic acid bacteria present either in the cultures or in the raw milk, any culture system must be based upon the premise that strict hygiene cannot eliminate phage entirely. Under present conditions the most important factor in preventing the build-up of phage appears to be the prudent selection of cultures.

Commercial systems of culture propagation

Traditionally three stages of culture propagation, which progressively increase in scale, are involved in cheese-making (Fig. 3). Commercial cultures can be divided into those whose composition is variable and usually unknown in terms of phage sensitivity and, in some cases, species identity; and those which are made up from pure defined cultures, and are therefore reproducible at will.

Cultures of undefined composition

In Holland, cultures for Gouda-type cheese manufacture are propagated without precautions against phage infection. Natural phage contamination results in constant infection of the starter cultures and up to 10^8 phage particles per ml may be present (Stadhouders, Bangma & Driessen, 1976). The cultures presumably contain the relatively stable balance of susceptible and resistant bacteria typical of the phage-carrier state, with sensitive cells being continuously infected by free phage particles. If these mixed cultures are transferred frequently in the laboratory, where phage contamination is prevented, they become susceptible to phage attack due to emergence of sensitive mutants. Recently the system has been modified by concentrating and freezing the cultures so that they reach the cheese plant with a minimum of transfers. The importance of freezing the mixed culture to stabilize the strain composition cannot be over-emphasized, since this enables the same mixed starter to be used continuously for long periods in any one cheese plant.

In some countries, particularly Northern Italy and Switzerland, whey starters are used. These are obtained by allowing residual whey from the previous day's manufacture to acidify overnight in open tanks. The predominant organism is *L. helveticus*, but the extent to which other bacilli and streptococci are present is determined largely by the temperature at which the whey is held. Whey starters have been found to function more successfully than normal cultures when raw milk is used, and are considered to give a more acceptable cheese (Bottazzi, 1966).

In the United States and Britain, a rotation of several 'different' mixed starters is often used (Fig. 3). This procedure is basically unsound, since it does not take into account the fact that the number of available starter strains significantly different in phage sensitivity is likely to be small. The use of a large number of supposedly different strains in a plant will increase the risk of phage problems, since there is a greater chance that more temperate phages will be released and more indicator strains will be present. The uncertainty of the system is shown by the fact that in many plants the bulk culture is ready for use at least 8 h before it is required. This enables its activity to be checked in the milk that is to be used in the vats. The use of phage-inhibitory media (Fig. 3), which contain phosphates to bind Ca^{2+}, is common in the United States. While the mechanism(s) by which Ca^{2+} affects phage growth in the lactic streptococci has yet to be determined, a study of *Lactobacillus* phages (Watanabe & Takasue, 1972) showed that Ca^{2+} was required for penetration of phage DNA into host cells but not for either the adsorption or phage-multiplication steps. The reliance placed on these phosphated media in some cheese-making areas is surprising in view of reports that some phages can grow in such media (Sandine, 1977).

Cultures of defined single strains

The commercial use of defined single strains in starter cultures has been limited to New Zealand and, to a lesser extent, the United States, Australia and Scotland. The importance of using strains which are sensitive to completely different groups of phages has been emphasized

(Collins, 1962; Lawrence & Pearce, 1972) but in practice these have proved difficult to find. Recently, however, it has been shown that strains suitable for cheese-making can be obtained relatively easily from traditional mixed cultures, where they have arisen as a result of natural selection, by culturing isolates repeatedly in the presence of cheese wheys (Heap & Lawrence, 1976). When the selected strains are introduced into the plants from which the cheese wheys came originally, phages eventually appear which attack each individual strain. These strains need not be replaced, however, until the rate of phage multiplication becomes very high. Mixtures of such strains, called multiple starters (Lawrence *et al.*, 1978) to distinguish them from undefined mixed starters, can be used continuously for long periods.

Recent innovations in culture technology

The most important development in the past 15 years has undoubtedly been the replacement of liquid stock cultures, traditionally prepared at the commercial plants, by frozen cultures (Fig. 3) provided either by dairy research establishments or starter supply companies. The cultures are grown in a suitable medium, preferably with pH control, frozen and then stored at temperatures no higher than −35 °C. The

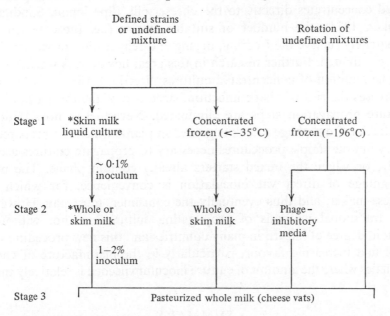

*Traditional procedure.

Fig. 3. Culture systems used in the manufacture of Cheddar and Gouda cheese, showing how the different stages of propagation are carried out.

survival of frozen cultures is influenced by the rate of freezing and temperature of storage, yet little information is available regarding the effect of these procedures on starter cultures. Some strains, particularly those of *L. bulgaricus*, are susceptible to freezing damage, but their survival in liquid nitrogen is improved by supplementation of the growth medium with Tween 80 prior to freezing (Tamime & Robinson, 1976).

The concentration of commercial cultures is now a common procedure, but offers little advantage unless the cultures are added directly to the vat (Fig. 3). Cell numbers in concentrates usually range from 10^{10} to 10^{11} cells per g, i.e. only 5 to 50 times more than in a culture where the pH has been kept constant, and the increase in acid-producing activity of such concentrates seldom relates to the increase in cell numbers. Centrifugation is the only concentration process being used commercially at present, but a pilot-scale diffusion culture technique, developed to remove inhibitory lactate from the medium, is reported to give a 20-fold concentrating effect during growth (Osborne, 1977). Much interest is being shown in the use of lyophilized culture concentrates to avoid the cost involved in transporting frozen cultures. Although it has been shown that cheese can be made by adding freeze-dried concentrates directly to the cheese milk (Speckman, Sandine & Elliker, 1974), the number of suitable strains (i.e. those which will adequately survive the freezing, drying and rehydration processes) is at present limited. Further research in this area, however, is warranted.

The addition of concentrated cultures directly to the vats obviously decreases the risk of phage infection, since two of the three traditional culture propagation steps are eliminated. Nevertheless no additional protection from phage attack is gained in plants which are prepared to carry out the simple procedures necessary to propagate cultures aseptically, or where the mixed starters already contain phage. The main advantage of direct vat inoculation is convenience, for which the cheese-maker, and thus eventually the consumer, must pay. However, the traditional methods of propagating cultures are not without a certain degree of hazard in many countries and this new procedure may well find increasing favour, particularly in the manufacture of cheese varieties where the amount of culture inoculum needed is relatively small.

SUMMARY

Only a limited understanding of many aspects of milk fermentations exists at present. The use in most countries of mixed cultures of undefined composition, and the marked changes that take place in the cheese curd during manufacture, often preclude the unambiguous interpretation of data. In addition the dominant role played by phages in commercial milk fermentations is almost without parallel in other microbial technologies. Vast numbers of cell divisions take place daily in the presence of phage, creating a unique opportunity for mutation and genetic exchange to occur.

Normally the primary microbial requirement is the fermentation of lactose to lactic acid. This is best carried out by lactic streptococci or the closely related lactobacilli, since use of bacteria with greater biochemical diversity would almost invariably give products with undesirable organoleptic characteristics. Rapid growth in milk requires an adequate proteinase system, and the PEP-dependent lactose – PTS is a prerequisite for homolactic fermentation by lactic streptococci. The limited range of metabolic activities and absence of endogenous energy reserve polymers in these organisms offer many advantages for metabolic studies, in particular the regulation *in vivo* of glycolysis and energy-dependent processes such as solute transport.

Acknowledgements

Grateful acknowledgement is made to our colleagues for their helpful advice and criticism during the preparation of this article, and particularly to J. Thompson and H. A. Heap for permission to quote unpublished results.

REFERENCES

In order to limit the number of references, a review or only the most recent of a series of papers on a particular topic has usually been quoted. We apologize to those authors whose work has not been referenced directly.

Accolas, J. P., Bloquel, R., Didienne, R. & Regnier, J. (1977). Acid producing properties of thermophilic lactic bacteria in relation to yoghurt manufacture. *Le Lait*, **57**, 1–23.

Anders, R. F., Hogg, D. M. & Jago, G. R. (1970). Formation of hydrogen peroxide by group N streptococci and its effect on their growth and metabolism. *Applied Microbiology*, **19**, 608–12.

Anderson, D. G. & McKay, L. L. (1977). Plasmids, loss of lactose metabolism, and appearance of partial and full lactose-fermenting revertants in *Streptococcus cremoris* B₁. *Journal of Bacteriology*, **129**, 367–77.

Babel, F. J. (1977). Antibiosis by lactic culture bacteria. *Journal of Dairy Science*, **60**, 815–21.

Bang, F. (1949). On the metabolism of *Streptococcus lactis*. 12*th International Dairy Congress, Stockholm*, **5**, 19–29.

BISSETT, D. L. & ANDERSON, R. L. (1974). Lactose and D-galactose metabolism in group N streptococci: presence of enzymes for both the D-galactose 1-phosphate and D-tagatose 6-phosphate pathways. *Journal of Bacteriology*, **117**, 318–20.

BOTTAZZI, V. (1966). Microbiology of natural starters used in cheese manufacture. *17th International Dairy Congress, Munich, D*, 509–13.

BOTTAZZI, V. & VESCOVO, M. (1969). Carbonyl compounds produced by yoghurt bacteria. *Netherlands Milk and Dairy Journal*, **23**, 71–8.

COGAN, T. M. (1975). Citrate utilization in milk by *Leuconostoc cremoris* and *Streptococcus diacetilactis*. *Journal of Dairy Research*, **42**, 139–46.

COLLINS, E. B. (1962). Behaviour and use of lactic streptococci and their bacteriophages. *Journal of Dairy Science*, **45**, 552–8.

COLLINS, E. B. & BRUHN, J. C. (1970). Roles of acetate and pyruvate in the metabolism of *Streptococcus diacetilactis*. *Journal of Bacteriology*, **103**, 541–6.

CROW, V. L. & PRITCHARD, G. G. (1977). Fructose 1,6-diphosphate-activitated L-lactate dehydrogenase from *Streptococcus lactis*: kinetic properties and factors affecting activation. *Journal of Bacteriology*, **131**, 82–91.

CUMMINGS, D. J. & BOLIN, R. W. (1976). Head length control in T4 bacteriophage morphologenesis: Effect of canavanine on assembly. *Bacteriological Reviews*, **40**, 314–59.

DEMKO, G. M., BLANTON, S. J. B. & BENOIT, R. E. (1972). Heterofermentative carbohydrate metabolism of lactose-impaired mutants of *Streptococcus lactis*. *Journal of Bacteriology*, **112**, 1335–45.

DESMAZEAUD, M. J. & GRIPON, J. C. (1977). General mechanism of protein breakdown during cheese ripening. *Milchwissenschaft*, **32**, 731–4.

DESMAZEAUD, M. J. & JUGE, M. (1976). Characterization of the proteolytic activity and fractionation of the dipeptidases and aminopeptidases from *Streptococcus thermophilus*. *Le Lait*, **56**, 241–60.

DE VRIES, W., KAPTEIJN, W. M. C., VAN DER BEEK, E. G. & STOUTHAMER, A. H. (1970). Molar growth yields and fermentation balances of *Lactobacillus casei* L3 in batch cultures and in continuous cultures. *Journal of General Microbiology*, **63**, 333–45.

DIRAR, H. & COLLINS, E. B. (1972). End-products, fermentation balances and molar growth yields of homofermentative lactobacilli. *Journal of General Microbiology*, **73**, 233–8.

DRIESSEN, F. M., UBBELS, J. & STADHOUDERS, J. (1977). Continuous manufacture of yoghurt. I. Optimal conditions and kinetics of the prefermentation process. *Biotechnology and Bioengineering*, **19**, 821–39.

EFSTATHIOU, J. D. & MCKAY, L. L. (1976). Plasmids in *Streptococcus lactis*: evidence that lactose metabolism and proteinase activity are plasmid linked. *Applied and Environmental Microbiology*, **32**, 38–44.

EXTERKATE, F. A. (1975). An introductory study of the proteolytic system of *Streptococcus cremoris* strain HP. *Netherlands Milk and Dairy Journal*, **29**, 303–18.

EXTERKATE, F. A. (1976). Comparison of strains of *Streptococcus cremoris* for proteolytic activities associated with the cell wall. *Netherlands Milk and Dairy Journal*, **30**, 95–105.

FUCHS, P. G., ZAJDEL, J. & DOBRZANSKI, W. T. (1975). Possible plasmid nature of the determinant for production of the antibiotic nisin in some strains of *Streptococcus lactis*. *Journal of General Microbiology*, **88**, 189–92.

HEAP, H. A. & LAWRENCE, R. C. (1976). The selection of starter strains for cheesemaking. *New Zealand Journal of Dairy Science and Technology*, **11**, 16–20.

HEAP, H. A., LIMSOWTIN, G. K. Y. & LAWRENCE, R. C. (1978). Contribution of

Streptococcus lactis strains in raw milk to phage infection in commercial cheese factories. *New Zealand Journal of Dairy Science and Technology*, **13**, 16–22.

HIRSCH, A. (1952). The evolution of the lactic streptococci. *Journal of Dairy Research*, **19**, 290–3.

HOFER, F. (1977). Involvement of plasmids in lactose metabolism in *Lactobacillus casei* suggested by genetic experiments. *FEMS Microbiology Letters*, **1**, 167–9.

HUGGINS, A. R. & SANDINE, W. E. (1977). Incidence and properties of temperate bacteriophages induced from lactic streptococci. *Applied and Environmental Microbiology*, **33**, 184–91.

JARVIS, A. W. (1977). The serological differentiation of lactic streptococcal bacteriophages. *New Zealand Journal of Dairy Science and Technology*, **12**, 176–81.

JONSSON, H. & PETTERSSON, H. E. (1977). Studies on the citric acid fermentation in lactic starter cultures with special interest in α-acetolactic acid. 2. Metabolic studies. *Milchwissenschaft*, **32**, 587–94.

KLETER, G. (1976). The ripening of Gouda cheese made under strict aseptic conditions. I. Cheese with no other bacterial enzymes than those from a starter streptococcus. *Netherlands Milk and Dairy Journal*, **30**, 254–70.

KNITTEL, M. D. (1965). Genetic homology and exchange in lactic acid streptococci. Ph.D. Thesis. Oregon State University, Corvallis, Oregon.

KOSIKOWSKI, F. V. (1977). *Cheese and Fermented Milk Foods*, 2nd edition. Ann Arbor, Michigan: Edwards Bros.

LAW, B. A. (1977). Dipeptide utilization by starter streptococci. *Journal of Dairy Research*, **44**, 309–17.

LAW, B. A., CASTANON, M. J. & SHARPE, M. E. (1976). The contribution of starter streptococci to flavour development in Cheddar cheese. *Journal of Dairy Research*, **43**, 301–11.

LAWRENCE, R. C. & PEARCE, L. E. (1972). Cheese starters under control. *Dairy Industries*, **37**, 73–8.

LAWRENCE, R. C., THOMAS, T. D. & TERZAGHI, B. E. (1976). Reviews of the progress of dairy science: cheese starters. *Journal of Dairy Research*, **43**, 141–93.

LAWRENCE, R. C., HEAP, H. A., LIMSOWTIN, G. K. Y. & JARVIS, A. W. (1978). Cheddar cheese starters: Current knowledge and practices of phage characterization and strain selection. *Journal of Dairy Science* (in press).

LEES, G. J. & JAGO, G. R. (1976). Acetaldehyde: an intermediate in the formation of ethanol from glucose by lactic acid bacteria. *Journal of Dairy Research*, **43**, 63–73.

LIMSOWTIN, G. K. Y., HEAP, H. A. & LAWRENCE, R. C. (1978). Heterogeneity among strains of lactic streptococci. *New Zealand Journal of Dairy Science and Technology*, **13**, 1–8.

LONDON, J. (1976). The ecology and taxonomic status of the lactobacilli. *Annual Review of Microbiology*, **30**, 279–301.

LOWRIE, R. J., LAWRENCE, R. C. & PEBERDY, M. F. (1974). Cheddar cheese flavour. V. Influence of bacteriophage and cooking temperature on cheese made under controlled bacteriological conditions. *New Zealand Journal of Dairy Science and Technology*, **9**, 116–21.

MCKAY, L. L. & BALDWIN, K. A. (1974a). Simultaneous loss of proteinase- and lactose-utilizing enzyme activities in *Streptococcus lactis* and reversal of loss by transduction. *Applied Microbiology*, **28**, 342–6.

MCKAY, L. L. & BALDWIN, K. A. (1974b). Altered metabolism in a *Streptococcus lactis* C2 mutant deficient in lactic dehydrogenase. *Journal of Dairy Science*, **57**, 181–6.

MCKAY, L., MILLER, A., SANDINE, W. E. & ELLIKER, P. R. (1970). Mechanisms of lactose utilization by lactic acid streptococci: enzymatic and genetic analysis. *Journal of Bacteriology*, **102**, 804–9.

McKay, L. L., Cords, B. R. & Baldwin, K. A. (1973). Transduction of lactose metabolism in *Streptococcus lactis* C2. *Journal of Bacteriology*, **115**, 810–15.

Mills, O. E. & Thomas, T. D. (1978). Release of cell wall-associated proteinase(s) from lactic streprococci. *New Zealand Journal of Dairy Science and Technology* (in press).

Mou, L., Sullivan, J. J. & Jago, G. R. (1975). Peptidase activities of group N streptococci. *Journal of Dairy Research*, **42**, 147–55.

O'Leary, V. S. & Woychik, J. H. (1976). Utilization of lactose, glucose, and galactose by a mixed culture of *Streptococcus thermophilus* and *Lactobacillus bulgaricus* in milk treated with lactase enzyme. *Applied and Environmental Microbiology*, **32**, 89–94.

Osborne, R. J. W. (1977). Production of frozen concentrated cheese starters by diffusion culture. *Journal of the Society of Dairy Technology*, **30**, 40–4.

Pearce, L. E., Skipper, N. A. & Jarvis, B. D. W. (1974). Proteinase activity in slow lactic acid-producing variants of *Streptococcus lactis*. *Applied Microbiology*, **27**, 933–7.

Postma, P. W. & Roseman, S. (1976). The bacterial phosphoenolpyruvate: sugar phosphotransferase system. *Biochimica et Biophysica Acta*, **457**, 213–57.

Premi, L., Sandine, W. E. & Elliker, P. R. (1972). Lactose-hydrolyzing enzymes of Lactobacillus species. *Applied Microbiology*, **24**, 51–7.

Reanney, D. (1976). Extrachromosomal elements as possible agents of adaptation and development. *Bacteriological Reviews*, **40**, 552–90.

Sandine, W. E. (1977). New techniques in handling lactic cultures to enhance their performance. *Journal of Dairy Science*, **60**, 822–8.

Schleifer, K. H. & Kandler, O. (1967). The chemical composition of the cell wall of streptococci. II. The amino acid sequence of mureins of *Str. lactis* and *cremoris*. *Archiv für Mikrobiologie*, **57**, 365–81.

Schmidt, R. H., Morris, H. A. & McKay, L. L. (1977). Cellular location and characteristics of peptidase enzymes in lactic streptococci. *Journal of Dairy Science*, **60**, 710–17.

Shankar, P. A. & Davies, F. L. (1977). Associative bacterial growth in yoghurt starters: initial observations on stimulatory factors. *Journal of the Society of Dairy Technology*, **30**, 31–2.

Sinha, R. P. (1977). Acriflavine-resistant mutants of *Streptococcus cremoris*. *Antimicrobial Agents and Chemotherapy*, **12**, 383–9.

Sozzi, T. & Maret, R. (1975). Isolation and characteristics of *Streptococcus thermophilus* and *Lactobacillus helveticus* phages from Emmental starters. *Le Lait*, **55**, 269–88.

Speck, M. L. (1976). Interactions among lactobacilli and man. *Journal of Dairy Science*, **59**, 338–43.

Speckman, R. A. & Collins, E. B. (1973). Incorporation of radioactive acetate into diacetyl by *Streptococcus diacetilactis*. *Applied Microbiology*, **26**, 744–6.

Speckman, C. A., Sandine, W. E. & Elliker, P. R. (1974). Lyophilized lactic acid starter culture concentrates. *Journal of Dairy Science*, **57**, 165–73.

Stadhouders, J. (1974). Dairy starter cultures. *Milchwissenschaft*, **29**, 329–37.

Stadhouders, J., Bangma, A. & Driessen, F. M. (1976). Control of starter activity and the use of starter concentrates. *Nordeuropaeisk mejeri-tidsskrift*, **42**, 190–208.

Tamime, A. Y. & Robinson, R. K. (1976). Recent developments in the production and preservation of starter cultures for yoghurt. *Dairy Industries International*, **41**, 408–11.

THOMAS, T. D. (1976). Regulation of lactose fermentation in group N streptococci. *Applied and Environmental Microbiology*, **32**, 474–8.

THOMAS, T. D. & LOWRIE, R. J. (1975). Starters and bacteriophages in lactic acid casein manufacture. 1. Mixed strain starters. *Journal of Milk and Food Technology*, **38**, 269–74.

THOMAS, T. D. & TURNER, K. W. (1977). Preparation of skim milk to allow harvesting of starter cells from milk cultures. *New Zealand Journal of Dairy Science and Technology*, **12**, 15–21.

THOMAS, T. D., JARVIS, B. D. W. & SKIPPER, N. A. (1974). Localization of proteinase(s) near the cell surface of *Streptococcus lactis*. *Journal of Bacteriology*, **118**, 329–33.

THOMPSON, J. & THOMAS, T. D. (1977). Phosphoenolpyruvate and 2-phosphoglycerate: endogenous energy source(s) for sugar accumulation by starved cells of *Streptococcus lactis*. *Journal of Bacteriology*, **130**, 583–95.

THOMPSON, J., TURNER, K. W. & THOMAS, T. D. (1978). Catabolite inhibition and sequential metabolism of sugars by *Streptococcus lactis*. *Journal of Bacteriology*, **133**, 1163–74.

TSANEVA, K. P. (1976). Electron microscopy of virulent phages for *Streptococcus lactis*. *Applied and Environmental Microbiology*, **31**, 590–601.

TURNER, K. W. & THOMAS, T. D. (1975). Uncoupling of growth and acid production in lactic streptococci. *New Zealand Journal of Dairy Science and Technology*, **10**, 162–7.

WATANABE, K. & TAKASUE, S. (1972). The requirement for calcium in infection with *Lactobacillus* phage. *Journal of General Virology*, **17**, 19–30.

WICKEN, A. J. & KNOX, K. W. (1975). Characterization of group N streptococcus lipoteichoic acid. *Infection and Immunity*, **11**, 973–81.

MICROBIAL ASPECTS OF WASTE TREATMENT WITH PARTICULAR ATTENTION TO THE DEGRADATION OF ORGANIC COMPOUNDS

J. HOWARD SLATER* AND HUGH J. SOMERVILLE†

*Department of Environmental Sciences,
University of Warwick, Coventry CV4 7AL, UK*

† *Shell Biosciences Laboratory, Sittingbourne Research Centre,
Sittingbourne ME9 8AG, UK*

INTRODUCTION

Originally, man was a minor component within the biosphere; his activities and wastes had little effect on the quality of the environment. The trend towards high-density community life in early societies brought the first significant waste-disposal problems, namely the effect of large quantities of domestic waste materials in a localized environment. As a result, initially of health considerations but latterly of the need to maintain an acceptable environmental quality, a controlled approach to waste collection and disposal was required. The early Assyrian and Babylonian civilizations developed advanced sewer systems (Gray, 1940) and the cities of the fertile crescent had waste-treatment processes based on digestion in cess pits (Hughes, 1977). The success of the Industrial Revolution in Britain was marred by a lack of concern for the consequences of rapid urbanization, which contributed to appalling public health standards and major disease outbreaks (Hawkes, 1965). The growth of human populations necessitated efficient waste collection and specific treatment processes since disposal by dilution and self-purification, for example direct discharge and biodegradation in natural waters, was no longer always practicable or desirable. Mainly by empirical means a variety of biological treatment systems were developed, ranging from cess pits, septic tanks and land effluent percolation (sewage farms) to gravel beds, percolating filters (trickle beds) and activated-sludge processes coupled with anaerobic digestion (Hawkes, 1965; Pike & Curds, 1971). The primary objectives of these biotreaters were to alleviate health hazards and to reduce the concentration of oxidizable organic compounds, yielding a final effluent which could be discharged into the natural environment without producing any adverse effects.

At first, biotreaters were designed to deal almost exclusively with domestic wastes, basically a simple operation since such wastes have a relatively standard composition and contain readily degradable materials. However, since the nineteenth century, waste treatment has increased in complexity, with a number of novel problems as well as the logistic difficulties associated with the large increase in waste production. The types of waste produced have altered radically because of the advances and expansion of modern industrial processes and agricultural practices. Effluents from industrial processes may be, for example, extremely acidic or alkaline or contain high concentrations of toxic compounds or ions and, as a result of the volume of waste waters produced, can, if directly discharged, seriously affect the efficiency of municipal treatment plants or damage the natural environment. For example, pulp and paper mills discharge large quantities of waste process waters containing many toxic compounds including resin acids, chlorinated resin acids, and chlorinated unsaturated fatty acids (Leach, Mueller & Walden, 1978). Tannery effluents contain high concentrations of sulphides (up to 2500 mg sulphide 1^{-1}) and chromium ions (up to 1500 mg Cr^{3+} ions 1^{-1}) (Bailey, 1977). In principle, the discharge of industrial wastes is subject to rigorous legal controls (Fisher, 1977) and, since the majority of industrial wastes are treated by local authority sewage works, the rate of discharge has to be controlled to maintain an efficient biological activity in the plant. Increasingly, local authorities demand preliminary factory pretreatment before discharge to minimize the detrimental effects of industrial wastes. In the United Kingdom, legislation controls nearly all discharges of trade and sewage effluent to rivers, the sea, or land through the appropriate water authority which may refuse the discharge or accept it subject to any reasonable conditions (Department of the Environment, 1976). For example, the pretreatment of tannery waste liquors involves vigorous aeration in the presence of manganese sulphate as catalyst to accelerate the rate of sulphide oxidation (Bailey & Humphreys, 1967).

One of the major contemporary problems has been the appearance of thousands of new chemicals released into the biosphere as products of the synthetic chemical industry, particularly the many novel compounds used in agriculture as pesticides and herbicides. Many of these compounds have not occurred in the biosphere prior to their synthesis by man and may be considered as environmentally foreign compounds (xenobiotic compounds). Some xenobiotic compounds are sufficiently similar to existing natural products to be degraded by existing mechanisms but others are not susceptible to biological transformation.

Consequently compounds in this latter category may persist in the environment for considerable periods of time and may become serious environmental hazards. The ability of micro-organisms to evolve the capacity to break down persistent xenobiotic compounds is of considerable interest and an important aspect of microbial waste utilization in the present-day biosphere. This topic is discussed in detail later.

A second modern waste disposal problem is concerned with the distribution of waste materials and materials which ultimately have to be considered as wastes in the environment. Traditionally waste treatment has been concerned with collected, bulk quantities of materials which are contained and treated *en masse*. However as a result of some present-day activities, many compounds which ultimately become wastes are widely dispersed into the biosphere either deliberately, for example agrochemicals, or accidentally, for example oil spillages. Clearly wastes of this nature cannot be collected for treatment and, although in most cases natural microbial activity may be sufficient to degrade these dispersed wastes, there may be occasions when novel treatments have to be effected *in situ*.

The biotreatment of waste depends almost entirely on microbial activity exploiting those sequences of the carbon cycle involved in the catabolism and mineralization of organic compounds. The success of any process depends on the capacity of existing micro-organisms to degrade the vast array of simple and complex organic compounds synthesized or derived within the biosphere. Accordingly, a fundamental feature of biotreaters is that they should contain a range of micro-organisms with the overall metabolic capacity to degrade any compound entering the system (Painter, 1978).

This paper is concerned with a broad outline of the principles of biotreatment, the potential advances in treatment strategy and aspects of microbial metabolism, biochemistry and genetics which are relevant to biodegradation. The design of efficient biotreaters has been extensively surveyed elsewhere (Bolton & Klein, 1971; Jones, 1976; Howell, 1978; Painter, 1978) and engineering considerations, operational characteristics and performance capabilities will not be considered here (Andrews, 1974; Andrews, Briggs & Jenkins, 1974; Everett, 1977; Forster, 1977).

BIOTREATMENT OF COLLECTED WASTE STREAMS

The treatment of collected wastes follows the generalized scheme outlined in Fig. 1. Different waste streams, with or without pretreatment, possibly by chemical oxidation (Teletzke, 1964), are collected

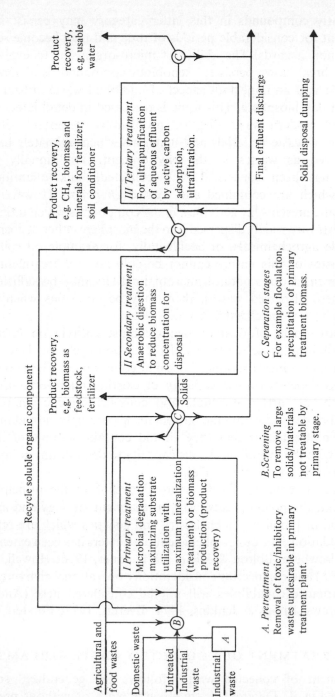

Fig. 1 General scheme for the treatment of collected aqueous effluents.

with the minimum of direct loss to the environment, mixed, and after the removal of large solids by mechanical screening passed to the primary treatment stage.

Primary treatment stage

In principle, the first stage ought to maximize the rate of biodegradation with maximum substrate utilization and minimum microbial biomass production. The latter aim is often important since in most treatment processes the biomass formed is itself an additional waste eventually requiring disposal (Ford, 1977), although the minimization of biomass production has to be balanced against the need for high concentrations in the primary stage to maximize the rate of clean effluent production (Jones, 1976).

The active biomass in biotreaters, in general, demonstrates the same physiological control mechanisms that operate in monocultures. On introduction of a biodegradable compound to the inflowing stream, a long lag may be observed before degradation starts (Painter, 1974). Activated sludge is also apparently subject to catabolite repression. For example, Gaudy (1962) demonstrated diauxic responses in activated sludge exposed to glucose, melibiose and lactose. Although sludge recycling contributes to the maintenance of catalytic potential, factors such as washout, enzyme decay and predation will contribute to the elimination of the relevant biodegradation capacity in the absence of continuous selection pressure exerted by a particular compound. In addition, the efficiency of activated sludge depends on relatively undefined properties of the floc such as its age, method of formation and the ratio of feed to biomass (Hawkes, 1978). Such overall parameters reflect the general physiological status of the sludge and may, to some extent, be used to monitor and optimize the efficiency of the treatment process.

Effluents, particularly those from industrial chemical complexes, are frequently of variable composition. In some cases this can be alleviated by providing upstream holding tanks to allow some control of the composition and uniformity of the medium. The basic nutrient composition, particularly the carbon:nitrogen:phosphorus ratio, can be variable and additions may be required for optimum performance and to control mineral levels in the final effluent (Barth, Brenner & Lewis, 1968).

The use of suspended micro-organisms in a floc form aerated by mixing is probably the most common system used for the primary treatment of aqueous effluents. Various alternatives to activated sludge plants involving suspended growth have been developed. For example, microalgae can be used either in unmixed ponds, where the top zone is oxygenated as a result of algal photosynthetic activity and the biomass sinks to an anaerobic zone where it is partially digested, or in mixed aerobic ponds where higher treatment rates may be achieved (Benemann *et al.*, 1977; Shelef *et al.*, 1977). Other high-rate suspended-solid biotreatment processes include the UNOX and FMC systems (see Howell, 1978) in which oxygen is used as the gas phase to obtain higher oxygen mass transfer rates (Boon, 1976) and the ICI Deep Shaft process (Bailey *et al.*, 1975) where a high dissolved oxygen concentration is achieved by a novel pressure lift process.

Immobilized microbial growth on percolating (trickling) filters (Painter, 1978) is another common primary treatment method and here too various alternative methods are available. These systems still result in biomass formation but have the advantages of less sensitivity to shock loadings and higher treatment rates than activated sludge plants. Some of these alternative approaches are discussed in detail by Howell (1978).

In the past, treatment processes have tended to ignore wastes as potential feedstocks for microbial fermentations and product formation, with the notable exception of methane generation from sludge. In many cases, waste treatment has been accepted as a charge on society without a detailed examination of all of the economic factors involved. Increasingly, however, a number of specific effluents (as opposed to mixed domestic and industrial effluents) have been examined as potential resources, although few effluent treatments coupled to product formation have been successfully operated (Table 1).

Whatever primary treatment process is used, biomass is a major product and various outlets have been investigated for its utilization, principally as an animal feed, as a fertilizer, or as a feedstock for anaerobic digestion to methane. An excellent example of the commercial possibilities comes from the use of waste starch in liquid waste streams from the processing of potatoes, wheat and other cereals. The Symba process is based on a mixed culture of *Candida utilis* and *Endomycopsis fibuliger* producing a high-quality single-cell protein product and simultaneously reducing the effluent's biological oxygen demand (BOD) by 90% (Skogman, 1976).

Vast quantities of biomass are produced in industrial and municipal treatment systems (Loll, 1976) and, as waste treatment becomes a more significant component of industrial costs, it is likely that new processes will be developed to exploit the available biomass. These are likely to

Table 1. *Some examples of proposed and operating microbial effluent treatments coupled to product formation*

Feedstock	Involved microbial species	Products (excluding dischargeable water)	References
Sulphate liquor from pulp mills	*Paecilomyces*	Biomass for animal food	Romantschuk & Lehtomaki (1978)
Residual tanker oil	Mixed culture producing emulsification	Dischargeable oil	Gutnick & Rosenberg (1977)
Sludge, cellulose, agricultural wastes, etc.	Bacteria degrading cellulose and organic acids, methanogens	Methane, sludge for fertilizer	Booram & Smith (1974); Buhr & Andrews (1977); Clausen, Sitton & Gaddy (1977)
Municipal, chemical wastes	Algae (*Spirulina, Chlorella*)	Biomass	Benemann *et al.* (1977) Shelef *et al.* (1977)
Sugar cane, casava	Yeast strains	Ethanol	Jackman (1976)
Brewery waste	Fungal strains	Citric acid	Hang *et al.* (1977)
Starch wastes	*Candida, Endomycopsis*	Biomass for animal food	Skogman (1976)
Sewage	*Nitrosomonas, Nitrobacter*	Ccoling water	Humphris (1977)

include the use of biomass for fuel generation, incorporation into utilizable solids and refinement to specific components such as lipids or vitamins.

Secondary treatment stage

Anaerobic digestion is normally used as an effective method of reducing sludge bulk derived from the primary biotreaters. However, some wastes, for example feed-lot waste (Bryant *et al.*, 1976), can be digested anaerobically without an aerobic primary treatment stage, but this is generally a slower process requiring longer residence times and a thermal input (Konstandt, 1976). Despite the heat input required, a thermophilic process can be operated to reduce the residence time (Bryant *et al.*, 1976; Buhr & Andrews, 1977). Anaerobic digestion is a complex fermentation producing principally fatty acids, carbon dioxide and methane from a wide range of substrates including some highly reduced organic compounds (McCarty *et al.*, 1976; Balba & Evans, 1977; Healy & Young, 1978*a*, *b*). Indeed, methane is usually produced in sufficient quantities to supply the complete energy needs of traditional activated sludge works (Konstandt, 1976). In many cases the final solid waste can be used as a high-quality fertilizer, provided that it contains acceptably low levels of toxic materials, particularly metals and pathogens (Shuval, 1977).

A number of alternative anaerobic fermentation mechanisms exist (Morris, 1975; Thauer, Jungermann & Decker, 1977) including production of hydrogen as an additional or alternative end-product to methane. However, hydrogen production has been discounted as a relatively inefficient method of energy conversion (Thauer, 1976).

TREATMENT OF DISPERSED WASTES

Inevitably much potential waste now reaches the environment through accidental spillages or direct discharge. Such events can range from small quantities of potentially harmful chemicals, for example dioxin and agrochemicals, usually at low concentrations, to the huge volumes of oil released as a result of blow-outs or tanker accidents. Clearly, these wastes cannot feasibly be collected and treated in the conventional manner and in the case of those compounds deliberately released into the biosphere a period of time must elapse for their function, for example herbicidal activity, to be expressed before degradation begins. Such a sequence of events is perfectly acceptable for those materials which are degradable, but poses major problems if they are less readily metabolized and liable to accumulate and interfere with the normal operation of the biosphere. For large volumes and concentration of wastes in the environment, the initial problem is one of dispersal and dilution but, in all cases, degradation ultimately depends on the presence of appropriate micro-organisms and on the supply of appropriate nutrients. For example, the degradation of crude oil may be limited in the marine environment by the supply of inorganic nitrogen and phosphorus, and an increase in the rate of degradation can be made if these nutrients are added (Gutnick & Rosenberg, 1977; van der Linden, 1978). The oxygen supply is also critical, since oil which reaches anoxic zones will remain undegraded, and techniques for forced aeration have been developed to enhance microbial degradation (Jamison, Raymond & Hudson, 1976).

Oxygen depletion is often a major consequence of waste metabolism and may have severe effects on local environments. For example, the discharge of pulp fibre into a sea loch has been shown to affect markedly sediment redox values and the associated microflora (Stanley, Pearson & Brown, 1978). The addition of formulations of micro-organisms able to degrade a spilled waste, with added nutrients including an electron acceptor where necessary, is a possibility for future action. This could eliminate the lag times involved in the adaptation and growth of any naturally occurring population. A number of such commercial products have been marketed but their efficacy remains to be rigorously established. The principle, however, has been exemplified by Daughton & Hsieh (1977a) who added a mixed culture adapted to degrade the insecticide Parathion to soil samples contaminated with the insecticide and showed considerably accelerated rates of degradation.

Recently, there has been speculation about the use of deliberately constructed bacterial strains carrying multiple degradative capacities (Williams, 1978). For example, the construction of a strain of *Pseudomonas* containing several degradative plasmids has led to the proposal that it should be exploited for the degradation of fuel oil (Friello, Mylroie & Chakrabarty, 1976). However, this approach is unlikely to be widely successful since this organism's degradative capacity is only sufficient to deal with a small proportion of crude oil. Furthermore, naturally occurring mixed populations which have evolved to grow in a particular environment are likely to be very competitive, growing more rapidly than a constructed strain. It is possible, however, that in future the addition of constructed strains may be successful in degrading specific compounds which are not metabolized by the existing natural microbial flora.

ANALYSIS OF EFFLUENTS AND THEIR EFFECTS ON MICRO-ORGANISMS IN THE ENVIRONMENT

It is a general concern of society to be aware of the extent to which the environment is contaminated with man-made products. Ideally, acquisition of this knowledge would involve specific detection methods for all of the compounds released into the environment and a comprehensive study of their subsequent effects. However, this is an impracticable goal at present and specific methods, such as the use of ^{14}C-labelled compounds, are only feasible for a limited number of situations, in particular where compounds with some known effect are being measured. The realistic alternative is to use general methods capable of quantifying gross contamination and of evaluating the extent to which microbial and other natural processes can recycle waste materials or are adversely affected by them.

Specific analytical techniques such as gas- and liquid-chromatography and gas analysis are excluded from this brief discussion. For more detailed discussion of the general methods described, reference should be made to texts and reviews (Painter, 1974; Higgins & Burns, 1975; Callely, Forster & Stafford, 1977; de Kreuk & Hansveit, 1978; Gledhill & Seager, 1978).

Analytical methods

Biological oxygen demand (BOD)

This method, of which there are several defined variations available, involves measurement of the amount of dissolved oxygen consumed for a given volume of waste under standard conditions. Normally, the oxygen concentration is measured before and after incubation of the waste at 20 °C for 5 d. BOD is expressed as mg O_2 consumed per ml of material. The method suffers from the disadvantages of being carried out in a closed system and it estimates only the extent of oxygen uptake. There is no estimation of undegraded components in the incubation mixture, and consequently the presence of recalcitrant or inhibitory compounds may be overlooked. BOD values can, however, give an approximation of the extent to which an effluent may remove oxygen from a water source and, if used in conjunction with TOC, can give an estimate of overall degradation.

Total Organic Carbon (TOC)

Significant recent advances have been made in TOC measurement and several instruments are available commercially, including some with automatic sampling. In these, the organic material, dissolved or suspended, is oxidized catalytically to carbon dioxide; this is either measured directly or reduced to methane before detection. This approach is of widespread application in the assessment of water quality and in biodegradation studies where, with appropriate sampling regimes, it results in an accurate and simple estimation of the rate of mineralization of a sample.

Chemical oxygen demand (COD)

This depends on chemical oxidation, and thus gives a measure of both organic and some inorganic material. It is particularly useful when a combination of chemical and biological treatment is envisaged and can give a maximum or 'theoretical' oxygen uptake value for complete biodegradation, although more detailed inorganic analyses may be necessary. This value is not however attainable because of conversion of some carbon into biomass. It is not unusual for COD values to be several times the BOD values for the same samples, indicating that the particular sample contains a large amount of organic material that is not degraded by micro-organisms.

Assessment of the effects of waste materials on micro-organisms

In assessing the impact of a particular waste it may be necessary to evaluate both its susceptibility to biotreatment and any effects on the existing biotreatment system. A number of methods are available to assess the effect of a certain waste, particularly when it is discharged into a conventional biotreatment system. These include the 'porous pot' (Stennet & Eden, 1971) and the OECD confirmatory test (OECD, 1971), both of which are models of activated sludge systems. Currently there is considerable activity in developing realistic laboratory-scale model systems to monitor the effect of waste materials on microbial ecosystems (see for example, Giddings *et al.*, 1978; Pritchard *et al.*, 1978). These are complex systems to operate, and are likely to be subject to numerous operational difficulties, but the results of such tests are likely to be more significant than simply testing the toxicity of a single compound against a single species. The appreciation of analytical techniques, such as gas analysis and determination of dissolved oxygen tension, gives useful information on the overall rates of respiration and may be used *in situ* to assess the general effect of specific effluents on water-treatment systems and on the ability of the microbial populations to accommodate and recover from novel challenges. A simple and effective initial test is to measure the respiration rates of activated sludge in dilutions of the suspect solution both before and after continuous culture in the presence of the waste (Swain & Somerville, 1978).

Chronic effects of diffuse wastes discharged in the environment have only been studied by taking each case individually. The marked effect of eutrophication on microbial populations of freshwater habitats has been studied, for example, by Daft, Stewart and co-workers (see Daft & Fallowfield, 1978) and effects of a particular chemical, kepone, on estuarine populations has been studied by Orndorff & Colwell (personal communication). Kimerle & Swisher (1977) have shown that the toxicity of linear alkylbenzenesulphonates (LAS) to fish can be eliminated as a result of partial biodegradation.

The identification of populations of particular microbial species can indicate both that waste or some change in carbon balance has reached the environment and that a hazard has arisen, for example the health hazards associated with pathogens in sludge biomass (Wielgolaski, 1975; Cairns, Dickson & Westlake, 1976; Elliott, 1977; Shuval, 1977). When large amounts of organic material are continuously discharged, the most serious general effect appears to be oxygen depletion which causes, in time, a permanent effect on both microbial and higher life forms (Mellanby, 1972; Stanley, Pearson & Brown, 1978).

Attempts are being made to regulate the use of pesticides by evaluation of soil-function tests; the tests involve the monitoring of microbial activities including CO_2 evolution, nitrification and nitrogen fixation. To date, reliable measurements have only been made in the laboratory and it is not clear to what extent observations can be extrapolated to the field. Although there are areas of specific concern, such as the effect of pesticides on blue-green algae in rice paddies, a recent review (Greaves et al., 1976) has concluded that no pesticide has been shown to have long-term detrimental effects on micro-organisms in soil when applied at the levels used in agriculture.

Some micro-organisms can be used in the study of particular environmental problems. For example, the survey of lichen populations has been used to measure SO_2 pollution of the atmosphere (Nash, 1976) and recently, monitoring of sulphate-reducing bacteria has been initiated in the oil industry to indicate potential corrosion problems (Herbert, 1976).

MICROBIAL BIODEGRADATION

Biodegradation is a general term applied to the partial or complete breakdown of organic compounds by micro-organisms. The term implies cleavage of covalent bonds with resulting transformation to one or more undefined products, with or without concomitant exploitation as a carbon- or energy-source for the growth of the transforming population. It can also describe complete mineralization where all of the original compound is converted to basic products such as carbon dioxide, water and ammonia. Typically for aerobic growth, 70% of the substrate is mineralized whilst 30% is utilized for new biomass production, although there may be some formation of other degradation products which are not utilized further by the growing population. For anaerobic growth, as little as 1% of the transformed substrates may be converted into new biomass. Studies on the fate of a compound in waste-treatment processes or in the natural environment have frequently centred on the overall rate of biodegradation, usually measured in terms of the rate of disappearance of the initial compound. Although a considerable body of information is available about the biochemistry of metabolism of many compounds degraded by pure cultures of micro-organisms, much less attention has been given to the ultimate fate of a compound in natural environments, or to the initial breakdown products and the biochemical details of degradation in complex multispecies habitats. The rate of biodegradation is, however, a useful parameter providing valuable information on the initial behaviour of a compound in a particular environment, but it must be stressed that its determination does not necessarily provide all of the relevant information. The initial compound may be degraded at a certain rate, generating products which may accumulate and become more important environmental hazards than the original compound, e.g. the generation of foam-stabilizing compounds during breakdown of some detergents (Patterson, Scott & Tucker, 1967; Cook, 1978). Furthermore, rates of biodegradation are often determined with little regard to the precise environmental conditions, and a value from one set of conditions may be very different in alternative circumstances. For example, 3,6-dichloropicolinic acid (the herbicide Lontrel) at an application concentration of 0.25 ppm and at 25 °C under non-leaching conditions has a half-life ranging from 14 to 309 d depending on the nature of the soil used. (A. J. Gilchrist, personal communication.)

For biodegradation to occur at all, the environment concerned must contain at least one population with the appropriate catabolic mechanism. The rate of biodegradation depends on the initial interaction between the compound and the organism, the kinetic properties of the metabolic process (largely determined by the concentration of the degrading population and the compound concentration), and the physicochemical conditions (which can be important in determining the fate of a compound in the biosphere (Plimmer, 1978)).

Micro-organisms have evolved an extensive range of enzymes, pathways and control mechanisms in order to be able to degrade a wide array of naturally occurring compounds (see for example, Dagley, 1971, 1975, 1978; Ornston, 1971; Kieslich, 1976; Evans, 1977). A striking feature of many enzyme reactions is their high degree of substrate specificity. Consequently, a large number of enzymes are required to degrade the range of natural products available. For example, *Pseudomonas* species are known to produce at least eight different dioxygenases catalysing similar *meta* ring-fission reactions, and many of these enzymes fail to attack other substrates (Dagley, 1975). However, substrate specificity is often not absolutely restricted to a single substrate and so, for example, protocatechuate 4,5-oxygenase will also attack gallate and 3-*o*-methylgallate, both substrates for other specific dioxygenases. Catechol 1,2-oxygenase catalyses oxidation of 3- and 4-methyl catechol besides catechol itself. Furthermore, complete pathways specific for single substrates may also be needed. *Pseudomonas putida* degrades catechols by two different pathways (Sala-Trepat, Murray & Williams, 1972; Collingworth, Chapman & Dagley, 1973) and, although there is a common initial reaction, i.e. the dioxygenase step, 4-methyl-catechol is only metabolized by the pathway involving the decarboxylation reaction, whilst 3-methyl catechol is metabolized by the alternative pathway (Dagley, 1975) (Fig 2).

In contrast, some enzymes show an apparently startling lack of specificity. For example, species of methane-oxidizing bacteria will also oxidize ammonia (Drozd, Godley & Bailey, 1978) with the cell-free enzyme system showing even wider specificity (Colby, Stirling & Dalton, 1977). Nitrogenase will reduce a number of structurally related compounds (Postgate, 1974) and phosphatases will hydrolyse a range of inorganic and organic phosphates (Flint & Hopton, 1976).

Each organism does not possess all the mechanisms available for biodegradation of every compound, and so the successful treatment of multicomponent wastes depends on the combined activities of a wide range of micro-organisms with diverse metabolic capabilities. Clearly, an understanding of the individual pathways and enzymes involved in the degradation of many different compounds is fundamentally important in describing biodegradation in natural environments. It is clear that shifts in metabolic pattern for a compound can occur as a result of relatively few mutational events, for example the shift of phenol metabolism from *meta* to *ortho* pathway in *Pseudomonas putida*

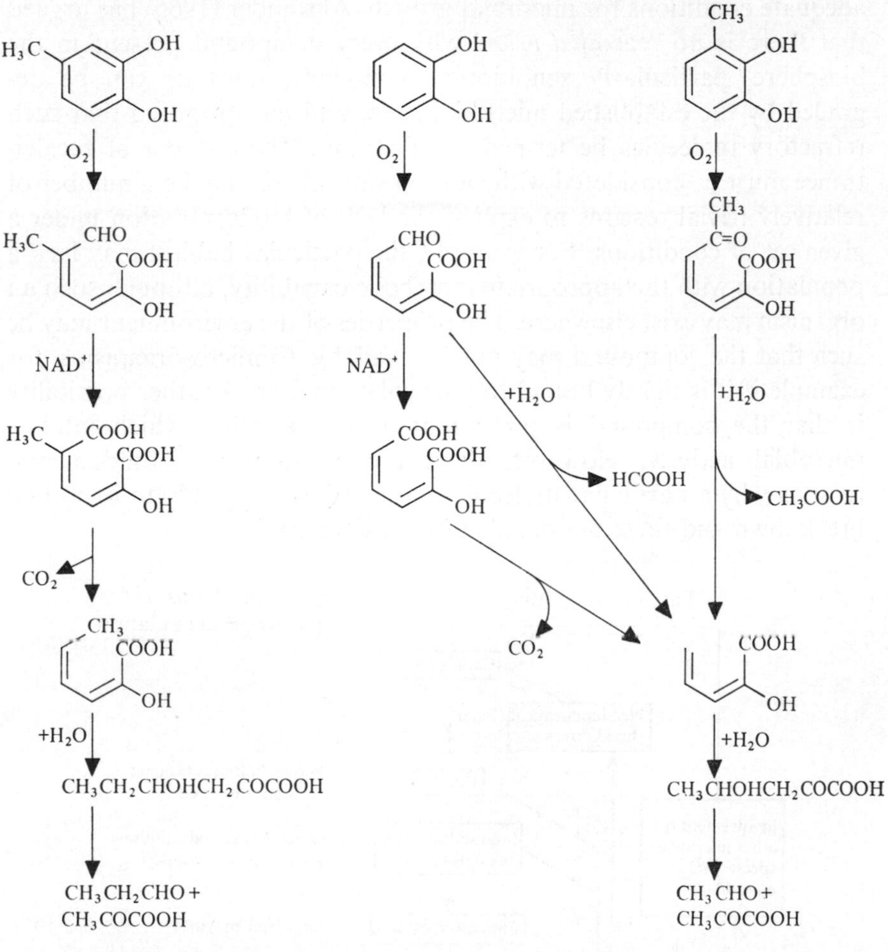

Fig. 2. The degradation pathways of catechol (*centre*), 3-methyl-catechol (*right*) and 4-methyl-catechol (*left*) by *Pseudomonas putida*.

(Feist & Hegeman, 1969) and the change in mannitol metabolism from mannitol to arabitol dehydrogenase (Tanaka, Lerner & Lin, 1967). There is, however, a great difference between the description of these processes in pure cultures of laboratory-adapted organisms, typically growing under nutrient-excess conditions, and the complex oxidations in natural habitats (Chapman, 1978).

Failure to transform compounds – recalcitrance

Despite the diversity of biodegradative mechanisms, it is a common observation that many compounds, especially environmentally foreign compounds, apparently are not degraded in a given habitat, despite adequate conditions for microbial growth. Alexander (1965) has argued that there is no reason *a priori* why every compound present in the biosphere, particularly xenobiotic compounds, must or can be degraded by the established microbial flora, and has proposed that such refractory molecules be termed 'recalcitrant'. The concept of recalcitrance must be considered with caution since there may be a number of relatively trivial reasons to explain the lack of biodegradation under a given set of conditions. For example, the particular habitat may lack a population with the appropriate metabolic capability, although such an organism may exist elsewhere. The properties of the environment may be such that the compound may not be available to micro-organisms, for example if it is tightly bound to particulate matter. Another possibility is that the compound is present at a concentration which inhibits microbial activity. However, there are a number of fundamental reasons why a particular molecule may not be susceptible to microbial breakdown and these are detailed below (Fig. 3).

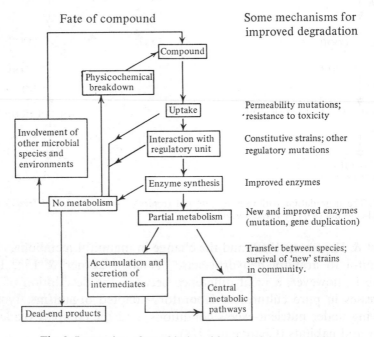

Fig. 3. Interaction of xenobiotics with microbial communities.

Uptake deficiencies

Although an organism may possess all the necessary enzymes and pathways to degrade a particular compound, degradation may not occur because the compound is unable to enter the cell. The classic example which illustrates this principle (but not absolute recalcitrance) is the inability of *Escherichia coli* to utilize citrate as a carbon-source when the organism is grown in a defined minimal medium, despite the fact that the organism contains a fully functional tricarboxylic acid cycle (Lara & Stokes, 1952). Under nutrient-rich conditions, citrate can enter the cell and is readily catabolized (Dagley, 1954). Similarly *cis*, *cis*-muconic acid, a common intermediate in breakdown of aromatic compounds, is not used as a growth substrate by fluorescent pseudomonads without a mutation that results in an alteration in permeability (Ornston, 1971). Specific transport mechanisms have evolved to facilitate the entry of naturally occurring compounds but on initial exposure to many unnatural molecules these uptake mechanisms are unlikely to function.

Catabolic enzyme deficiencies

Similarly, organisms may lack an appropriate enzyme or set of enzymes to convert the compound to intermediates of central metabolism. Thus, molecules with novel configurations are unlikely to be metabolized because of the generally high specificity of enzyme catalysis and the lack of an appropriate catalytic activity. Apparently minor changes to the molecular structure of compounds which are readily metabolized can produce an analogue which is recalcitrant: many substitutions to aromatic compounds, particularly halogens, nitro, amino and sulphonate groups, sufficiently alter the structure to reduce the rate of transformation and may even completely prevent metabolism (Alexander & Lustigman, 1966; Alexander, 1967; Cripps, 1975; Chapman, 1976; Furukawa, Tonomura & Kamibayashi, 1978). The gentisate dioxygenase from *Moraxella osloensis* also cleaves gentisates substituted with halogen or methyl groups at the C_3 or C_4 positions at rates which are as low as 25% of the original unsubstituted rate (Crawford, Hutton & Chapman, 1975). A *Pseudomonas* sp. which can grow on diphenylethane (Francis *et al.*, 1976) cannot degrade a number of analogues including 2,2-diphenylethanol and *p,p'*-dichlorodiphenylethane. Wild-type strains of *Pseudomonas aeruginosa* contain an amidase which allows acetamide and propionamide to be used as sole carbon-energy- and nitrogen-sources, but a number of structurally related amides including valeramide and phenylacetamide are not utilized (Clarke, 1974). Although these molecules may be catabolized by other organisms, the principle to be established here is that other molecules may not serve as the substrates for any existing enzyme mechanisms.

Control mechanism deficiencies

A compound may be recalcitrant because it is unable to induce the required catabolic enzymes although suitable enzymes, inducible by other substrates, may be present. For example, butyramide is a rather poor substrate for the amidase of *Pseudomonas aeruginosa* mentioned above, but even though it can be utilized for growth, it is unavailable because it cannot induce the enzyme (Clarke, 1974). *Pseudomonas putida* S3 is unable to grow on 2,2'-dichloropropionic acid (22DCPA) but can grow slowly on 2-monochloropropionic acid (2MPCA) (Senior, Bull & Slater, 1976). 22DCPA does not induce dehalogenase activity but 2MPCA can induce both 2MCPA and 22DCPA dehalogenase activity under certain growth conditions, attaining rather low specific activities of approximately 0.5 μmole 22DCPA converted per mg protein per h (Slater, 1978a). One possible explanation is that despite the potential for 22DCPA dehalogenase activity, 22DCPA cannot be metabolized because it is unable to act as an inducer.

Transformations uncoupled from microbial energy production and growth – co-metabolism

Biodegradation of organic compounds has largely been considered from the standpoint of utilization as carbon- and energy-sources for microbial growth. However, an important principle has been firmly established (Horvath, 1972) that many micro-organisms growing at the expense of one substrate may also have the capacity to transform one or more compounds without deriving any direct benefit from the process, either in terms of energy production or as a source of carbon for biomass production. This phenomenon was first described by Leadbetter & Foster (1959) who showed that cultures of *Pseudomonas methanica* (*Methylomonas methanica*), growing on methane as the sole carbon- and energy-source, could concomitantly oxidize a number of other alkanes, including ethane to acetic acid and propane to propionic acid. Neither hydrocarbon alone supported the growth of *P. methanica* and, since the aliphatic acids were not further metabolized, they accumulated in the culture. Initially this type of metabolism was associated with oxidation reactions and was therefore termed 'co-oxidation' (Foster, 1962). However, Jensen (1963) widened the scope of the phenomenon in describing bacterial dehalogenation of certain chlorinated aliphatic acids in reactions which failed to produce unchlorinated products able to support the growth of the dehalogenating organism. Jensen proposed the broader term of 'co-metabolism' which has found general acceptance despite some arguments against its use (Hulbert & Krawiec, 1977). Co-metabolism, which should be considered as a general term including co-oxidation as a specific process, occurs widely in nature, effecting the transformation of many naturally occurring compounds, but it is probably most significant in the degradation of a number of xenobiotic compounds (Horvath, 1972; Alexander, 1978).

The general phenomenon is probably the result of a number of slightly different mechanisms. In its original conception, co-metabolism was the result of a simultaneous attack on the growth-promoting metabolite and the non-utilized co-metabolite by the same enzyme or sequence of enzymes. This form of co-metabolism depended on the activity of enzymes of broad specificity, in contrast to the highly specific mechanism discussed earlier. For example, it has been shown (Ooyama & Foster, 1965; Beam & Perry, 1973; 1974; de Klerk & van der Linden, 1974) that organisms growing at the expense of an alkane, co-metabolize cycloalkanes to the corresponding cycloalcohol or cycloalkanone. Moreover, cyclohexane alone (Beam & Perry, 1974) was unable to induce the oxygenase system responsible for its co-metabolism. Similar results were obtained with a pure culture by de Klerk & van der Linden (1974).

Focht & Alexander (1970a, b) isolated from sewage a species of *Hydrogenomonas* capable of growth on diphenylmethane by two successive *meta* ring-fission reactions. The isolate could also grow on benzyhydrol and p-chlorobenzhydrol (Fig. 6, p. 246) and a number of analogues were co-metabolized by washed-cell suspensions of organisms grown on diphenylmethane. p,p'-Dichlorodiphenylmethane was degraded by a co-metabolic process which initially yielded p-chlorophenylacetic acid which in time was cleaved to yield a chlorinated unsaturated keto-enol. Chloride was not released at any stage, indicating that the chlorinated cleaved intermediates could not serve as substrates for the later enzymes involved in diphenylmethane metabolism. The nature of the substitution on the carbon atom linking the two benzene rings, i.e. a carbonyl carbon or a trichloromethyl group, prevented complete degradation but not co-metabolism. However, the combined influence of *para*, *para*-dichlorosubstitution and the presence of carbonyl group or trichloromethyl substitution sufficiently altered the molecule to preclude even co-metabolism.

Co-metabolism may also occur through the activity of enzymes not directly associated with the catabolism of the growth substrate. For example, Jamison *et al.* (1969) isolated a species of *Nocardia* which grew on n-hexadecane and co-metabolized p-xylene to 2,5-dimethyl-*cis*,

cis-muconic acid. This co-metabolite was the product of an *ortho* ring-fission enzyme not involved in the metabolism of n-hexadecane and presumably induced by the presence of *p*-xylene. Its inability to be metabolized further suggests that subsequent enzymes were unable to utilize the methylated muconic acid as a substrate. The partial break-down of several groups of chlorinated compounds (Jensen, 1963; Beynon, Hutson & Wright, 1973) provides a further example of this form of co-metabolism. Organisms growing on succinate, for example, will synthesize the enzyme dehalogenase, which is clearly not involved in succinate metabolism, and will metabolize chlorinated acetic acids without being able to utilize the products for growth (Slater *et al.*, 1976).

POPULATION AND COMMUNITY STRUCTURE

Natural habitats are heterogeneous in terms of environmental resources and conditions and so are able to sustain the growth of a wide range of micro-organisms with a broad spectrum of metabolic activities occupying all the available niches. Biotreater environments come within this category, and considerable attention has focussed on the composition of the microbial flora in these systems (Banks, 1977; Hawkes, 1978). Generally there is a greater heterogeneity to the flora of percolating filters compared with activated-sludge processes, but both are dominated by heterotrophic bacteria, especially species of *Achromobacter*, *Alcaligenes*, *Bacillus*, *Chromobacter*, *Flavobacterium*, *Micrococcus* and *Pseudomonas*, with chemolithotrophs predominantly represented by *Nitrosomonas* and *Nitrobacter* species and thiobacilli (Hawkes, 1965, 1978; Painter, 1970; Pike & Curds, 1971). Fungal populations are important components under certain conditions in percolating filters, as are algal surface-growths (Hawkes, 1965). The higher trophic levels are represented by a diverse protozoan population, tending to be dominated by ciliates, but including rotifers and nematodes. These have an important role in reducing the bacterial biomass concentration and clarifying the final effluent (Hawkes, 1978).

Much is known about the microbial composition of different waste-treatment plants, how these vary from plant to plant (Banks, 1977) and the compositional variations which lead to changes in the overall conditions or properties of the microbial biomass, for example during bulking (Pike, 1975; Hawkes, 1978). A great deal is also known about the overall capacity of a biotreater's microbial flora to degrade different effluents and compounds and the individual capabilities of different isolates to grow on different compounds.

However, even though it is obvious that most natural habitats support the growth of mixed assemblages of micro-organisms, very little is known about the consequences of growth and coexistence of mixed microbial populations, particularly with respect to biodegradative processes. Although environmental heterogeneity enables different microbial populations to grow comparatively independently, it seems highly likely that many beneficial relationships between different microbial populations exist, resulting in the formation of structured microbial communities which, in some cases, constitute units which are better equipped to exploit a given set of environmental conditions (Slater, 1978a, b). Many different types of microbial community have been isolated, in many cases using continuous-flow culture enrichment procedures, but at the present time little can be said about their general distribution or importance in natural habitats.

In the following sections an outline is given of the various categories of microbial community which have been described so far, with particular reference to their potential in the degradation of complex natural products and xenobiotic compounds.

Interactions based on the provision of specific nutrients or the removal of toxic end-products of metabolism

A number of mixed microbial cultures have been described in which the growth of one auxotrophic population depends on the presence of a second population which synthesizes and excretes a compound, usually a growth factor or an amino acid, required by the first population. Furthermore, the interaction may be reciprocal as, for example, in the case of a two-membered mixed culture isolated by Nurmikko (1954) in which a phenylalanine-requiring strain of *Lactobacillus plantarum* provided folic acid required by the second population, *Streptococcus faecalis*, which reciprocated by furnishing the required amino acid. Associations of this character seem to be common (Slater, 1978b) and, indeed, their frequency is probably underestimated since the methods chosen to isolate pure cultures discriminate against nutritionally fastidious organisms which in natural growth situations depend on the activity of other organisms. Herbert, Fairbairn & Davies (1976) showed that such interactions may be common between component populations of activated sludge. They tested a large number of isolates in an extensive pair-wise cross-feeding programme and concluded that in 14% of the cases some growth stimulation was observed. Such microbial interactions could be significant in contributing to the stability of a mixed microbial flora, particularly under changing environmental conditions, and may ensure the continuing presence of a particular species with an important biodegradative capacity.

The continuing growth of an organism may be dependent on the removal of a waste product of metabolism. Wilkinson and his colleagues (Wilkinson, Topiwala & Hamer, 1974) demonstrated the importance of such a relationship in a four-membered mixed culture growing on methane. The primary pseudomonad population oxidized methane, as a result methanol was excreted and, in pure culture, inhibited the growth of the pseudomonad. A species of *Hyphomicrobium* was present in the community, at about 4% of the total population, growing on excreted methanol and thereby ensuring continued growth of the primary population.

Microbial communities based on combined metabolic attack

Although complex organic compounds are degraded in nature, often at reasonable rates, it may be difficult to isolate individual populations with the capacity to effect a complete degradation or achieve a rate of degradation equal to that in the natural environment. Furthermore, the inability to demonstrate the breakdown of an apparently recalcitrant compound may be due to screening for pure cultures which could lack the capability to utilize the compound, whereas a mixture of micro-organism may be able to degrade the compound in question.

Gunner & Zuckerman (1968) described a synergistic relationship between species of *Arthrobacter* and *Streptomyces* which accounted for the degradation of the insecticide Diazinon (*o,o*-diethyl-*o*-2-isopropyl-4-methyl-6-pyrimidyl thiophosphate) in soil, which suggested some form of combined metabolic attack. Application of the insecticide resulted in the selective enrichment of these two organisms, although neither organism as a pure culture could use Diazinon as the sole carbon- and energy-source. Alone, the *Arthrobacter* sp. could degrade the ethyl substituents but neither organism metabolized the pyrimidine ring, whereas in a mixed culture the insecticide was completely degraded after 21 d growth, with resulting accumulation of two unidentified breakdown products. Unfortunately, the precise nature of the combined ring-cleavage mechanism has not been elucidated but these results were among the first to indicate the potential importance of interactions between different microbial populations in the degradation of complex molecules. One possibility is that the pyrimidine ring-cleavage mechanism required different components separately synthesized by the two organisms. Bates & Liu (1963) have shown that such a cooperative mechanism is indeed possible in a mixed culture of two pseudomonads which produce lecithinase activity. Separately, neither culture contained this activity and the interpretation was that the two strains produced different enzyme subunits which together resulted in the formation of a fully functional enzyme.

Studies on the biodegradation of surfactants provide a further indication that, in nature, microbial community activity may be more significant than the activity of individual populations. Johanides & Hršak (1976) using continuous-flow culture enrichment techniques isolated a bacterial community growing on linear alkyl benzenesulphonates (LAS) as the limiting carbon- and energy-source. The enrichment produced a stable mixed culture which was dominated by species of *Pseudomonas* and *Alcaligenes* and whose complexity, in terms of species composition, may have been partially due to the substrate containing at least twenty isomers and homologues. The proportions of the component bacteria depended on the growth conditions, being dominated by *Alcaligenes* species at low specific growth rates and by the pseudomonads at higher growth rates. The significant observation, however, was that individually none of the isolates was able to grow on LAS, and it was suggested that the degradation of the surfactant depended on the concerted activity of the complete community. Again, the mechanism has not been determined and it cannot be completely excluded that the inability to grow the pure cultures on LAS reflected the fact that other nutritional interactions, for example specific nutritional requirements, were not accounted for. Similar observations have been made by G. Hollis (personal communication) on a four-membered community growing on LAS with none of the purified organisms capable of LAS degradation alone. Moreover, the four-membered community grew as a floc, a characteristic only exhibited by the complete community. The mechanism of floc formation remains to be elucidated but must depend on the combined activity of all the organisms and may be particularly significant in dealing with the metabolism of a surface-active agent.

Non-ionic surfactants of the alkylphenol ethoxylate type have been considered resistant to microbial attack, although they are slowly degraded in nature (Osburn & Benedict, 1966; Cain, 1977) and there is some evidence to suggest that their complete degradation is the result of the activity of several species. For example, Baggi *et al.* (1978) isolated a *Nocardia* species able to grow on hexaethoxylate phenylether as a single carbon- and energy-source. The *Nocardia* species was only able to metabolize the ethoxylate side-chain by successive β-oxidation steps leaving the aromatic ring uncleaved, and ethoxylate phenylether accumulated. Further metabolism of this product depended on a species of *Cylindrocarpin* which completed the degradation via phenol and *ortho* ring-fission.

$$CH_3—O—\text{(ring)}—(CH_2 CH_2 O)_6 H$$

Nocardia sp.

$$CH_3—O—\text{(ring)}—CH_2 CH_2 OH$$

Phenol

Cylindrocarpin sp.

Intermediary
metabolites

Microbial communities based on co-metabolism

One consequence of co-metabolism may be that complete mineralization depends on the activity of a microbial community. Recently, some examples of this type of interaction have been described. For example, Hsieh and his colleagues have succeeded in adapting mixed culture to grow on the toxic insecticide Parathion (*o-o*-diethyl-*o-p*-nitrophenol phosphorothionate) using continuous-flow cultures. (Munnecke & Hsieh, 1974, 1975). Daughton & Hsieh (1977*b*) have characterized one such stable multispecies community which had been continuously grown on Parathion as the growth-limiting substrate for over two years (Fig. 4). The isolated community contained four organisms, none of which could grow on Parathion in monoculture, but a slow-growing strain of *Pseudomonas stutzeri* was identified which had an active Parathion-hydrolysis capability. The pseudomonad was unable to grow on the two hydrolysis products, diethyl thiophosphate and *p*-nitrophenol, but a second member of the community, *P. aeruginosa*, grew on *p*-nitrophenol. Thus in the complete community the growth of all the organisms depended on the co-metabolic activity of one organism whose own growth requirements depended on excretion by the *P. aeruginosa* population. No role was ascribed to the two other stable members of the mixed culture and none of the organisms could utilize the second accumulated product, diethyl thiophosphate. The enrichment regime resulted in the selection of a small community with a limited genetic complement which lacked the capacity to metabolize diethyl thiophosphate. It would be interesting to know if this compound can be used as a growth substrate and whether or not such an organism(s) could be integrated into the established community.

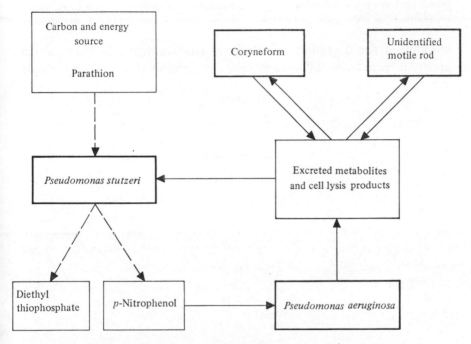

Fig. 4. Degradation of *o,o*-diethyl-*o-p*-nitrophenol phosphorothionate by a microbial community isolated by continuous-flow culture enrichment. Co-metabolism, broken line; growth-associated utilization or provision, solid line.

Despite the ready metabolism of cyclohexane in natural environments, the difficulty in obtaining pure cultures of cyclohexane utilizers has led to the suggestion that cycloalkane degradation may be the result of a commensal relationship between two organisms including a co-metabolic interaction (Beam & Perry, 1974; de Klerk & van der Linden, 1974). A mixed culture growing on a mixture of propane and cyclohexane was established (Beam & Perry, 1974) and shown to assimilate cyclohexane carbon into biomass and evolve carbon dioxide derived from cyclohexane (Fig. 5). One member of the community, *Mycobacterium vaccae* JOB5, co-metabolized cyclohexane yielding cyclohexanone, the growth substrate of the second member of the community, an unidentified organism (strain CY6) which could not utilize propane or cyclohexane. Although this could be a major

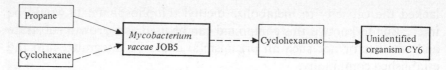

Fig. 5. The growth of a two-membered microbial community on propane and cyclohexane; growth includes a co-metabolism stage. Co-metabolism, broken line; growth-associated utilization, solid line.

mechanism for degradation in natural environments, in view of the apparent resistance of cyclohexane to microbial attack by single

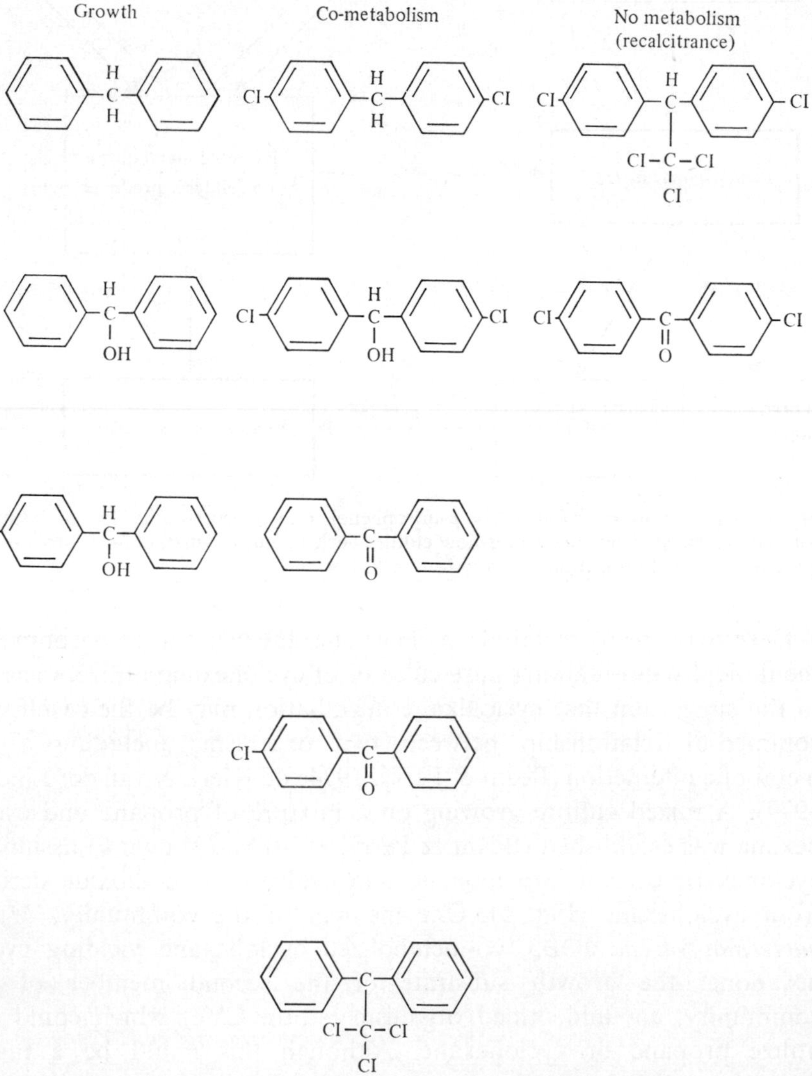

Fig. 6. The metabolism of diphenylmethane and analogues by *Hydrogenomonas* sp. (Focht & Alexander, 1970*a*, *b*).

organisms, other mechanisms cannot be excluded. For example, Stirling, Watkinson & Higgins (1977) have recently isolated from estuarine mudflats another stable mixed culture growing solely on cyclohexane, and a *Nocardia* species was identified as the primary cyclohexane utilizer. However, growth only occurred in the presence of an unidentified pseudomonad which provided the cyclohexane utilizer with a number of growth factors including biotin. Subsequently growth of a monoculture of the primary utilizer on cyclohexane was established in the presence of several growth factors.

The slow degradation of DDT (1,1-*bis* (*p*-chlorophenyl)2,2,2-trichloroethane) may also be achieved by a number of organisms using a co-metabolic step. Wedemeyer (1967) found that DDT could be dechlorinated under anaerobic conditions to produce *p,p'*-dichlorodiphenylmethane which is known to be co-metabolized by at least one organism growing in the presence of diphenylmethane (Focht & Alexander, 1970*b*) (Fig. 6). This proposed model has yet to be demonstrated conclusively in natural environments as the two organisms fulfil these activities under different conditions (i.e. aerobic and anaerobic environments). Nevertheless this work illustrates that the genetic potential exists in the biosphere to effect such a transformation.

Ring-cleaved chlorinated keto-enol product

Microbial communities in anaerobic environments –
interactions based on hydrogen (electron) transfer

The fermentation of organic compounds as well as the degradation of aromatic compounds occurs readily in anaerobic environments (Evans, 1977). It is also clear that microbial community activity is significant in these processes and a number of microbial associations have been isolated from several anoxic environments, including the rumen, anaerobic sludge digesters and aquatic sediments.

Although single species have been shown to degrade aromatic compounds in the presence of nitrate (Williams & Evans, 1975), or light (Dutton & Evans, 1969; Guyer & Hegeman, 1969), a number of microbial communities have been characterized which degrade aromatic compounds in association with methane production. For example, Balba & Evans (1977) isolated a four-membered microbial community from rumen liquor which grew on benzoate and contained an unidentified long rod, a shorter rod similar to *Methanobacterium ruminantium*, a spiral-shaped organism similar to *Methanospirillum hungati* and a facultative anaerobe. Moreover, cultures obtained from sewage sludge showed a remarkably similar composition to the rumen microbial communities. Healy & Young (1978b) have isolated a stable community growing on ferulic acid (a lignin monomer) and have shown it to be morphologically similar to the benzoate microbial community obtained by Ferry & Wolfe (1976). Whether or not these separate communities contain exactly the same organisms is unknown (a detailed characterization of the isolates has not yet been reported), nevertheless they appear to be similar functionally. That is, some members of the community are involved in degradation of aromatic compounds producing fatty acids, carbon dioxide and hydrogen, which in turn are exploited as carbon- and energy-sources by the associated methanogens. Ferry & Wolfe (1976) have shown that these two activities can be partially separated by observing methane production from acetate in the presence of *o*-chlorobenzoate, which inhibited the degradation of benzoate. The exact nature of the metabolic structure of the community has yet to be elucidated but it is possible that there is a parallel with other methanogenic communities growing on fermentable carbon sources, originally isolated from the rumen.

Methanobacillus omelianskii (Bryant *et al.*, 1967) was shown to be a tight association between two bacteria able to use ethanol as a single carbon- and energy-source in a fermentation which also yielded methane. A detailed examination showed that one species, the 'S' organism, oxidized ethanol to acetate and generated hydrogen which in the mono-culture caused growth to cease. In the mixture, however, the hydrogen was utilized as an energy-source for the growth of the methanogen, *Methanobacterium* strain MOH. This association, which maximized the amount of biomass produced from the oxidation of ethanol, also ensured that a toxic waste-product did not accumulate. Similar com-munities, for example composed of *Desulfovibrio vulgaris* and *Methano-bacterium formicum*, have been isolated from anaerobic sewage digesters in the absence of sulphate (Bryant *et al.*, 1977). In the presence of sulphate, however, methane was no longer formed, since sulphate acted as an alternative hydrogen sink producing an equivalent amount of hydrogen sulphide.

Recently Bakker (1977) has presented evidence showing that an-aerobic degradation of a number of aromatic compounds, including phenol, benzoic acid, hydroxybenzoic acids and cresols, could be effectively linked to the reduction of nitrate by a stable mixed-bacterial culture. The community was isolated by anaerobic growth on phenol from a mixed inoculum which included soil from a phenol-plant site and sludge from a phenol-oxidation ditch. At least three different Gram-negative bacteria were isolated as pure cultures capable of anaerobic growth on phenol and nitrate, but individually each showed a very slow rate of phenol degradation, which suggested some form of unspecified interaction within the community to accelerate the rate of phenol degradation.

MICROBIAL ADAPTATION TO GROWTH ON NOVEL COMPOUNDS

A major contemporary interest concerns the capacity of micro-organisms to adapt to degrade new compounds. Micro-organisms have had geological time-scales in which to evolve catabolic sequences to effect the breakdown of naturally occurring compounds. When new natural products arose it is likely that the evolution of degradative capacity closely followed. In contrast, the novel challenge presented to micro-organisms by man-made chemicals has almost entirely arisen within the last one hundred years; an extremely short period of time on an evolutionary scale. The presence of such compounds provides an appro-priate selection pressure for evolution of degradative mechanisms and this raises the intriguing questions of how rapidly and by what mech-anisms do micro-organisms acquire new catabolic activities, and whether or not there are any limitations to the potential of microbial catabolism.

There are two categories of adaptation phenomenon which can be considered. Firstly, the adaptation of existing catabolic enzymes, including associated mechanisms such as uptake and regulatory processes, to the degradation of the novel compounds. Secondly, the problem of the evolution of whole metabolic pathways.

The evolutionary capabilities of micro-organisms have been extensively examined in laboratory experimental systems, principally using pure cultures, and several elegant reviews have summarized this work and drawn attention to the potential of such systems (Hegeman & Rosenberg, 1970; Ornston, 1971; Hegeman, 1972; Hartley *et. al.*, 1972; Clarke, 1974; Hartley, 1974).

Adaptation through alteration in enzyme activities

Studies with the aliphatic amidases of *Pseudomonas aeruginosa* (Clarke, 1974) have illustrated the importance of point mutations in altering the activity towards novel compounds related to the original substrate. Regulatory mutations, including selection of strains constitutive for enzyme production, were involved in the step-wise derivation of new degradative capacities. However, most degradative pathways are apparently closely regulated in nature and such events may represent a transient step in the evolution of new pathways. Experiments on pentose utilization by *Klebsiella aerogenes* have shown the importance of enzyme concentrations and changes in specific activity, including the influence of gene duplication and the selection of constitutive mutants (Hartley, 1974; Rigby, Burleigh & Hartley, 1974; Hansche, 1975).

Adaptation by one member, a strain of *P. putida*, in a multispecies community growing on 2,2-dichloropropionic acid (22DCPA) has been demonstrated by Senior, Bull & Slater (1976). The original strain, *P. putida* S3, was unable to utilize 22DCPA as a growth substrate, although under certain growth conditions it was shown that there was very low 22DCPA-specific dehalogenase activity. Extracts of a mutant, *P. putida* P3, which arose spontaneously in the community, contained a sixteen-fold increase in dehalogenase activity (Slater *et al.*, 1976) and, as suggested above (p. 238), the likely explanation is that this reflects an alteration in the regulatory mechanism.

Further evidence has indicated that totally distinct enzymes can arise from the breakdown of natural substrates, for example lactose (Campbell, Lengyel & Langridge, 1973), where the starting point may be an enzyme with extremely low specificity for the substrate (Hart & Hall, 1974; Hall, 1975). It is not only in the evolution of degradative enzymes that micro-organisms may need to adapt to new challenges. Prior to metabolism, chemicals must penetrate the cell and active transport mechanisms may be required. As already described, the metabolic intermediates citric acid (Lara & Stokes, 1952) and *cis,cis*-muconic acid (Ornston, 1971) are two examples where permeability mutations must occur, at least in some species, before metabolism can take place.

Evolution of metabolic pathways

Discussion of the evolution of whole pathways centres around two opposing hypotheses. Horowitz (1945) proposed a step-wise retrograde mechanism, in which gene duplication is followed by modification of one copy to produce an altered protein which has the required catalytic function for the next step in the pathway. Although supported by some experimental evidence (Hartley, 1974), this may be criticized on several grounds (see Hegeman, 1972; Clarke, 1974). By operation of this mechanism, adjacent enzymes in a pathway should show considerable homology, even though the enzymes may catalyse reactions which are not related mechanistically and examples where this does not seem to be the case have been demonstrated (see Clarke, 1974 for examples). The second proposal (Wu, Lin & Tanaka, 1968) is that new pathways evolve from existing pathways by alterations of enzymes with slight activity towards the metabolites of the new pathway. By this hypothesis, enzymes in different pathways fulfilling similar functions should be closely related. This is supported, for example, by comparison of the lactonizing enzymes of two related β-ketoadipate pathways (Meagher & Ornston, 1973; Patel, Meagher & Ornston, 1973).

We wish to propose an additional mechanism, similar in principle to the Wu, Lin & Tanaka model, but operating at the level of metabolic interactions between microbial communities in the first instance. A particular compound may be degraded because a number of enzymes (evolved for other, different pathways) exist in two or more individuals within a community. Clearly, the chance of locating an existing enzyme with gratuitous activity to the novel substrate is considerably greater in the larger genetic pool of many different organisms. Furthermore, the chance of locating a second enzyme to transform the product of the first reaction is also high, although this enzyme is unlikely to be coded for by the same organism as the first. Thus degradation, as we have seen in a number of cases mentioned earlier, is the result of the collective activity of a metabolically structured community. It is now quite clear that many mechanisms exist for genetic transfer (Wheelis 1975; Cohn, 1976; Reanney, 1976; Atherton, Byrom & Dart, this volume pp. 379–405), and at the extreme may allow a substantial portion of the information within different populations to exist in a state of genetic flux between the populations. Thus, genetic transfer, mediated by non-specific phage, or plasmids, or transformation, could result in the eventual consolidation of all the relevant genetic information of the new pathway within a single organism. For example, the gene which codes for the enzyme that hydrolyses Parathion in *P. stutzeri* could be transferred to *P. aeruginosa* (Fig. 4) to evolve an organism with the capacity to grow on Parathion in pure culture. For such an event the single organism would need to be at a selective (growth) advantage over the microbial community and it may be argued that, under certain conditions, a stable microbial community is more competitive than a single organism with the same capability (Slater, 1978a). Nevertheless in the contemporary environment this mechanism could be of major importance.

FUTURE DEVELOPMENTS

Current activity in translating public concern with environmental quality into legislation (King, 1978) allows firm prediction that effluent treatment will be a field of interest in the next few years. It is more difficult to predict the areas in which advances will be made. However, it seems likely that integrated physical, chemical and biological processes will be required to treat some of the more complex wastes and effluents. Where it can be directly applied, biotreatment is more economical than chemical methods in the oxidation of industrial wastes (Pearson, 1978).

Within biological treatment, advances are likely to be made in systems designed to control the composition of inflowing medium and in understanding the role of the biocatalyst. A multidisciplinary approach is required in which the microbial physiologist is involved in the planning and design stages. It is possible to envisage systems where more detailed on-line analyses contribute to automation of process control and where specific biocatalysts, possibly in immobilized form, are added to deal with specific components of effluent. A major challenge in this area of microbiology is to exploit the vast amounts of cellulose and similar wastes that are treated by man's activities, particularly where they are available in suitably concentrated quantities (Humphrey, 1975; Mandels & Andreotti, 1978). One of the most exciting areas for research in the microbiology of wastes is in the evolution of new degradative capacities in individual strains and in communities. It remains to be established, to what extent developments in this area will contribute to improvements in the disposal and utilization of wastes.

REFERENCES

ALEXANDER, M. (1965). Biodegradation: problems of molecular recalcitrance and microbial fallibility. *Advances in Applied Microbiology*, 7, 35–80.

ALEXANDER, M. (1967). The breakdown of pesticides in soils. In *Agriculture and the Quality of Our Environment*, ed. N. C. Brady, pp. 331–42. Washington: American Association for the Advancement of Science.

ALEXANDER, M. (1978). Role of co-metabolism. In *Microbial Degradation of Pollutants in Marine Environments*, eds P. H. Pritchard & A. W. Bourquin. Washington: United States Environmental Protection Agency. (In press.)

ALEXANDER, M. & LUSTIGMAN, B. K. (1966). Effect of chemical structure on microbial degradation of substituted benzenes. *Journal of Agricultural and Food Chemistry*, 14, 410–13.

ANDREWS, J. F. (1974). Dynamic models and control strategies for wastewater treatment processes. *Water Research*, 8, 261–89.

ANDREWS, J. F., BRIGGS, R. & JENKINS, S. H. (eds) (1974). Instrumentation control and automation for waste-water treatment systems. *Progress in Water Technology*, vol. 6. Oxford: Pergamon Press.

BAGGI, G., BERETTA, L., GALLI, E., SCOLASTICO, C. & TRECCANI, V. (1978). Biodegradation of polyoxyethylene alkylphenols. In *The Oil Industry and Microbial Ecosystems*, eds K. W. A. Chater & H. J. Somerville, pp. 129–36. London: Heyden and Sons.

BAILEY, D. A. (1977). The origin, treatment and disposal of tannery effluents. *Process Biochemistry*, 12(1), 13–25.

BAILEY, D. A. & HUMPHREYS, F. E. (1967). The removal of sulphide from limeyard wastes by aeration. *Journal of the Society of Leather Trades Chemists*, 51, 154–72.

BAILEY, M., HINES, D. A., OUSBY, J. C. & ROESLAR, F. C. (1975). The I.C.I. Deep Shaft aeration process for effluent treatment. *Institute of Chemical Engineers Symposium*, 41, D1–10.

BAKKER, G. (1977). Anaerobic degradation of aromatic compounds in the presence of nitrate. *FEMS Microbiology Letters*, 1, 103–8.

BALBA, M. J. & EVANS, W. C. (1977). The methanogenic fermentation of aromatic substrates. *Biochemical Society Transactions*, 5, 302–4.

BANKS, C. J. (1977). Methodology for bacterial population studies on activated sludge. PhD Thesis, University of York.

BARTH, E. F., BRENNER, R. L. & LEWIS, R. F. (1968). Chemical–biological control of nitrogen and phosphorus in wastewater effluent. *Journal of the Water Pollution Control Federation*, 40, 2040–54.

BATES, J. L. & LIU, P. V. (1963). Complementation of lecithinase activities by closely related pseudomonads: its taxonomic implication. *Journal of Bacteriology*, 86, 585–92.

BEAM, H. W. & PERRY, J. J. (1973). Co-metabolism as a factor in microbial degradation of cycloparaffinic hydrocarbons. *Archiv für Mikrobiologie*, 91, 87–90.

BEAM, H. W. & PERRY, J. J. (1974). Microbial degradation of cycloparaffinic hydrocarbons via co-metabolism and commensalism. *Journal of General Microbiology*, 82, 163–9.

BENEMANN, J. R., WEISSMAN, J. C., KOOPMAN, B. L. & OSWALD, W. J. (1977). Production of biomass from fresh water aquatic systems – concepts of large-scale bioconversion systems using microalgae. *Nature, London*, 268, 19–23.

BEYNON, K. I., HUTSON, D. H. & WRIGHT, A. N. (1973). The metabolism and degradation of vinyl phosphate insecticides. *Residue Reviews*, 47, 55–142.

BOLTON, R. L. & KLEIN, L. (1971). *Sewage Treatment: Basic Principles and Trends*, London: Butterworth.

BOON, A. G. (1976). Technical review of the use of oxygen in the treatment of waste water. *Water Pollution Control*, **75**, 206–13.

BOORAM, C. V. & SMITH, R. J. (1974). Manure management in a 700-head swine-finishing unit in the American Midwest. An integrated system incorporating hydraulic manure transport with recycled anaerobic lagoon liquor and final effluent use by corn (*Zea mays*). *Water Research*, **8**, 1089–97.

BRYANT, M. P., WOLIN, E. A., WOLIN, M. J. & WOLFE, R. S. (1967). *Methanobacillus omelianskii*, a symbiotic association of two species of bacteria. *Archiv für Mikrobiologie*, **59**, 20–31.

BRYANT, M. P., VAREL, V. H., FROBISH, R. A. & ISAACSON, H. R. (1976). Biological potential of thermophilic methanogenesis from cattle wastes. In *Microbial Energy Conversion*, eds H. G. Schlegel & J. Barnea, pp. 347–59. Göttingen: Erich Goltze.

BRYANT, M. P., CAMPBELL, L. L., REDDY, C. A. & CRABILL, M. R. (1977). Growth of *Desulfovibrio* in lactate or ethanol media low in sulphate in association with hydrogen-utilising methanogenic bacteria. *Applied and Environmental Microbiology*, **33**, 1162–9.

BUHR, H. O. & ANDREWS, J. F. (1977). The thermophilic anaerobic digestion process. *Water Research*, **11**, 129–43.

CAIN, R. B. (1977). Surfactant biodegradation in waste waters. In *Treatment of Industrial Effluents*, eds A. G. Callely, C. F. Forster & D. A. Stafford, pp. 283–327. London: Hodder & Stoughton.

CAIRNS, J., DICKSON, K. L. & WESTLAKE, G. F. (1976). *Biological Monitoring of Water and Effluent Quality*. Special Technical Publication 607. Philadelphia: American Society for Testing and Materials.

CALLELY, A. G., FORSTER, C. F. & STAFFORD, D. A. (eds) (1977). *Treatment of Industrial Effluents*. London: Hodder and Stoughton.

CAMPBELL, J. H., LENGYEL, J. A. & LANGRIDGE, J. (1973). Evolution of a second gene for β-galactosidase in *Escherichia coli*. *Proceedings of the National Academy of Sciences, USA*, **70**, 1841–5.

CHAPMAN, P. (1976). Microbial degradation of halogenated compounds. *Biochemical Society Transactions*, **4**, 463–546.

CHAPMAN, P. J. (1978). Mechanisms of biodegradation. In *Microbial Degradation of Pollutants in Marine Environments*, eds P. H. Pritchard & A. W. Bourquin. Washington: United States Environmental Protection Agency. (In press.)

CLARKE, P. H. (1974). The evolution of enzymes for the utilization of novel substrates. In *Evolution in the Microbial World*, eds M. J. Carlile & J. J. Skehel, pp. 183–217. *24th Symposium of the Society for General Microbiology*. Cambridge University Press.

CLAUSEN, E. C., SITTON, O. C. & GADDY, J. L. (1977). Bioconversion of crop materials to methane. *Process Biochemistry*, **12**(7), 5–7.

COHN, S. N. (1976). Transposable genetic elements and plasmid evolution. *Nature, London*, **263**, 731–8.

COLBY, J., STIRLING, D. I. & DALTON, H. (1977). The soluble methane monooxygenase of *Methylococcus capsulatus* (Bath). *Biochemical Journal*, **165**, 395–402.

COLLINGWORTH, W. L., CHAPMAN, P. J. & DAGLEY, S. (1973). Stereospecific enzymes in the degradation of aromatic compounds by *Pseudomonas putida*. *Journal of Bacteriology*, **113**, 922–31.

COOK, K. A. (1978). Degradation of the non-ionic surfactant Dobanol 45–7 by activated sludge. *Water Research* (in press).

CRAWFORD, R. L., HUTTON, S. W. & CHAPMAN, P. J. (1975). Purification and properties of gentisate 1, 2-dioxygenase from *Moraxella osloensis. Journal of Bacteriology*, **121**, 794–9.

CRIPPS, R. E. (1975). The microbial metabolism of acetophenone. *Biochemical Journal*, **152**, 233–41.

DAFT, M. J. & FALLOWFIELD, H. J. (1978). Seasonal variations in algae and bacterial populations in Scottish Freshwater habitats. In *The Oil Industry and Microbial Ecosystems*, eds K. W. A. Chater & H. J. Somerville, pp. 41–50. London: Heyden and Sons.

DAGLEY, S. (1954). Dissimilation of citric acid by *Aerobacter aerogenes* and *Escherichia coli. Journal of General Microbiology*, **11**, 218–27.

DAGLEY, S. (1971). Catabolism of aromatic compounds by micro-organisms. *Advances in Microbial Physiology*, **6**, 1–46.

DAGLEY, S. (1975). A biochemical approach to some problems of environmental pollution. *Essays in Biochemistry*, **11**, 81–138.

DAGLEY, S. (1978). Microbial catabolism, the carbon cycle and environmental pollution. *Naturwissenschaften*, **65**, 85–95.

DAUGHTON, C. G. & HSIEH, D. P. H. (1977*a*). Accelerated Parathion degradation in soil by inoculation with Parathion-utilizing bacteria. *Bulletin of Environmental Contamination and Toxicology*, **18**, 48–56.

DAUGHTON, C. G. & HSIEH, D. P. H. (1977*b*). Parathion utilization by bacterial symbionts in a chemostat. *Applied and Environmental Microbiology*, **34**, 175–84.

DE KLERK, H. & VAN DER LINDEN, A. C. (1974). Bacterial degradation of cyclohexane. Participation of a cooxidation reaction. *Antonie van Leeuwenhoek, Journal of Microbiology and Serology*, **40**, 7–15.

DE KREUK, J. F. & HANSVEIT, A. O. (1978). Assessment of biodegradation. In *The Oil Industry and Microbial Ecosystems*, eds K. W. A. Chater & H. J. Somerville, pp. 155–76. London: Heyden and Sons.

DEPARTMENT OF THE ENVIRONMENT. Central Unit on Environmental Pollution (1976); *Pollution Control in Britain: How it works*, Pollution paper No. 9. London: Her Majesty's Stationery Office.

DROZD, J. W., GODLEY, A. & BAILEY, M. L. (1978). Ammonia oxidation by methaneoxidizing bacteria. *Proceedings of the Society for General Microbiology*, **5**, 66–7.

DUTTON, P. L. & EVANS, W. C. (1969). Metabolism of aromatic compounds by *Rhodopseudomonas palustris*. A new reductive method of aromatic ring cleavage. *Biochemical Journal*, **113**, 525–36.

ELLIOTT, L. F. (1977). Bacterial and viral pathogens associated with land application of organic wastes. *Journal of Environmental Quality*, **6**, 245–51.

EVANS, W. C. (1977). Biochemistry of the bacterial catabolism of aromatic compounds in anaerobic environments. *Nature, London*, **270**, 17–22.

EVERETT, J. G. (1977). Sludge treatment and disposal. In *Treatment of Industrial Effluents*, eds A. G. Callely, C. F. Forster & D. A. Stafford, pp. 103–28. London: Hodder & Stoughton.

FEIST, C. F. & HEGEMAN, G. D. (1969). Phenol and benzoate metabolism by *Pseudomonas putida* – regulation of tangential pathways. *Journal of Bacteriology*, **100**, 869–77.

FERRY, J. G. & WOLFE, R. S. (1976). Anaerobic degradation of benzoate to methane by a microbial consortium. *Archiv für Mikrobiologie*, **107**, 33–40.

FISHER, N. S. (1977). Legal aspects of pollution. In *Treatment of Industrial Effluents*, eds A. G. Callely, C. F. Forster & D. A. Stafford, pp. 18–29. London: Hodder & Stoughton.

FLINT, K. & HOPTON, J. W. (1976). Substrate specificity and iron inhibition of bacterial and particle-associated alkaline phosphatases of wastes and sewage sludges. *European Journal of Applied Microbiology*, **4**, 195–204.

FOCHT, D. D. & ALEXANDER, M. (1970*a*). Bacterial degradation of diphenyl-methane, a DDT model substrate. *Applied Microbiology*, **20**, 608–11.

FOCHT, D. D. & ALEXANDER, M. (1970*b*). DDT metabolites and analogues: ring fission by *Hydrogenomonas*. *Science*, **170**, 91–2.

FORD, J. (1977). Handling of waste stream sludges. *Process Biochemistry*, **12**(5), 16–17.

FORSTER, C. F. (1977). Bio-oxidation. In *Treatment of Industrial Effluents*, eds A. G. Callely, C. F. Forster & D. A. Stafford, pp. 65–87. London: Hodder & Stoughton.

FOSTER, J. W. (1962). Hydrocarbons as substrates for micro-organisms. *Antonie van Leeuwenhoek, Journal of Microbiology and Serology*, **28**, 241–74.

FRANCIS, A. J., SPANGGARD, R. J., OUCHI, G. I., BRAMHALL, R. & BOHNOS, N. (1976). Metabolism of DDT analogues by a *Pseudomonas* sp. *Applied and Environmental Microbiology*, **32**, 213–16.

FRIELLO, D. A., MYLROIE, J. R. & CHAKRABARTY, A. M. (1976). Use of genetically engineered multi-plasmid micro-organisms for rapid degradation of fuel hydrocarbons. In *Proceedings of the Third International Biodegradation Symposium*, eds J. M. Sharpley & A. M. Kaplan, pp. 205–14. London: Applied Science.

FURUKAWA, K., TONOMURA, K. & KAMIBAYASHI, A. (1978). Effect of chlorine substitution in the biodegradability of polychlorinated biphenyls. *Applied and Environmental Microbiology*, **35**, 223–7.

GAUDY, A. F. (1962). Studies on induction and expression in activated sludge systems. *Applied Microbiology*, **10**, 264–71.

GIDDINGS, J. M., WALTON, B. T., EDDLEMON, G. K. & OLSON, K. G. (1978). Transport and fate of anthracene in aquatic microcosms. In *Microbial Degradation of Pollutants in Marine Environments*, eds P. H. Pritchard & A. W. Bourquin. Washington: United States Environmental Protection Agency (in press).

GLEDHILL, W. E. & SAEGER, V. W. (1978). Microbial degradation in the environmental hazard evaluation process. In *Microbial Degradation of Pollutants in Marine Environments*, eds P. H. Pritchard & A. W. Bourquin. Washington: United States Environmental Protection Agency (in press).

GRAY, H. F. (1940). Sewerage in ancient and medieval times. *Sewerage Works Journal*, **12**, 939–46.

GREAVES, M. P., DAVIS, H. A., MARSH, J. A. P. & WINGFIELD, G. I. (1976). Herbicides and Soil and the Micro-organisms. *CRC Critical Reviews in Microbiology*, **5**, 1–38.

GUNNER, H. B. & ZUCKERMAN, B. M. (1968). Degradation of 'Diazinon' by synergistic microbial action. *Nature, London*, **217**, 1183–4.

GUTNICK, D. L. & ROSENBERG, E. (1977). Oil tankers and pollution: a microbiological approach. *Annual Review of Microbiology*, **31**, 379–96.

GUYER, M. & HEGEMAN, G. D. (1969). Evidence for a reductive pathway for the anaerobic metabolism of benzoate. *Journal of Bacteriology*, **99**, 906–7.

HALL, B. G. (1975). Experimental evolution of a new enzymatic function. Kinetic analysis of the ancestral (ebg⁰) and evolved (ebg⁺) genes. *Journal of Molecular Biology*, **107**, 71–84.

HANG, Y. D., SPITTSTOERER, D. S., WODDAMS, E. E. & SHARMAN, R. N. (1977). Citric acid fermentation of brewing wastes. *Journal of Food Science*, **42**, 383–4.

HANSCHE, P. E. (1975). Gene duplication as a mechanism of genetic adaptation in *Saccharomyces cerevisiae*. *Genetics*, **79**, 661–74.

HART, D. L. & HALL, E. G. (1974). Second naturally occurring β-galactosidase in *Escherichia coli. Nature, London,* **248**, 152–3.

HARTLEY, B. S. (1974). Enzyme families. In *Evolution in the Microbial World,* eds M. J. Carlile & J. J. Skehel, pp. 151–82. *24th Symposium of the Society for General Microbiology.* Cambridge University Press.

HARTLEY, B. S., BURLEIGH, B. D., MIDWINTER, G. G., MOORE, C. H., MORRIS, H. R., RIGBY, P. W. J., SMITH, M. J. & TAYLOR, S. S. (1972). In *Enzymes: Structure and Function,* eds J. Drenth, R. A. Oosterbaan & C. Velgar, pp. 151–76. Amsterdam: North-Holland.

HAWKES, H. A. (1965). The ecology of sewage bacteria beds. In *Ecology and the Industrial Society, Symposium of the British Ecological Society,* eds G. T. Goodman, R. W. Edwards & J. M. Lambert, pp. 119–48. Oxford: Blackwell Scientific.

HAWKES, H. A. (1978). The ecology of activated sludge. In *The Oil Industry and Microbial Ecosystems,* eds K. W. A. Chater & H. J. Somerville, pp. 217–33. London: Heyden and Sons.

HEALY, J. B. & YOUNG, L. Y. (1978a). Catechol and phenol degradation by a methanogenic population of bacteria. *Applied and Environmental Microbiology,* **35**, 216–18.

HEALY, J. B. & YOUNG, L. Y. (1978b). Methanogenic biodegradation of aromatic compounds. In *Microbial Degradation of Pollutants in Marine Environments,* eds P. H. Pritchard & A. W. Bourquin. Washington: United States Environmental Protection Agency (in press).

HEGEMAN, G. D. (1972). The evolution of metabolic pathways in bacteria. In *The Degradation of Synthetic Organic Molecules in the Biosphere,* pp. 56–72. Washington: National Academy of Sciences.

HEGEMAN, G. D. & ROSENBERG, S. L. (1970). The evolution of bacterial enzyme systems. *Annual Review of Microbiology,* **24**, 429–59.

HERBERT, B. N. (1976). Quality of seawater for injection into North Sea oil wells. In *The Genesis of Petroleum and Microbiological Means for its Recovery,* pp. 47–56. London: Institute of Petroleum.

HERBERT, S. D., FAIRBAIRN, S. R. & DAVIES, M. (1976). Syntrophic interactions between bacteria isolated from activated sludge. *Proceedings of the Society for General Microbiology,* **4**(1), 31–2.

HIGGINS, I. J. & BURNS, R. G. (1975). *The Chemistry and Microbiology of Pollution.* London: Academic Press.

HOROWITZ, N. H. (1945). On the evolution of biochemical syntheses. *Proceedings of the National Academy of Sciences, USA,* **31**, 153–7.

HORVATH, R. S. (1972). Microbial co-metabolism and the degradation of organic compounds in nature. *Bacteriological Reviews,* **36**, 146–55.

HOWELL, J. A. (1978). Alternative approaches to activated sludge and trickling filters. In *The Oil Industry and Microbial Ecosystems,* eds K. W. A. Chater & H. J. Somerville, pp. 199–216. London: Heyden and Sons.

HUGHES, D. E. (1977). Microbes and effluent treatment. In *Treatment of Industrial Effluents,* eds A. G. Callely, C. F. Forster & D. A. Stafford, pp. 1–6. London: Hodder & Stoughton.

HULBERT, M. H. & KRAWIEC, S. (1977). Co-metabolism: a critique. *Journal of Theoretical Biology,* **69**, 287–91.

HUMPHREY, A. E. (1975). Economical factors in the assessment of various cellulosic substances as chemical and energy resources. *Biotechnology and Bioengineering Symposium,* **5**, 49–65.

HUMPHRIS, T. H. (1977). The use of sewage effluent as power station cooling water. *Water Research,* **11**, 217–23.

JACKMAN, E. A. (1976). Brazil's national alcohol programme. *Process Biochemistry*, 11(5), 29–30.

JAMISON, V. W., RAYMOND, R. L. & HUDSON, J. O. (1969). Microbial hydrocarbon co-oxidation, III. Isolation and characterization of an α, α'-dimethyl-*cis*, *cis*-muconic acid-producing strain of *Nocardia corallina*. *Applied Microbiology*, 17, 853–6.

JAMISON, V. W., RAYMOND, R. L. & HUDSON, J. O. (1976). *Biodegradation of high octane gasoline*. In *Proceedings of the Third International Biodegradation Symposium*, eds J. M. Sharpley & A. M. Kaplan, pp. 187–96. London: Applied Science.

JENSEN, H. L. (1963). Carbon nutrition of some micro-organisms decomposing halogen-substituted aliphatic acids. *Acta Agriculturae Scandinavica*, 13, 404–12.

JOHANIDES, V. & HRŠAK, D. (1976). Changes in mixed bacterial cultures during linear alkybenzenesulphonate (LAS) biodegradation. In *Fifth International Symposium on Fermentation*, ed. H. Dellweg, p. 426. Berlin: Westkreuz.

JONES, G. L. (1976). Microbiology and activated sludge. *Process Biochemistry*, 11(1), 3–5.

KIESLICH, K. (1976). *Microbial Transformations of Non-steroid Cyclic Compounds*. New York, London: Wiley.

KIMERLE, R. A. & SWISHER, R. D. (1977). Reduction of aquatic toxicity of linear alkylbenzene sulfonate (LAS) by biodegradation. *Water Research*, 11, 31–7.

KING, N. J. (1978). Industry and the environment. In *The Oil Industry and Microbial Ecosystems*, eds K. W. A. Chater & H. J. Somerville, pp. 1–11. London: Heyden and Sons.

KONSTANDT, H. G. (1976). Engineering, operation and economics of methane gas production. In *Microbial Energy Conversion*, eds H. G. Schlegel & J. Barnea, pp. 379–98. Göttingen: Erich Goltze.

LARA, F. J. S. & STOKES, J. L. (1952). Oxidation of citrate by *Escherichia coli*. *Journal of Bacteriology*, 63, 415–20.

LEACH, J. M., MUELLER, J. C. & WALDEN, C. C. (1978). Biological toxification of pulp with effluents. *Process Biochemistry*, 13(1), 18–22.

LEADBETTER, E. R. & FOSTER, J. W. (1959). Oxidation products formed from gaseous alkanes by the bacterium *Pseudomonas methanica*. *Archives of Biochemistry and Biophysics*, 82, 491–2.

LOLL, U. (1976). Engineering, operation and economics of biodigesters. In *Microbial Energy Conversion*, eds H. G. Schlegel & J. Barnea, pp. 361–78. Göttingen: Erich Goltze.

MCCARTY, P. L., YOUNG, L. Y., STUCKEY, D. L. & HEALEY, J. B. (1976). Heat treatment for increasing methane yields from organic materials. In *Microbial Energy Conversion*, eds H. G. Schlegel & J. Barnea, pp. 179–200. Göttingen: Erich Goltze.

MANDELS, M. & ANDREOTTI, R. E. (1978). Problems and challenges in the cellulose to cellulase fermentation. *Process Biochemistry*, 13(5), 6–13.

MEAGHER, R. B. & ORNSTON, L. N. (1973). Relationships among enzymes of the β-ketoadipate pathway. I. Properties of *cis*, *cis*-muconate lactonizing enzyme from *Pseudomonas putida*. *Biochemistry*, 12, 3523–30.

MELLANBY, K. (1972). *The Biology of Pollution*. London: Edward Arnold.

MORRIS, J. G. (1975). The physiology of obligate anaerobiosis. *Advances in Microbial Physiology*, 12, 169–246.

MUNNECKE, D. M. & HSIEH, D. P. H. (1974). Microbial decontamination of Parathion and *p*-nitrophenol in aqueous medium. *Applied Microbiology*, 28, 212–17.

MUNNECKE, D. M. & HSIEH, D. P. H. (1975). Microbial metabolism of a Parathion–xylene pesticide. *Applied Microbiology*, 30, 575–80.

NASH, T. H. (1976). Lichens as indicators of air pollution. *Naturwissenschaften*, **63**, 364–7.

NURMIKKO, V. (1954). Symbiosis experiments concerning the production and biosynthesis of certain amino acids and vitamins in associations of lactic acid bacteria. *Annals Academiae Scientianim Fenniciae*, **54**, 7–58.

OECD (1971). *Pollution by Detergents. Determination of the Biodegradability of Anionic Synthetic Surface-Active Agents*. Paris: OECD.

OOYAMA, J. & FOSTER, J. W. (1965). Bacterial oxidation of cycloparaffinic hydrocarbons. *Antonie van Leeuwenhoek, Journal of Microbiology and Serology*, **31**, 45–65.

ORNSTON, L. N. (1971). Regulation of catabolic pathways in *Pseudomonas*. *Bacteriological Reviews*, **35**, 87–116.

OSBURN, Q. W. & BENEDICT, J. H. (1966). Polyethoxylated alkyl phenols: relationship of structure to biodegradation mechanism. *Journal of the American Oil Chemists' Society*, **43**, 141–54.

PAINTER, H. A. (1970). A review of literature in inorganic nitrogen metabolism in micro-organisms. *Water Research*, **4**, 393–450.

PAINTER, H. A. (1974). Biodegradability. *Proceedings of the Royal Society of London, Series B*, **185**, 149–58.

PAINTER, H. A. (1978). Biotechnology of wastewater treatment. In *The Oil Industry and Microbial Ecosystems*, eds K. W. A. Chater & H. J. Somerville, pp. 178–98. London: Heyden and Sons.

PATEL, R. N., MEAGHER, R. B. & ORNSTON, L. N. (1973). Relationships among enzymes of the β-keto adipate pathway. II. Properties of crystalline β-carboxycis, cis-muconate lactonizing enzyme from *Pseudomonas putida*. *Biochemistry*, **12**, 3531–7.

PATTERSON, S. J., SCOTT, C. C. & TUCKER, K. B. E. (1967). Detergent and polyglycol foaming in river waters. *Journal of the Institute of Water Pollution Control*, **3**, 3–10.

PEARSON, C. R. (1978). Technical and economic problems in the design of biological treatment plant for chemical works effluents. In *The Oil Industry and Microbial Ecosystems*, eds K. W. A. Chater & H. J. Somerville, pp. 171–7. London: Heyden and Sons.

PIKE, E. B. (1975). Aerobic Bacteria. In *Ecological Aspects of Used Water Treatment*, eds C. R. Curds & H. A. Hawkes, pp. 1–63. London, New York: Academic Press.

PIKE, E. B. & CURDS, C. R. (1971). The microbial ecology of the activated sludge process. *Society for Applied Bacteriology Symposium*, **1**, 123–48.

PLIMMER, J. (1978). In *Microbial Degradation of Pollutants in Marine Environments*, eds P. H. Pritchard & A. W. Bourquin. Washington: United States Environmental Protection Agency (in press).

POSTGATE, J. R. (1974). New Advances and future potential in biological nitrogen fixation. *Journal of Applied Bacteriology*, **37**, 185–202.

PRITCHARD, P. H., MAZIARZ, J. L., FREDRICKSON, H. & BOURQUIN, A. W. (1978). System design factors affecting methyl parathion degradation in microcosms. In *Microbial Degradation of Pollutants in Marine Environments*, eds P. H. Pritchard & A. W. Bourquin. Washington: United States Environmental Protection Agency. (In press.)

REANNEY, D. (1976). Extrachromosomal elements as possible agents of adaptation and development. *Bacteriological Reviews*, **40**, 552–90.

RIGBY, P. W. J., BURLEIGH, B. D. & HARTLEY, B. S. (1974). Gene duplication in experimental enzyme evolution. *Nature, London*, **251**, 200–4.

ROMANTSCHUK, H. & LEHTOMAKI, M. (1978). Operational aspects of the first full-scale Pekilo SCP mill application. *Process Biochemistry*, **13**(3), 16–18.

SALA-TREPAT, J. M., MURRAY, K. & WILLIAMS, P. A. (1972). The metabolic divergence in the *meta*-cleavage of catechols by *Pseudomonas putida* NCIB 10015. Physiological significance and evolutionary implications. *European Journal of Biochemistry*, **28**, 347–56.

SENIOR, E., BULL, A. T. & SLATER, J. H. (1976). Enzyme evolution in a microbial community growing on the herbicide Dalapon. *Nature, London*, **263**, 476–9.

SHELEF, G., MORAINE, R., MEYDAN, A. & SANDBANK, E. (1977). Combined algae production – wastewater treatment and reclamation systems. In *Microbial Energy Conversion*, eds H. G. Schlegel & J. Barnea, pp. 427–42. Göttingen: Erich Goltze.

SHUVAL, H. I. (1977). Public health considerations in wastewater and excreta reuse for agriculture. In *Water, Wastes and Health in Hot Climates*, eds R. Feacham, M. McGarry & D. Mara, pp. 365–81. New York: Wiley.

SKOGMAN, H. (1976). Production of Symba-yeast from potato wastes. In *Food from Waste*, eds G. G. Birch, K. J. Parker & J. T. Worgan, pp. 167–79. London: Applied Science.

SLATER, J. H. (1978a). Microbial community structure. In *Microbial Degradation of Pollutants in Marine Environments*, eds P. H. Pritchard & A. W. Bourquin. Washington: United States Environmental Protection Agency (in press).

SLATER, J. H. (1978b). The role of microbial communities in the natural environment. In *The Oil Industry and Microbial Ecosystems*, eds K. W. A. Chater & H. J. Somerville, pp. 137–54. London: Heyden and Sons.

SLATER, J. H., WEIGHTMAN, A. J., SENIOR, E. & BULL, A. T. (1976). The dehalogenase from *Pseudomonas putida*. *Proceedings of the Society of General Microbiology*, **4**, 39–40.

STANLEY, S. O., PEARSON, T. H. & BROWN, C. M. (1978). Marine microbial ecosystems and the degradation of organic pollutants. In *The Oil Industry and Microbial Ecosystems*, eds K. W. A. Chater & H. J. Somerville, pp. 178–98. London: Heyden and Sons.

STENNET, G. V. & EDEN, G. E. (1971). Assessment of biodegradability of synthetic surfactants by tests simulating sewage treatment. *Water Research*, **5**, 601–9.

STIRLING, L. A., WATKINSON, R. J. & HIGGINS, I. J. (1977). The microbial metabolism of alicyclic hydrocarbons: isolation and properties of a cyclohexane-degrading bacterium. *Journal of General Microbiology*, **99**, 119–25.

SWAIN, H. M. & SOMERVILLE, H. J. (1978). Microbial metabolism of methanol in a model activated sludge system. *Journal of Applied Bacteriology* (in press).

TANAKA, S., LERNER, S. A. & LIN, E. C. C. (1967). Replacement of a phosphoenolpyruvate-dependent phosphotransferase by a nicotinamide adenine dinucleotide-linked dehydrogenase for the utilization of mannitol. *Journal of Bacteriology*, **93**, 642–8.

TELETZKE, G. H. (1964). Wet air oxidation. *Chemical Engineering Progress*, **60**, 33–8.

THAUER, R. K. (1976). Limitation of microbial H_2-formation via fermentation. In *Microbial Energy Conversion*, eds H. G. Schlegel & J. Barnea, pp. 201–4. Göttingen: Erich Goltze.

THAUER, R. K., JUNGERMANN, K. & DECKER, K. (1977). Energy conservation in chemotrophic anaerobic bacteria. *Bacteriological Reviews*, **41**, 100–80.

VAN DER LINDEN, A. C. (1978). Degradation of oil in the marine environment. In *Developments in Biodegradation of Hydrocarbons*, vol. I, ed. R. J. Watkinson, pp. 165–200. London: Applied Science.

WEDEMEYER, G. (1967). Dechlorination of 1,1,1-trichloro-2,2-*bis* (*p*-chlorophenyl) ethane by *Aerobacter aerogenes*. *Applied Microbiology*, **15**, 569–74.

WHEELIS, M. L. (1975). The genetics of dissimilatory pathways in *Pseudomonas*. *Annual Review of Microbiology*, **29**, 505–24.

WIELGOLASKI, F. E. (1975). Biological indicators on pollution. *Urban Ecology*, **1**, 63–79.

WILKINSON, T. G., TOPIWALA, H. H. & HAMER, G. (1974). Interactions in a mixed bacterial culture growing in methane in continuous culture. *Biotechnology and Bioengineering*, **16**, 41–7.

WILLIAMS, P. A. (1978). Microbial genetics relating to hydrocarbon formation. In *Developments in Biodegradation of Hydrocarbons*, vol. I, ed. R. J. Watkinson, pp. 135–64. London: Applied Science.

WILLIAMS, R. J. & EVANS, W. C. (1975). The metabolism of benzoate by *Moraxella* species through anaerobic nitrate respiration – evidence for a reductive pathway. *Biochemical Journal*, **148**, 1–10.

WU, T. T., LIN, E. C. C. & TANAKA, S. (1968). Mutants of *Aerobacter aerogenes*, capable of utilising xylitol as a novel carbon source. *Journal of Bacteriology*, **96**, 447–56.

MICROBIOLOGICAL METHODS FOR THE EXTRACTION AND RECOVERY OF METALS

DON P. KELLY,* PAUL R. NORRIS* AND CORALE L. BRIERLEY†

Department of Environmental Sciences, University of Warwick, Coventry CV4 7AL, UK

† *New Mexico Bureau of Mines and Mineral Resources, Socorro, New Mexico 87801, USA*

INTRODUCTION

Our purpose in this review is to explore a field of microbiology which is of considerable ecological, biochemical and economic significance, but with which many microbiologists and those hydrometallurgists concerned with metal extraction and recovery are poorly acquainted.

The interactions of metals and micro-organisms are diverse but can be considered in three major categories:

(1) The roles of metals in essential metabolism, or the interruption of metabolism in the case of toxic metals.

(2) Accumulation of metals by organisms, including binding at organism surfaces, and intracellular uptake of metals.

(3) Biochemical transformations of metals which can involve solubilization or precipitation, valency changes through oxidative or reductive processes, and the interconversion of inorganic and organic metal compounds.

We are not concerned with the first category nor with the biochemistry and genetics of mineral nutrient transport, but with those areas having actual or potential biotechnological application. The economically significant interrelation of organisms and metals can be largely covered by two headings. First, the *extraction* of metals from insoluble minerals by various organisms, principally through leaching by acidophilic iron-oxidizing and sulphur-oxidizing bacteria, which involves processes in category (3) above; and secondly, the *recovery* of metals from solution by organisms, which can involve processes in categories (1) and (2). It is under these two headings that this review is presented.

METAL EXTRACTION: THE LEACHING OF MINERALS

What is bacterial leaching?

The term 'bacterial leaching' is used to describe the solubilization of metals (such as copper and uranium) from their ores in an operation which is usually relatively crude, consisting of the percolation of acidified water through heaps of broken, low-grade ore (or the 'waste' from conventional treatment of richer ones). Such dumps or heaps may contain from 10^5 to 4×10^9 tons of rock (C. L. Brierley, 1978a). Within the dumps, bacterial activity results in mineral sulphide oxidation and release of metal. The solution is collected when it drains from the dump and is processed to recover the metal. The residual liquid, containing sulphuric acid and ferrous/ferric iron, is recycled to the dump. Economically, bacterially-assisted leaching is a valuable means of extracting some metals from their ores. In the case of copper the industrial processes have been run principally by chemical engineers with minimal knowledge of microbiology, while much of the microbiological study has been carried out by microbiologists with little grasp of the possible engineering problems or applications. This situation is improving with realization that the effective exploitation of leaching processes to obtain metals which have a declining resource base needs much cross-fertilization between fundamental microbiological and industrial requirements. While leaching may be effected by various organisms under acid or alkaline conditions, it is acid leaching that has been most studied and is undoubtedly of the greatest present use and future potential.

Acid leaching of mineral sulphides

This is a natural phenomenon occurring since geologically ancient times, but accelerated following man's exposure of mineral-rich materials in historical times through the mining of copper, iron, coal and many sulphide minerals. The phenomenon of the drainage of dissolved metals and acid from mine workings, mine waste dumps and coal heaps has been known for many years but it is only in the past thirty years that the essential role of bacteria has been proved. Copper released in consequence of bacterial dissolution of its sulphide minerals was probably first recovered by the Romans; then by the 16th-century Welsh in Anglesey; at Rio Tinto, Spain, in the 18th century; and in the very large-scale operations developed in the USA during this century (Sheffer & Evans, 1968; Fletcher, 1970; Sasson, 1975; C. L. Brierley, 1978a). The importance of these operations in the recovery of copper is shown by the fact that leaching of low-grade copper waste material in the western USA accounts for about 11.5% of that country's total annual copper production (C. L. Brierley, 1978a). An earlier estimate (Burkin, 1971) gave leaching a 15% share of the total US primary copper production.

The principles, practices and applications of bacterial leaching have been subject to a thorough coverage in the literature (Duncan & Trussell, 1964; Le Roux, 1969; Fletcher, 1970; Duncan & Bruynesteyn, 1971; Tuovinen & Kelly , 1972, 1974a; Kelly, 1976; Karavaiko, Kuznetsov & Golomzik, 1977; Schwartz, 1977; Torma, 1977; Murr, Torma & Brierley, 1978) and is the subject of a critical review (C. L. Brierley, 1978a) to which reference should be made.

In the following sections we summarize the important basic processes and micro-organisms involved: we are mainly concerned with evaluating the relative importance of bacterial and chemical mechanisms in the bacterially-assisted leaching systems; with the possible mechanisms of bacterial reaction with a mineral surface; and with recent developments in the microbiology of leaching. Progress in understanding leaching processes has shown relatively little fundamental advance during the 1970s, in that many of the problems outlined by Tuovinen & Kelly (1972) are still inadequately solved. Two areas have, however, seen special developments: the demonstration that *mixed* populations of bacteria existing in natural leaching systems can be as effective as, or better than, pure cultures in mineral solubilization; and, particularly, the demonstration that various types of *thermophilic* bacteria exist in leaching systems and can effect mineral breakdown.

What is the microbiology of the process?

Several specialized types of bacteria occurring in waters associated with exposed sulphides or sulphur-rich coals can derive energy from oxidizing inorganic substances such as ferrous iron, sulphur and soluble or insoluble sulphides. They are generally capable of fixing carbon dioxide, with at least some being fully autotrophic (Whittenbury & Kelly, 1977) whereas others also require organic substances for growth. The relations of these organisms are summarized in Fig. 1.

The 'classical' view that *Thiobacillus ferrooxidans* is the only organism of importance in leaching (Kelly, 1976; Torma, 1977) is being modified by recent observations. The types of organisms now known to be involved are as follows:

(a) *Thiobacillus ferrooxidans* An immense amount of information has

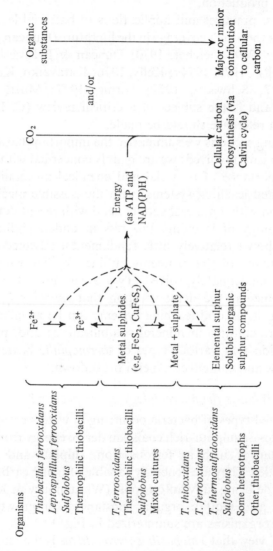

Fig. 1. The acidophilic leaching bacteria: their activities and basic metabolism.

been accumulated on this organism which is capable of chemolithotrophic growth in acid environments using energy from the oxidation of
ferrous iron, soluble and insoluble sulphides, sulphur and soluble
sulphur compounds, such as thiosulphate and tetrathionate, and
possibly cuprous and stannous ions (Nielsen & Beck, 1972; Tuovinen &
Kelly, 1972, 1973, 1974 b, c, d, e; Kelly & Tuovinen, 1972, 1975; Kelly,
1976; Lewis & Miller, 1977; Schwartz, 1977; Murr et al., 1978;
Eccleston & Kelly, 1978).

(b) Thiobacillus thiooxidans This organism cannot oxidize iron; it is
best known for its ability to grow on elemental sulphur and some
soluble sulphur-compounds. It may have a role in the oxidation of
sulphur generated in leaching systems and has been shown to be effective
in catalysing zinc sulphide oxidation (Khalid & Ralph, 1977). Mineral
leaching by T. ferrooxidans could be enhanced by T. thiooxidans (Karavaiko & Moshniakova, 1974).

(c) Leptospirillum ferrooxidans and mixed cultures of acidophiles
L. ferrooxidans was isolated from copper deposits in Armenia (Balashova
et al., 1974) and in pure culture can grow only on soluble ferrous iron;
it cannot attack sulphur or mineral sulphides. A new, facultatively
chemolithotrophic species of thiobacillus, T. organoparus, was isolated
from the same deposits as L. ferrooxidans and from other locations
(Markosyan, 1973). This organism could only oxidize sulphur, not
soluble iron or mineral sulphides. Mixed cultures of L. ferrooxidans and
T. organoparus would, however, grow on and degrade both pyrite and
chalcopyrite, which neither organism alone can metabolize. Subsequently it has been shown that mixtures of L. ferrooxidans with the other
sulphur-oxidizing acidophiles, T. thiooxidans or T. acidophilus (Guay &
Silver, 1975), will also degrade pyrite rapidly (Kelly, 1978; Norris &
Kelly, 1978b), and that apparently stable natural mixed populations of
organisms (including types probably similar to L. ferrooxidans and T.
organoparus as well as T. ferrooxidans) can be enriched on pyrite as a
substrate (Norris & Kelly, 1978b). It is therefore likely that in natural
and contrived leaching systems operating at temperatures below 35 °C,
mineral breakdown is not due exclusively to T. ferrooxidans, but is
effected at least in part by the cooperative action of physiologically
distinct species including L. ferrooxidans, T. thiooxidans, T. organoparus
and T. acidophilus. Heterotrophic bacteria may also play a stimulatory
role in mineral leaching, as witnessed by the demonstration that the
leaching of a copper–nickel sulphide concentrate by T. ferrooxidans was

accelerated by *Beijerinckia* sp. (Tsuchiya, Trivedi & Schuler, 1974). This was thought to be due to stimulation of bacterial development using atmospheric nitrogen fixed by *Beijerinckia* (Tsuchiya *et al.*, 1974), but this was not conclusively proved, and a more subtle stimulatory mechanism, such as removal of autoinhibitory organic secretions by the heterotroph or even stimulation by some mutual interaction with the mineral, might have to be sought. It should be noted that *T. ferrooxidans* itself is capable of fixing dinitrogen from the atmosphere (Mackintosh, 1978).

(*d*) *Thermophilic thiobacilli*   These have been isolated from thermal springs (Emoto, 1929; Egorova & Deryugina, 1963; Zavarzin, 1966; Williams & Hoare, 1972; Le Roux, Wakerley & Hunt, 1977; C. L. Brierley, 1978*a*), but it is only very recently that moderately thermophilic thiobacilli capable of degrading minerals such as pyrite, copper sulphides and pentlandite have been demonstrated. Several strains are now known. These include *Thiobacillus* TH1 (Brierley & Le Roux, 1977; Brierley, Norris, Kelly and Le Roux, 1978; J. A. Brierley, 1978), which can grow on ferrous iron, pyrite, $CuFeS_2$, CuS or pentlandite, but requires yeast extract, glutathione or cysteine for growth on any of these materials; *Thiobacillus* TH2 and TH3 (Brierley & Lockwood, 1977; J. A. Brierley, 1978) isolated respectively from a test leach facility and a copper leach dump, both of which also require organic supplements; and *T. thermosulfidooxidans* (Golovacheva & Karavaiko, 1977), which is also facultatively heterotrophic on sugars. The existence of bacteria which are quite distinct from *T. ferrooxidans* (as indicated by distinct temperature optima and widely different DNA composition (Brierley *et al.*, 1978; Tuovinen, Kelly, Dow & Eccleston, 1978) indicates that attempts to select for mutations to thermophily in *T. ferrooxidans* (Karavaiko *et al.*, 1977) are not the only courses of action in any industrial interest in high-temperature leaching.

(*e*) *Extreme thermophiles*   A number of strains of spherical bacteria, lacking cell walls and resembling *Sulfolobus acidocaldarius*, have been isolated from hot-spring and other environments (Darland, Brock, Samsonoff & Conti, 1970; Brock, Brock, Belly & Weiss, 1972; Brierley & Brierley, 1973; de Rosa, Gambacorta & Bu'Lock, 1975; C. L. Brierley, 1978*a*) and have been shown to be able to grow at temperatures up to about 80 °C using sulphur or iron oxidation for energy. Such bacteria are able to leach recalcitrant minerals such as chalcopyrite and molybdenite at 60 °C apparently more effectively than mesophilic bacteria (Brierley & Murr, 1973; C. L. Brierley, 1974, 1975, 1977, 1978; Brierley & Brierley, 1978).

(*f*) *Anaerobic bacteria*   No anaerobic iron-oxidizing bacteria have yet been found, but denitrifying bacteria, including a sulphur-oxidizing thiobacillus, have been found in copper–molybdenum ore deposits (Karavaiko, Shchetinina, Pivovarova & Mubarakova, 1973). Anaerobically or microaerophilically, *Sulfolobus* and *T. ferrooxidans* (and *T. thiooxidans* aerobically) can oxidize sulphur with the coupled reduction of ferric to ferrous iron as an alternative to the reduction of oxygen (Brock & Gustafson, 1976). *Sulfolobus*-like bacteria can also use sulphur to reduce molybdenum(VI) to molybdenum(V) blue (Brierley & Brierley, 1975; C. L. Brierley, 1978*a*).

(*g*) *Other bacteria*   Heterotrophic bacteria and fungi also occur in leaching environments, but their contribution to leaching, if any, is little known. Other thiobacilli may also occur: *T. delicatus* and *T. rubellus* with pH optima in the range pH 5–7 were isolated from copper-mine water and shown to have slight activity in releasing sulphur, but not copper, from an ore sample (Mizoguchi, Sato & Okabe, 1976). *T. thioparus* showed poor ability to attack zinc sulphide minerals, but did release some zinc from the synthetic sulphide (Khalid & Ralph, 1977). The occurrence of thiobacilli of this type, and species such as *T. neapolitanus* and *T. novellus*, in the early stages of a leach system could be beneficial in lowering the pH from the pH 4–7 range to the pH 1–4 range in which the significant leaching organisms function (Tuovinen & Kelly, 1974*a*). Goroll (1976) reported large numbers of *T. neapolitanus* in mine waters and indicated that good growth of the organism occurred on FeS_2, FeS, Sb_2S_3, Ag_2S_3, PbS, ZnS and $CuFeS_2$. This deserves further study since it indicates an important role for this organism in initiating leaching and acidification, although *T. neapolitanus* (pH optimum around pH 6.6) could not be significantly active at the pH values required by *T. ferrooxidans* and other acidophiles. It is noteworthy that in a microbiological study of 1210 samples of Brazilian mine waters, Sayão Lobato (1974*a*) found thiobacilli in 776, but *T. ferrooxidans* in only 14. Other autotrophs which were found included *Nitrosomonas* in 15 samples, *Nitrobacter* (6), *Hydrogenomonas* (103) and *Methanomonas* (4), as well as 'unclassified autotrophs' in 283 samples.

Significance of thermophilic and anaerobic organisms

In leach dumps (and in an experimental controlled leach tower (Brierley

& Lockwood, 1977)) the internal temperature rises as a result of bacterial and chemical mineral oxidation to between 54 °C and 80 °C (Beck, 1967; Brierley & Lockwood, 1977; Murr & Brierley, 1978). It is likely that the interior of dumps or even of column percolation systems become anaerobic, since the oxygen demand for sulphide mineral oxidation is considerable. These facts led to the view that, as *T. ferrooxidans* cannot grow and function at temperatures above 35–40 °C, the reactions within a dump at 45 °C or more must be entirely chemical. The discovery of thermophilic leaching thiobacilli, and more recently of spherical organisms morphologically like *Sulfolobus* in leach systems (Murr & Brierley, 1978), indicates that bacterial breakdown of minerals can take place even at 50–60 °C. Similarly, the discovery of anaerobic sulphur oxidation at the expense of ferric iron means that the anaerobic interior of leaching mineral heaps may still generate sulphuric acid (and remove from mineral surfaces sulphur coatings produced by chemical reaction with ferric iron) even in the absence of oxygen. All these processes would enhance the efficiency of an operational leach system.

The reactions of bacterial leaching: How are the organisms involved?

Bacterially-assisted leaching is usually discussed in terms of *direct* bacterial attack on mineral sulphides and *indirect* attack via ferric iron and sulphuric acid, which can be generated by bacterial oxidation of pyrite (FeS_2)), soluble ferrous iron and elemental sulphur. We consider that the terms 'direct' and 'indirect' can be inadequate or misleading in dealing with bacterially-assisted leaching of solid substrates. We will consider this further, initially dividing the chemical and biologically-catalysed reactions in leaching systems into three classes.

(*a*) *Oxidations undoubtedly catalysed by bacteria* Substrates which constitute the principal oxidizable components of mineral sulphides are involved:

$$4FeSO_4 + O_2 + 2H_2SO_4 \rightarrow 2Fe_2(SO_4)_3 + 2H_2O \tag{1}$$
$$S_8 + 12O_2 + 8H_2O \rightarrow 8H_2SO_4 \tag{2}$$
$$H_2S + 2O_2 \rightarrow H_2SO_4 \tag{3}$$

(*b*) *Chemical reactions independent of bacteria* Ferric iron is well known as a very effective chemical solubilizing agent for numerous minerals (Stumm-Zollinger, 1972; Dutrizac & MacDonald, 1974; Kelly, 1976; Derry, Garrett, Le Roux & Smith, 1976) and can be generated together with sulphuric acid from the bacterial oxidations indicated above.

$$2FeS_2 + 2Fe_2(SO_4)_3 \rightarrow 6FeSO_4 + 4S^0 \tag{4}$$
(pyrite)
$$CuFeS_2 + 2Fe_2(SO_4)_3 \rightarrow CuSO_4 + 5FeSO_4 + 2S^0 \tag{5}$$
(chalcopyrite)
$$Cu_2S + 2Fe_2(SO_4)_3 \rightarrow 2CuSO_4 + 4FeSO_4 + S^0 \tag{6}$$
(chalcocite)
$$CuS + Fe_2(SO_4)_3 \rightarrow CuSO_4 + 2FeSO_4 + S^0 \tag{7}$$
(covellite)
$$Bi_2S_3 + 3Fe_2(SO_4)_3 \rightarrow Bi_2(SO_4)_3 + 6FeSO_4 + 3S^0 \tag{8}$$
$$UO_2 + Fe_2(SO_4)_3 \rightarrow UO_2SO_4 + 2FeSO_4 \tag{9}$$
(uraninite)
$$UO_3 + H_2SO_4 \rightarrow UO_2SO_4 + H_2O \tag{10}$$

(c) Reactions reported to be 'directly' effected by bacteria

$$4FeS_2 + 15O_2 + 2H_2O \rightarrow 2Fe_2(SO_4)_3 + 2H_2SO_4 \tag{11}$$
$$4CuFeS_2 + 17O_2 + 2H_2SO_4 \rightarrow 4CuSO_4 + 2Fe_2(SO_4)_3$$
$$+ 2H_2O \tag{12}$$
$$2Cu_2S + O_2 + 2H_2SO_4 \rightarrow 2CuS + 2CuSO_4 + 2H_2O \tag{13}$$
$$CuS + 2O_2 \rightarrow CuSO_4 \tag{14}$$

Reaction (14) has been regarded as typical of the overall reaction of 'direct' attack of *T. ferrooxidans* on divalent metal sulphides, i.e.

$$MS + 2O_2 \rightarrow MSO_4 \tag{15}$$

where M could be Zn, Pb, Co, Ni etc.

Further examples of apparent direct attack by bacteria are:

$$2AsFeS + 7O_2 + 2H_2O \rightarrow 2FeAsO_4 + 2H_2SO_4 \tag{16}$$
(arsenopyrite)
$$2MoS_2 + 9O_2 + 6H_2O \rightarrow 2H_2MoO_4 + 4H_2SO_4 \tag{17}$$
(molybdenite)
$$Sb_2S_3 + 6O_2 \rightarrow Sb_2(SO_4)_3 \tag{18}$$
(stibnite)
$$2CuSe + O_2 + 2H_2SO_4 \rightarrow 2CuSO_4 + 2H_2O + 2Se^0 \tag{19}$$
$$(Ni, Fe_9)S_8 + 17\tfrac{5}{8}O_2 + 3\tfrac{1}{4}H_2SO_4 \rightarrow 4\tfrac{1}{2}NiSO_4 + 2\tfrac{1}{4}Fe_2(SO_4)_3$$
(pentlandite)
$$+ 3\tfrac{1}{4}H_2O \tag{20}$$

It is immediately apparent that in any leaching system in which iron is present, a combination of 'direct' (Fe^{3+}-independent) and 'indirect' (chemical, non-enzymic) attack can occur. Some authorities on bacterial leaching (see C. L. Brierley, 1978a) suggest that all leaching can be explained in terms of chemical solubilization by ferric iron (Stumm-Zollinger, 1972) and sulphuric acid (reactions 4–10) supplied by bacterial oxidations (reactions 1 and 2), i.e. by 'indirect' attack. However, ferric leaching of sulphides can generate sulphur which coats or passivates the potentially reactive surface of the mineral. The inability of *Leptospirillum ferrooxidans* to oxidize pyrite (FeS_2) in acid media despite its ability to maintain soluble iron in the ferric state suggests that bacteria, or mixed cultures of bacteria, which can oxidize pyrite must contribute more than simple regeneration of ferric iron to the successful leaching systems. This extra contribution appears to be the 'direct' involvement of the bacteria, and has been considered as oxidation of the sulphur moiety and, if an iron-containing sulphide is involved, oxidation of ferrous iron in the crystal lattice structure. Clearly, whether the sulphide or sulphur moiety is attacked while it is part of the crystal lattice or when it has been deposited as a reaction layer at the mineral surface after some ferric iron leaching is of terminological interest; while in practical terms, the fact that an important part in efficient leaching could depend on the bacteria solubilizing a solid substrate (sulphur) seems to merit the description of 'direct' attack. Evidence of bacterial oxidation, independently of ferric iron, of the sulphur moiety of mineral sulphides which contain iron (Duncan, Landesman & Walden, 1967) and the oxidation of synthetic mineral sulphides in the absence of iron (Torma, 1971; Silver & Torma, 1974; Sakaguchi, Torma & Silver, 1976) has, therefore, provided the basis for the proposed 'direct' attack by bacteria on mineral sulphides. We wish to consider further the evidence for real direct bacterial attack on mineral sulphides, i.e. oxidation of constituents in the crystal lattice, and also the significance of ferric iron in leaching systems.

Are there mechanisms for direct bacterial attack on minerals?

Some controversy surrounds this topic and a final conclusion cannot yet be presented (C. L. Brierley, 1978a).

Reaction (15) represented the oxidation of metal sulphides as:

$$MS + 2O_2 \rightarrow MSO_4$$

Torma (1971) demonstrated that the microbiological oxidation of synthetic cobalt, nickel and zinc sulphides was preceded by a lag phase during which the pH of the leach solutions tended to rise, so that reaction (15) should be considered initially as an acid-requiring reaction:

$$MS + H_2SO_4 + \tfrac{1}{2}O_2 \rightarrow MSO_4 + H_2 + S^0$$

followed by bacterial oxidation of the sulphur:

$$S + 1\tfrac{1}{2}O_2 + H_2O \rightarrow H_2SO_4$$

Clearly, a cycle of acid leaching of the solid sulphide could be established once the pH was stabilized in the range suitable for bacterial oxidation of elemental sulphur. Leaching during the lag phase depended on the solubility product of the particular sulphide (Torma, 1971). Torma *et al.* (1974) further suggested a relation between the rate of sulphide (CoS, CdS, NiS and ZnS) oxidation by *T. ferrooxidans* and the solubility product of the respective sulphide. The bacterially-assisted leaching of 'iron-free' galena (PbS) followed the above pattern with the role of the bacteria apparently the oxidation of elemental sulphur formed by chemical oxidation (Tomizuka & Yagisawa, 1978). At pH values lower than 1.3 the oxidation of galena was equivalent in the presence or absence of bacteria. Solubility-dependent leaching rather than direct attack on the crystal lattice would appear to be confirmed by the absence of significant oxygen consumption by *T. ferrooxidans* on synthetic ZnS and CdS in buffered suspensions (Silver & Torma, 1974). Other studies have failed to demonstrate bacterial growth and cadmium release from CdS unless excess elemental sulphur was added as the primary substrate for *T. thiooxidans* (Brissette, Champagne & Jutras, 1971) and for *T. thiooxidans* or *T. ferrooxidans* (P. R. Norris, unpublished). The latter study showed that the rate of cadmium leaching was proportional to acid produced and was more rapid by *T. thiooxidans* than *T. ferrooxidans* (which grew more slowly on elemental sulphur); both studies showed that the addition of acid equivalent to that produced by the bacteria from sulphur was less effective in releasing cadmium when added to sterile controls. Further experiments in our laboratory have shown that when extra sulphur was added *T. thiooxidans* also solubilized ZnS more rapidly than did *T. ferrooxidans*, but that without pH control neither organism could become established on ZnS in the absence of excess sulphur or iron. Khalid & Ralph (1977) also found *T. thiooxidans* more efficient than *T. ferrooxidans* during leaching of synthetic zinc sulphide in pH-controlled cultures. It appears, therefore, that these sulphides might be leached following oxidation of elemental sulphur or free sulphide rather than of the sulphur moiety of the solid substrate *in situ* and that any acidophile capable of oxidizing the 'free' sulphur is capable of promoting the leaching.

A different sequence of events operates in the bacterially-assisted leaching of copper sulphides (CuS) and iron pyrite. Here the metals are not released by *T. thiooxidans* even when extra sulphur is added and *T.*

thiooxidans growth is profuse (P. R. Norris, unpublished). Covellite and chalcocite (Cu_2S) oxidation by *T. ferrooxidans* has been studied in detail and readily demonstrated in iron-free conditions (Beck, 1977; Imai, 1978). The ability of *T. ferrooxidans* to leach these substrates suggests, therefore, either the involvement of the *metal*-oxidizing as well as the sulphur-oxidizing system or that only *T. ferrooxidans* can abstract sulphide from the mineral lattice. The likelihood of oxidation of the mineral's metal and sulphur constituents is clearly governed by the sulphur 'availability' and the chemical form and valence state of the metal in the lattice.

Most sulphide crystal structures have been related to type structures, forming the basis of a classification with two main groups based on cubic or hexagonal structures and further subdivisions with a limited number of basic structures (Ross, 1957). The minerals of interest in bacterial leaching are dispersed among these groups, so precluding any simple correlation of structure with susceptibility to bacterial attack. Moreover, the basic structural classification can conceal the nature of the coordination in the crystals. Thus both pyrite, essentially a cubic structure, and covellite, a 'miscellaneous' structure with a hexagonal two-part cell, contain S_2 groups in critical positions. In contrast, ZnS (zincblende) and CdS generally have each metal or sulphur atom surrounded by four equidistant atoms of the opposite sort at the corners of a regular tetrahedron (Wyckoff, 1963). The possibility arises that the sulphur pairs in pyrite and covellite may be a site of bacterial attack, allowing their direct solubilization, in contrast, for example, to the attack on CdS. Furthermore, chalcopyrite, which is generally regarded as refractory to dissolution, exhibits polymeric isomorphism with zincblende, where (if the distinction between Fe and Cu is ignored) the structures are identical. However, a mechanism of attack initially on S_2 groups would have to explain the unique ability of *T. ferrooxidans* in this direction compared with *T. thiooxidans* when both organisms can utilize elemental sulphur. Thus the crystal structures of the sulphides and the ability of *T. ferrooxidans* to oxidize metals should be examined more closely in seeking the key to bacterial leaching mechanisms.

Covellite (CuS), although apparently cupric sulphide (Cu^{II}), contains monovalent copper (Cu^I), but it is difficult to assess the relative valence states of the copper in a number of leachable copper minerals (including covellite, bornite and cubanite) to give a consistent stereochemistry of Cu^I or Cu^{II}. If covellite, for example, is regarded as a covalent structure, it can be represented as $Cu_4^I Cu_2^{II}(S_2)_2 S_2$ (Wells, 1975). Just as pyrite contains a metal (Fe^{II}) oxidizable by *T. ferro-*

oxidans, so covellite (and of course chalcocite, which is predominantly cuprous sulphide) contains not only S_2 pairs but an oxidizable metal (Cu^I) probably available to the energy-yielding metabolism of *T. ferrooxidans* (Nielsen & Beck, 1972). Chalcopyrite seems to provide the same possibilities. The valence state of the metals in $CuFeS_2$ is not simply described as Cu^{II}/Fe^{II} as the formula might suggest but has been shown to comprise both $Cu^I Fe^{III} S_2$ and $Cu^{II} Fe^{II} S_2$ (Gmelin, 1955). The contribution of these two forms could be variable over a wide range, and such a probability has been proposed (Wyckoff, 1970) to explain the different degrees of recalcitrance to leaching observed with different chalcopyrite samples. Similarly, some contribution from Cu^I/Cu^{II} oxidation could occur during copper selenide oxidation by *T. ferrooxidans* (Torma & Habashi, 1972) since CuSe has a covellite-type structure and also contains Cu^I (Gmelin, 1955). Other possibilities for metal oxidations in reactions catalysed by *T. ferrooxidans* could be in Sb_2S_3 and Sb_2O_3, in which both antimony and sulphur can be oxidized (Lyalikova, 1972; Torma & Gabra, 1977); and ores of other metals with more than one possible valence state but clearly excluding such metals as zinc and cadmium. In future studies of the probability of such oxidation mechanisms, caution must be exercised in assessing results because of the need to consider complicating reactions between free metals and the bacteria: it is particularly important to consider the high toxicity of metals such as mercury, molybdenum and silver which could be released during leaching of some minerals.

What is the physicochemical relationship between the bacterium and a mineral surface?

If leaching is to be explained exclusively in terms of chemical attack on minerals by soluble substances (Fe^{3+}, H_2SO_4) generated by bacteria from soluble iron and sulphides or elemental sulphur, no hypothesis under this heading is necessary. The indication that S_2 groups or metals may, however, be directly 'recognized' within minerals, at least by *T. ferrooxidans,* implies a more intimate physical association between the organism and mineral surface. Whether or not direct reaction with the surface occurs, there is clearly an advantage to the organism to be attached to the surface, since it is then proximal to the oxidizable products of chemical leaching (Fe^{2+} and sulphur).

Electron microscope observations (Brierley, Brierley & Murr, 1973; Karavaiko & Pivovarova, 1977; Berry & Murr, 1978) have shown that physical attachment (using *T. ferrooxidans, T. thiooxidans* and *Sulfolobus*) to surfaces occurs and that attachment appears to be selectively to sulphide phases (i.e. FeS_2, $CuFeS_2$) rather than to the silicate matrix of prepared mineral specimens. Attachment can, however, occur even with dead cells (Beck, 1977). In the case of sulphur, recent cinephoto-micrography studies (M. Silver, personal communication, 1978) show that *T. acidophilus* attaches more readily to sulphur than does *T. ferrooxidans*. Cultures of *T. thiooxidans* show more rapid use of sulphur, with a much greater proportion of the population attached to sulphur than *T. ferrooxidans* treated under otherwise identical conditions (Norris, unpublished). Using scanning electron microscopy, Bennett & Tributsch (1978) demonstrated characteristic patterns of bacterially-etched pits in the surface of pyrite crystals exposed to *T. ferrooxidans* and proposed that bacterial distribution on the surface (and hence leaching sites) were critically dependent on deviations in the crystal structure, such as fracture lines or dislocations. The problem of under-standing the relation between leaching bacteria and the surface of insoluble but metabolizable minerals is analogous to that, as yet un-resolved, of the oxidation of elemental sulphur by thiobacilli and may, in fact, involve similar mechanisms. The cell envelope of the thiobacilli precludes direct contact of the cytoplasm with the mineral (unless a membrane extrusion-vacuolation system, as proposed for sulphur, is involved (Karavaiko & Pivovarova, 1977)) but 'contact' could be effected by external membrane layers, secreted enzymes or secreted metabolites such as phospholipids (Agate, Korczinski & Lundgren, 1969). The formation of sulphur by the ferric leaching of mineral sul-phides (equations 4–8) could result in subsequent inhibition of chemical leaching by passive coating of the mineral surface. Bacterial oxidation of this layer could counteract such an effect and could be assisted by sul-phur solubilization into secreted phospholipids. Sulphur oxidation can, however, proceed under conditions where extracellular product con-centrations (including enzymes) must be low (e.g. in Warburg experi-ments with washed suspensions) so intimate attack on the sulphur layer could also be occurring. The combined attack on pyrite by *Lepto-spirillum ferrooxidans* and sulphur-oxidizing thiobacilli could result from the removal of the passive sulphur layers on the surface of the pyrite by the thiobacilli, but preliminary experiments (P. R. Norris, unpublished) failed to obtain growth of *L. ferrooxidans* plus *T. thio-oxidans* on CuS under conditions where *T. ferrooxidans* alone would have developed. The mixture, however, did develop on pyrite. Iron availability might be critical, but the general indication is that, while chemical leaching and bacterial sulphur oxidation do contribute, there are as yet unresolved mechanisms by which *T. ferrooxidans* may attack at least some minerals. Among these is the electrochemical oxidation system postulated by Karavaiko & Pivovarova (1977). In this hypothesis, the mineral sulphide resembles a natural electrode with electron transport either to a mineral of higher electrode potential or the bacteria. One consequence of this is prior oxidation, in a mixture of minerals, of a mineral with a lower electrode potential; thus arsenopyrite was oxidized in advance of chalcopyrite when both were available (G. I. Karavaiko, personal communication). Since the significant oxidation reactions are thus considered to be at the electronic structure-level of the minerals, deviations from ideal electronic structure in the mineral will influence leaching. Such deviations can include regional cation or anion deficiency in the sulphide surface providing positive (hole) or negative (electron) conductivity. Comparing the leaching of pyrite with such deficiences showed that *T. ferrooxidans* oxidized specimens with hole conductivity more efficiently than those with electron conductivity (Karavaiko & Pivovarova, 1977).

Returning to the problem of whether all sulphide minerals are susceptible to direct attack, determining the effect of small amounts of iron on leaching rates becomes important. Thus Sakaguchi, Torma & Silver (1976) demonstrated that copper release from covellite and chalcocite by *T. ferrooxidans* was doubled or trebled by 0.01M ferric iron. Even in reports where no effect of added iron is observed, iron in the mineral, even at low levels, might assist by ferric leaching. This is discussed elsewhere (C. L. Brierley, 1978a). Khalid & Ralph (1977) showed that marmatite (ZnS containing 3% (w/w) iron) was leached more rapidly than sphalerite or wurtzite (ZnS containing 0.03–0.08 and 0.06% Fe respectively). 'Iron-free' (0.06% Fe) synthetic zinc sulphide was in fact leached more rapidly by *T. thiooxidans* than *T. ferrooxidans* in the absence of iron, but was leached much faster by *T. ferrooxidans* when iron(II) was provided (100–4000 ppm). These experiments again indicate different degrees of recalcitrance by different crystal structures, and the ease of leaching of marmatite could be due in part to iron but perhaps more significantly to the 'opening up' of the structure by the substitution of iron for zinc. Added ferric iron has in some cases proved inhibitory to mineral leaching (see C. L. Brierley, 1978a), but the concentrations tested (0.27–0.81 M with chalcopyrite) would certainly have been somewhat inhibitory to iron oxidation (Kelly & Jones, 1978).

Clearly, sulphide mineral leaching can be due to a combination of direct bacterial attack and indirect chemical solubilization. The relative contributions of the two processes could depend on the types of minerals and bacteria and on the physical and chemical conditions of the leaching.

Applications of bacterially-assisted leaching

There are what can be described as 'Contact' and 'Remote' applications for bacterial oxidations applied to leaching.

(a) *Contact leaching systems*   These include the large-scale heap and dump systems in which liquid percolates through an ore and the bacteria as well as the acid ferric lixiviant solutions are in intimate contact with the minerals. This has been the most important system to date in terms of metal recovered. Developments of this type of system can include the heap or vat leaching of high-grade copper ores, uranium oxides or mineral concentrates; contact leaching has been investigated on the laboratory and pilot-plant scales (see publications by Torma and others, 1970–1974; Barbic & Krajincanic, 1972; Le Roux, North & Wilson, 1973; Guay *et al.*, 1975, 1976; Pinches, 1975; Barbic *et al.*, 1976; Sakaguchi, Silver & Torma, 1976; Tomizuka *et al.*, 1976; Bruynesteyn & Duncan, 1977; C. L. Brierley, 1978*b*; McElroy & Bruynesteyn, 1978; Tomizuka & Yagisawa, 1978). In some cases, the further development of bacterial processes could prove economic for the extraction of high-grade concentrates as an alternative to conventional methods. Some selectivity can be achieved also in bacterial leaching. For example, the bacterial oxidation of a lead–zinc sulphide concentrate or ore will solubilize the zinc, which will be removed in acid solution, but the lead sulphate produced is insoluble and will thus be separated from the zinc and can be chemically removed from the solid residue. Possibly a great potential application of contact leaching is in the removal of pyrite from high-sulphur coal. At present high-sulphur coals are not burnt because of the concomitant SO_2 pollution, but continued use of coal makes the exploitation of reserves such as the coals of the US Appalachian fields (containing 3.0–5.5% total sulphur) a growing necessity (see Dugan & Apel, 1978, for source references). Pyrite in coal (which can be 70% of the total sulphur) is amenable to bacterial leaching and rapid, almost complete, removal of it can be effected in the laboratory with pure or mixed cultures (Kuznetsov, Ivanov & Lyalikova, 1963; Silverman, 1963; Ponsford, 1966; Dugan & Apel, 1978). Mixed cultures (including *T. ferrooxidans* and *T. thiooxidans*), particularly those from coal enrichments, were more effective than pure cultures in removing pyrite, and particle size was also an important factor in determining absolute leaching rates (Dugan & Apel, 1978) as had been shown previously for other sulphides (Torma *et al.*, 1972; Kelly, 1976). The work of Capes *et al.* (1973) showed that after the initiation of pyrite oxidation by thiobacilli on the coal surface, pyrite particles could be separated from the coal by flotation. Coal treatment could thus employ a variety of methods for the separation of sulphur.

Also in its infancy is the use of contact leaching to remove valuable metals from industrial wastes such as slags, ashes and the 'red mud' by-product of alumina production (Ebner, 1977, 1978; Szolnoki, 1977; Zajic, Jack & Sullivan, 1977). Such processes will depend on the availability of suitable ferrous or sulphidic bacterial substrates and the wastes may in fact be more amenable to remote leaching as discussed below.

(b) *Remote leaching systems* While direct attack by bacteria can dissolve minerals, it is well documented that the generation of ferric iron by bacterial oxidation is the 'key to the solubilization of metal sulfides and reduced uranium minerals' (C. L. Brierley, 1978a). The very wide range of minerals amenable to dissolution in acid ferric media (Dutrizac & MacDonald, 1974) includes uraninite (UO_2) which is not amenable to direct attack by acidophilic bacteria and whose leaching depends entirely on oxidative ($U^{IV} \rightarrow U^{VI}$) dissolution by chemical oxidants such as ferric iron, MnO_2, H_2O_2, chlorates or nitric acid. The role of bacteria in uranium leaching is thus exclusively the generation of ferric iron (i.e. equations (1) and (11)) for indirect leaching (equation (9)). This means that the two processes of ferric iron generation and uranium leaching can be physically separated from each other since the bacteria do not need to come into contact with the ore. In some cases uranium leaching by 'conventional' heap or vat systems using bacteria could be necessary if the oxidation of pyrite in the uranium ore (a common mixture) is necessary to generate the ferric iron. Advantages of separating the two stages include the fact that ferric leaching can be operated at elevated temperatures unfavourable for bacterial growth, and that the bacteria do not have to oxidize iron in the presence of uranium. This is significant, since uranyl ion is one of the few metals toxic at low concentrations, at least to *T. ferrooxidans* (Tuovinen & Kelly, 1974a, b, c), although tolerance can be developed by some strains (Tuovinen & Kelly, 1972, 1974a). *T. ferrooxidans* and other mineral leaching acidophiles are in general remarkably tolerant of metals (Tuovinen, Niemela & Gyllenberg, 1971; Tuovinen & Kelly, 1972; C. L. Brierley, 1978a), but *T. ferrooxidans* is inhibited also by low concentrations of silver (Hoffman & Hendrix, 1976; Norris & Kelly, 1978a), mercury and gold (C. L. Brierley, 1978a; Norris & Kelly, 1978a), thallium and rubidium

(Tuovinen & Kelly, 1974c), and some anions (C. L. Brierley, 1978a), particularly molybdate (Tuovinen & Kelly, 1972). Molybdate is also toxic to *Thiobacillus* TH1 (Brierley *et al.*, 1978) but not to the extremely thermophilic *Sulfolobus*-like organisms (Brierley, 1974). Sayão Lobato (1974b) reported *T. ferrooxidans* from mine waters, contacting a uranium mineral (0.125% U_3O_8) containing 0.53% (w/w) MoO_3, to be tolerant of molybdenum, although the level of resistance was not quoted. These metal sensitivities indicate that, in any process involving uranium, mercury, molybdenum or silver, a two-stage system with separate bacterial generation of ferric iron would be advisable if a feasible leaching operation is to be operated.

Successful uranium extraction in two-stage systems has been reported (Kelly, 1976; Derry *et al.*, 1976; C. L. Brierley, 1978a) and is currently being operated on a production scale in Canada (Gow *et al.*, 1971; Kelly, 1977). The further development of such two-stage processes is obviously the application of bacterial ferric iron generation to the leaching of minerals *in situ*. The ultimate application is underground solution mining, in which the acid ferric lixiviant is pumped into the ore material (which must either be porous and supported on an impervious bed-rock layer or previously shattered with explosives to allow liquor percolation) and withdrawn to the surface after reaction and solubilization of metals by the ferric iron. Mining *in situ* is currently receiving greater attention because the method increases the feasibility of extracting deeper and lower-value mineral reserves economically and, of course, has less adverse environmental impact than conventional mining. For uranium, deep solution mining could be effected solely with acid ferric solutions generated in oxidation vats or fermenters on the surface (see Derry *et al.*, 1976; Livesey-Goldblatt, Tunley & Nagu, 1977) and would not require underground aeration. For cheap copper extraction, deep solution mining of a disintegrated ore body could be effected in aqueous sulphur–acid systems with oxygen introduced (under hydrostatic pressure) into the system, thus allowing direct bacterial attack underground as well as ferric leaching. In such cases, further studies would be needed on the tolerance of the organisms (which would probably have to be externally introduced to the newly shattered ore) to pressure and increased dissolved oxygen concentrations. *T. ferrooxidans* is apparently barotolerant (Torma, 1972; C. L. Brierley, 1978a) but chalcopyrite oxidation by it was slightly affected after exposure to 6–8 atm for 2 h (K. Bosecker, A. E. Torma & J. A. Brierley, personal communication). Deep solution-mining employing underground bacterial activity would probably have to resort to the use of thermophiles, as temperatures might be expected to rise considerably once ferric leaching was under way.

Potential for genetic improvement and strain selection in applied leaching systems

Little work has yet been done to seek improvement of practical leaching systems by genetic manipulation of the organisms. In general the leaching systems currently operating are doing so with the natural population of organisms that developed on the sulphidic ores available. As previously pointed out (Tuovinen & Kelly, 1972), the natural leaching environment will itself be the most potent selective system in the production of strains best suited to the prevailing conditions. Introducing a genetically 'marked' strain of *T. ferrooxidans* did not, in fact, increase copper leaching, although repeated inoculation of the dumps with the strain over a two-year period eventually led to 55% of the detectable 'ferrobacilli' exhibiting the genetic marker (Groudev, Genchev & Gaidarjiev, 1978). In deep solution-mining of virgin disintegrated ores, however, it could prove necessary to 'seed' with suitable strains if underground bacterial activity is desired. Similarly, rapid through-put fermenters or vats for ferric iron regeneration will probably ultimately need pure cultures of iron-bacteria for guaranteed operational characteristics. In the latter case, selection for decreased sensitivity to ferric iron (Kelly & Jones, 1978) could be desirable, although no such selection has yet been observed in a chemostat system (M. Eccleston and D. P. Kelly, unpublished work).

Genetic selection for increased leaching rates for specific minerals and greater tolerance to toxic metals would also seem desirable aims but it is obvious that in our present state of poor knowledge of the actual reactions between bacteria and minerals, it is not clear what kind of selection methods, other than empirical testing, should be employed. There is a complete lack of 'classical' genetic information about thiobacilli and, of course, even fewer basic data have been accumulated about other potentially highly important leaching organisms, such as *Leptospirillum, Sulfolobus* and *T. organoparus*. If it can be found that acidophilic thiobacilli carry resistance factors for metals on transmissible plasmids, as occurs in heterotrophs (Summers & Silver, 1978), then the potential for strain improvement by selective development of multiple metal tolerance could possibly be beneficial. Among problems in working on the genetics of such organisms are the small growth yields available on ferrous iron (Kelly, Eccleston & Jones, 1977), the difficulty of using solid substrates (e.g. pyrite) in genetic selection techniques such as replica plating and the difficulty, in some cases at least, even of obtaining growth of single colonies on solidified media (Tuovinen & Kelly, 1973; C. L. Brierley, 1978a). The slowness with which one can expect to advance in the field of genetics of leaching bacteria is partly indicative of the great amount of fundamental biochemical and physiological work that is still needed to understand the activity of the bacteria.

Leaching by other organisms

Thiobacilli and the metabolically similar mineral-degrading lithotrophs are the only organisms that have been obviously exploited industrially and whose activities are sufficiently great for their natural attack on minerals to have been recognized in the field. Heterotrophic organisms (bacteria and fungi) are, however, effective to some extent in releasing metals from various materials including copper–nickel concentrates (Le Roux, Wakerley & Perry, 1978), low-grade copper ore (Kiel, 1977), uranium from granites (Berthelin, Belgy & Magne, 1977), manganese ore (Agate & Deshpande, 1977) and potassium from leucite (Rossi, 1978). As outlined elsewhere (Tuovinen & Kelly, 1974a), leaching by heterotrophic organisms is due not to direct attack of any metabolic significance by the organisms on the minerals but entirely to chemical reaction of excreted microbial metabolites with the rock material. In many cases, a combination effect is important, for example when the organism secretes organic acids (e.g. glutamic, lactic, citric or glycollic acids) which may have the dual effect of increasing metal dissolution by lowering pH and increasing the load of soluble metal by complexing metals such as U or Al into soluble organic combinations. Heterotrophic leaching could occur in environments where pH falls (e.g. to pH 2–4) because of organic acid production (e.g. Le Roux et al., 1978); in environments of pH 6–9 with no acid production, if suitable complexing or 'leaching' by-products are excreted (e.g. Agate & Deshpande, 1977; Berthelin et al., 1977); or in environments where pH is significantly increased by the production of ammonia. Ammonium hydroxide is a well known complexing agent for some metals, including copper, cobalt and zinc, and could, for example, be used to solubilize cobalt minerals. Some bacteria (ammonifiers) produce large amounts of ammonia from the degradation of proteinaceous wastes and it has been suggested that their activities could be exploited in alkaline leaching.

In assessing the technical feasibility of leaching by heterotrophs, the basic question to be answered is 'what does it cost in material terms?'. In contrast to 'thiobacillus-type' acid-leaching (in which the organisms obtain carbon from CO_2, CO_3^{2-} or HCO_3^-, or other nutrients in inorganic form, and can obtain energy exclusively from the mineral-breakdown reactions), the heterotrophic bacteria and fungi require *organic* carbon for growth and energy-generation and do not derive any known benefit from the degradation of the minerals which occurs because of their presence. Consequently a large external supply of organic matter is needed. For example, Kiel's (1977) data showed effective chemical leaching of copper at pH 4 with 0.05 M lactic acid (releasing 81.25% of the copper from a 10% (w/w) suspension of a 1.2% (w/w) copper ore in 10 days). For a micro-organism to generate a 0.05 M solution of lactic acid (assuming the use of homofermentative bacteria, converting glucose quantitatively to lactate, while using about 5.5% of the substrate glucose for new bacterial growth) it would need to metabolize the equivalent of a 0.026 M solution of glucose, or about 4.7 g glucose l^{-1}. So to treat 10^8 t copper ore to effect 81% efficient leaching with lactic acid at pH 4, the microbial production of lactic acid from about 4.7 Mt glucose would be required. This is only feasible if a fermentable carbohydrate waste is available as a substrate. Similarly, to exploit ammonia-producing organisms, simply to generate a litre of a 1% (w/v) solution of NH_4OH would require the quantitative deamination of about 60 g protein. In industrial terms, a million gallons of 1% ammonia would thus come from 273 t of protein. Again it would seem economically realistic to do this only if a suitable nitrogenous waste were available and suitable organisms were chosen.

Uranium is frequently extracted from its 'alkaline' ores by carbonate–bicarbonate leaching carried out either in vat processes or heap leaching or by underground solution-mining *in situ*. It has been suggested (Zajic, 1969) that the oxidation of uranium oxides (e.g. UO_2 to UO_3) is thermodynamically sufficiently exergonic to be permissive of microbial growth if organisms exist capable of oxidizing such substrates, although as yet no such microbes have been conclusively isolated (B. Czeglédi, personal communication). A patent exists claiming leaching by thiobacilli of zinc–lead and copper sulphide ores under alkaline conditions (pH 7.1–9.0) with subsequent ammonia extraction of the metals (Mayling, 1969) but the biochemistry of this system has not yet been studied.

METAL RECOVERY: UPTAKE AND ACCUMULATION

Investigation of the possible use of organisms to remove useful or harmful metals from solution has been suggested, following discussion of future mineral resources (Cloud, 1968) and many observations of metal accumulation by organisms (Broda, 1972). As this area of study does not have the historical or practical background of bacterial leaching, we propose to review the main interactions of metals and micro-organisms which are the potential basis for microbial metal recovery schemes.

Mg^{2+} and K^+ are accumulated to high concentrations for the maintenance of normal cellular processes by all micro-organisms. Numerous other metals, including iron, zinc, manganese, copper, cobalt, nickel, molybdenum and vanadium are required by various organisms but, as pointed out by Trudinger (1976), the contribution of these metals to the biomass is normally very low despite their concentration, often with great enrichment factors, from the environment. However, such trace elements can be accumulated to abnormal concentrations from solutions which contain more of the metals than normally encountered in the environment and, under such conditions, significant quantities of generally toxic metals, such as cadmium, silver, lead and thallium without metabolic functions can also be accumulated. The accumulation of metals is considered in two sections – intracellular uptake and binding to organism surfaces, sections which reflect two phases which generally occur when metals and organisms interact. Finally, the removal of metals from solution by microbial activity which does not necessarily result in accumulation of metals by organisms is considered.

Intracellular uptake of metals

(a) *Uptake via divalent cation transport systems*   Micro-organisms have evolved highly specific 'micronutrient' transport systems which enable selective concentration of trace metals. *Alcaligenes eutrophus*, which requires nickel for chemolithotrophic growth, accumulates Ni^{2+} via a temperature- and energy-dependent process, with only slight inhibition of Ni^{2+} uptake by Zn^{2+}, Co^{2+}, Mn^{2+} or Cu^{2+} (Tabillion & Kaltwasser, 1977) but growth of the organism is inhibited by greater than 0.3 μM Ni. Highly specific temperature- and energy-dependent transport systems for Mn^{2+} accumulation have been found in *B. subtilis*, *R. capsulata* and *E. coli* (Silver & Jasper, 1977). In *E. coli*, high-affinity Mn^{2+} uptake (K_m, 0.2 μM) was unaffected by a 100 000-fold excess of Mg^{2+} or Ca^{2+}. Fe^{2+} and Co^{2+} appeared to be competitive substrates but only with K_i values 100 times the K_m for Mn^{2+} uptake. The system did not transport Ni^{2+}, Cu^{2+} or Zn^{2+}. However, Silver *et al.* (1972) calculated that the high-affinity Mn^{2+} uptake system had the major role in Mn^{2+} uptake only when the external Mn^{2+} concentration was less than 40 μM. Above this concentration, Mn^{2+} uptake would proceed via a less specific Mg^{2+} transport system. Therefore, it appears that the specific micronutrient pathways would be of little significance in accumulation of metals by bacteria from solutions containing economically significant quantities of most metals.

The magnesium concentration of bacteria is generally between 15 and 35 mmole per kg wet weight even when the external Mg^{2+} concentration varies by a factor of 100 000 (Jasper & Silver, 1977). This magnesium is accumulated via temperature- and energy-dependent transport systems by various micro-organisms. We do not propose to describe in detail the studies which have led to the present understanding of the uptake systems but refer the reader to excellent recent reviews of the transport of magnesium, manganese, zinc, calcium and iron (Weinberg, 1977). These transport systems will be considered only in so far as they are involved in uptake of some potentially toxic or economically important metals. Although Mg^{2+} transport by $E.$ $coli$ is specific, in that Ca^{2+}, Sr^{2+} or K^+ do not inhibit Mg^{2+} uptake, there is evidence that Mn^{2+}, Ni^{2+}, Co^{2+} and perhaps Zn^{2+} may be accumulated by the same transport system (see Jasper & Silver, 1977). First, the K_m for Mg^{2+} uptake is similar to the K_i for Mg^{2+} in its inhibition of Co^{2+} uptake, while the K_m for Co^{2+} uptake is similar to the K_i for Co^{2+} in its inhibition of Mg^{2+} uptake. Secondly, the influx and efflux of Mg^{2+} appear to be coupled, and Mn^{2+} and Co^{2+}, as well as extracellular Mg^{2+}, can promote Mg^{2+} efflux. Thirdly, Co^{2+}-resistant mutants are deficient in Co^{2+}, Ni^{2+} and Mn^{2+} uptake. Mg^{2+} uptake by $E.$ $coli$ occurs by at least two distinct systems: a high-affinity system, which also transports Co^{2+}, Mn^{2+} and Ni^{2+}, and a system which is repressed during growth at high Mg^{2+} concentrations and for which Co^{2+} is possibly an inhibitor but not a substrate (see Jasper & Silver, 1977). Thus, the first system may be the principal mechanism of Co^{2+}, Ni^{2+} and Mn^{2+} uptake from high solution concentrations of the metals, but whether this is true for all strains or mutants of the organism and for all experimental conditions is not yet clear. Fujiwara et $al.$ (1977) have shown that a Co^{2+}-resistant $E.$ $coli$ was resistant as a result of reduced uptake of Co^{2+} but retained sensitivity to Ni^{2+} and showed, in Mg^{2+}-deficient conditions, Ni^{2+} uptake equivalent to that of the sensitive parent strain. Bucheder & Broda (1974) have described Zn^{2+} uptake by $E.$ $coli$ as a temperature- and energy-dependent process. Cd^{2+} at 0.1 mM inhibited Zn^{2+} uptake from a 10 μM solution by 50%, indicating considerable selectivity for Zn^{2+} against Cd^{2+}, but interaction with the Mg^{2+} transport system was not reported and, in view of the low Zn^{2+} concentrations, it is not clear if a specific micronutrient pathway or Mg^{2+} transport systems were involved. Webb (1970) has concluded that Ni^{2+}, Co^{2+} and Zn^{2+} were probably transported by the same system in *E. coli*, *Klebsiella aerogenes* and *B. megaterium*. Co^{2+}, Mn^{2+} and Ni^{2+} are also accumulated by *B. subtilis* via a citrate-inducible, energy-dependent system which is primarily concerned with Mg^{2+} uptake as a citrate-divalent cation complex (Willecke, Gries & Oehr, 1973). The active uptake systems for Mg^{2+} in this organism and in another Gram-positive organism, *Staphylococcus aureus*, appear to interact differently with the transition metals and Zn^{2+} compared with the equivalent systems in *E. coli*. Mg^{2+} uptake in *B. subtilis* and *S. aureus* was inhibited by Zn^{2+}, Mn^{2+} and probably Co^{2+}, but only Mg^{2+} and Mn^{2+} promoted Mg^{2+} efflux (Jasper & Silver, 1977).

The extent to which these metals which may substitute for Mg^{2+} in uptake systems are accumulated is evident from some of the data obtained with Co^{2+} and *E. coli*. Fujiwara *et al.* (1977) demonstrated that Co^{2+} uptake from 50 mg Co^{2+} l^{-1} reached the equivalent of 0.3% of the organism dry weight. *E. coli* (0.5 mg dry weight ml^{-1}) accumulated Co^{2+} from 10 mg Co^{2+} l^{-1} during 2 h at 25 °C in the presence of glucose to approximately 0.5% of the organism dry weight (P. R. Norris, unpublished) and Nelson & Kennedy (1971) found the maximum level of Co^{2+} uptake by *E. coli* was equivalent to 2.3% by weight of cell protein.

The divalent cation transport systems of micro-organisms other than bacteria have been less well studied. Zn^{2+} uptake by a filamentous fungus, *Neocosmospora vasinfecta*, was inhibited by Mg^{2+}, except at high Zn^{2+} concentrations where the Zn^{2+} uptake rate was increased, and was competitively inhibited by Mn^{2+}. There was some inhibition of Zn^{2+} uptake by Co^{2+}, Ni^{2+} and Cu^{2+} (Paton & Budd, 1972). Energy-dependent uptake of Zn^{2+} by the green alga, *Chlorella*, was inhibited by Ca^{2+} (Matzku & Broda, 1970) whereas the inhibition of Mg^{2+} uptake in *Euglena gracilis* by Co^{2+} and Mn^{2+} (see Jasper & Silver, 1977) and competitive inhibition of Co^{2+} uptake by Mg^{2+} in *Neurospora crassa* (Venkateswerlu & Sastry, 1970) suggest at least some interactions familiar from the studies with bacteria. As with bacteria, *Saccharomyces cerevisiae* possesses a Mg^{2+} transport system for which Co^{2+}, Ni^{2+}, Mn^{2+} and Zn^{2+} are probably competitive substrates (Fuhrmann & Rothstein, 1968). Co^{2+} uptake by *S. cerevisiae* was inhibited most strongly by cations with similar ionic radii to Co^{2+} and decreased across the series $Zn^{2+} > Ni^{2+} > Mg^{2+} > Mn^{2+} > Cd^{2+} > Ca^{2+}$ (Norris & Kelly, 1977). Similarity of cation size also appeared to influence the extent of inhibition by other cations of Cd^{2+} uptake by *S. cerevisiae* (Norris & Kelly, 1977) and of Ca^{2+} uptake by *Schizosaccharomyces pombe* (Boutry, Foury & Goffeau, 1977). However, this pattern is not followed with all combinations of metal cations or with all yeasts. For example, Co^{2+} uptake by *S. cerevisiae* was most strongly inhibited by Zn^{2+} (Norris & Kelly, 1977) but the rate of Ni^{2+} uptake by the same strain and by another strain of *S. cerevisiae* was only slightly inhibited by Zn^{2+} although Ni^{2+}, Co^{2+} and Zn^{2+} are all of similar ionic radii and were taken up at similar rates from solution in the absence of other cations (Norris & Kelly, 1979). Furthermore, Ni^{2+} uptake by non-growing suspensions of *Saccharomycopsis* (*Candida*) *lipolytica* was greater than that of Co^{2+} and was strongly inhibited by Mn^{2+} and Zn^{2+} but less by Co^{2+} or Mg^{2+} (Norris & Kelly, 1979). Zn^{2+} uptake was not strongly inhibited by Ni^{2+}, possibly indicating that Zn^{2+} shows the highest affinity for the cation transport system. A highly specific Zn^{2+} uptake system has been described in *Candida utilis* in which Zn^{2+} uptake was inhibited by Cd^{2+} but not by Ca^{2+}, Cr^{3+}, Mn^{2+}, Co^{2+} or Cu^{2+} (Failla, Benedict & Weinberg, 1976). This specificity, and the low Zn^{2+} concentration (1.1 μM) used in the study, suggest that the uptake process possibly involved a micronutrient transport system of similar function to those responsible for trace element accumulation in the bacteria. Zn^{2+} uptake from higher solution concentrations of the metal (200 μM) by a different strain of *C. utilis* was reduced to a similar extent by equimolar concentrations of Ni^{2+}, Mg^{2+} and Cd^{2+} (P. R. Norris, unpublished). Clearly, more work is required to establish the number and specificity of divalent metal cation uptake systems in yeasts. The extent of metal accumulation appears to be slightly higher than with bacteria when the organisms are compared on an equal dry weight basis. For example, Zn^{2+} uptake by *C. utilis* reached 0.7–0.85% of the organism's dry weight (Failla, 1977) and Co^{2+} uptake during 1 h at 25 °C in the presence of glucose and 0.5 mM $CoSO_4$ by *S. cerevisiae* (0.6 mg dry weight ml^{-1}) reached 1.5% of the organism dry weight (P. R. Norris, unpublished). Higher concentrations of accumulated Mn^{2+} (Okorokov, *et al.*, 1977) and Mg^{2+} (see Jasper & Silver, 1977) have been reported in the vacuoles than in the cytoplasm of some fungi.

(*b*) *Uptake via monovalent cation transport systems* The high intracellular concentration of K^+ as well as Mg^{2+} in micro-organisms has encouraged many detailed studies of monovalent cation uptake (Armstrong, 1972; Harold & Altendorf, 1974). The complexity and organisation of K^+ accumulation is shown by the presence of four kinetically and genetically distinct K^+ transport systems in *E. coli* (Rhoads, Waters & Epstein, 1976). Other alkali metal ions can be competitive substrates for the K^+ transport systems, with relative affinities apparently determined by the similarity of the radius of the ion to that of K^+. Thus an

affinity series for uptake by *S. cervisiae* has been given as $K^+ > Rb^+$
$> Cs^+ > Na^+ > Li^+$ (Armstrong & Rothstein, 1964) and the characteristics of the uptake of Tl^+ by *S. cerevisiae* suggest that it should be
included close to K^+ in the affinity series (Norris *et al.*, 1976). Tl^+ uptake also occurred, probably via K^+ transport systems, in *E. coli*
(Norris *et al.*, 1976) and *Chlorella* (Solt, Paschinger & Broda, 1971).
Rubidium, accumulated via K^+ and, probably, NH_4^+ uptake systems,
substituted for K^+ in supporting growth of *C. utilis* and, in NH^+_4-
limited chemostat culture, accounted for almost 8% of the organism
dry weight (Aiking & Tempest, 1977).

(*c*) *Uptake via undefined processes*   There are many reports of metal
cation uptake by organisms involving toxic metals which appear to
enter cells other than via recognized cation transport systems. Temperature-dependent Cd^{2+} uptake by *Staphylococcus aureus* from 1 mM
$CdCl_2$ reached about 1% of the organism dry weight (Tynecka, Zajac
& Gos, 1975). With lower Cd^{2+} concentrations, a plasmid-bearing
strain of the organism did not accumulate and was resistant to the metal
compared with a Cd^{2+}-accumulating plasmid-negative strain. Cd^{2+} is
accumulated by energy-dependent uptake in *S. cerevisiae* (Heldwein,
Tromballa & Broda, 1977; Norris & Kelly, 1977) but it is not known
whether its tendency to form similar complexes to Zn^{2+} or its similar
ionic size to Ca^{2+} enables substitution for these cations in transport
systems. Pb^{2+} uptake by *S. cerevisiae* can reach a similar level to that of
Co^{2+} and Cd^{2+} but does not require an energy source (Heldwein *et al.*,
1977). Cd^{2+}, Pb^{2+}, Ag^+, Hg^{2+} and some other heavy metals can affect
yeast cell membrane permeability, with Cu^{2+}, for example, causing loss
of K^+ and Mg^{2+} from a variety of yeasts (Norris & Kelly, 1979).
Similar loss of K^+ and Mg^{2+} from *Thiobacillus ferrooxidans* has been
observed concurrently with Ag^+ uptake by the organism at pH 1.5
under conditions in which most metals, including Cu^{2+}, are not accumulated (Norris & Kelly, 1978*a*). With heavy metals that bind to cell
membranes with high affinity, it seems possible that access to and
binding at intracellular sites could follow alteration of membrane permeability rather than utilization of cation transport systems. However,
as with Hg^{2+} and *S. cerevisiae* (Murray & Kidby, 1975), greater amounts
of these metals may be retained bound to organism surface components
than accumulated intracellularly.

(*d*) *Uptake via anion transport pathways*   The existence of some metals
in oxoanion configurations in aqueous solution indicates some potential

for uptake via microbial anion instead of cation transport systems. Energy-dependent uptake of chromate (CrO_4^{2-}) in *Neurospora crassa* has been described as via the sulphate (SO_4^{2-}) permease systems. Chromate was a competitive inhibitor of sulphate uptake with a K_i (3×10^{-5} M) for inhibition of sulphate uptake similar to the K_m (4×10^{-5} M) for chromate uptake. Tungstate (WO_4^{2-}) and molybdate (MoO_4^{2-}) showed weaker inhibition of sulphate uptake (Roberts & Marzluf, 1971). The binding and transport of sulphate by *Salmonella typhimurium* was inhibited by anions structurally similar to the tetrahedral sulphate ion with an inhibition sequence of $CrO_4^{2-} \gg SeO_4^{2-} > MoO_4^{2-} > WO_4^{2-} > VO_4^{2-}$ (Pardee *et al.*, 1966). Vanadate inhibited binding of sulphate more strongly than its subsequent transport, so the degree to which these oxoanions may be competitive substrates for sulphate transport systems remains in doubt. Springer & Huber (1973) reported that selenate uptake by *E. coli* occurred by diffusion rather than by active uptake and selenate-tolerant strains incorporated selenium less readily into cell proteins and produced a sulphate permease resistant to selenate inhibition. The extent to which organisms can accumulate such oxoanions is not known, and the site of their accumulation may vary. Tellurium accumulated by *Streptococcus faecum* is predominantly intracellular (Tucker *et al.*, 1966) and selenium granules have been observed in the cytoplasm of *E. coli* and other bacteria (Silverberg, Wong & Chau, 1976) but these metals may also be deposited at cell surfaces (see below).

Organism surface-binding of metals

(a) Cation adsorption In contrast to the intracellular uptake of metal cations, their adsorption to negatively charged sites at organism surfaces is typically rapid, reversible and independent of temperature and energy metabolism. The anionic ligands responsible for cation adsorption include phosphoryl, carboxyl, sulphydryl and hydroxyl groups of membrane proteins and lipids and of cell wall structural components such as bacterial peptidoglycans and associated polymers. The latter includes peptides and teichoic acids of Gram-positive bacteria. The electrophoretic behaviour of organisms at different pH values has indicated that dissociation of phosphodiester groups probably gives rise to the negative charge at the surface of *Saccharomyces cerevisiae* even at low pH, and that carboxyl groups appear to be the main source of anionic sites at the surface of *Escherichia coli*, *Micrococcus lysodeikticus* and, together with some phosphoryl groups, *Bacillus megaterium* (Neihof & Echols, 1973).

The cation-binding behaviour of Gram-positive bacteria has been likened to that of commercial carboxylic ion-exchange resins with an affinity series of $H^+ \gg La^{3+} \gg Cd^{2+} > Sr^{2+} > Ca^{2+} > Mg^{2+} > K^+ > Na^+$ shown by *B. megaterium, M. lysodeikticus* and *Streptococcus mutans* (Marquis, Mayzel & Cartensen, 1976). Cell wall fragments of *B. subtilis* also bound a variety of metals but showed extreme binding of a few, with retention of 200 μg of Mg^{2+}, Fe^{3+} and Cu^{2+} per mg dry weight of wall fragments from 5 mM metal solutions (Beveridge & Murray, 1976). Substantial amounts of Na^+ and K^+, intermediate amounts of Mn^{2+}, Zn^{2+}, Ca^{2+}, Au^{3+} and Ni^{2+} and small amounts of Hg^{2+}, Pb^{2+}, Sr^{2+} and Ag^+ were also retained, whereas there was no adsorption of Li^+, Ba^{2+}, Co^{2+} or Al^{3+}. Partial lysozyme digestion of the walls greatly reduced Mg^{2+} retention but not that of Ca^{2+}, Fe^{3+} or Ni^{2+}, indicating diverse mechanisms or sites of metal binding and retention. The difference in binding of Co^{2+} and Ni^{2+} by the wall fragments has not been seen with whole cells of *B. megaterium* which bound equivalent amounts of cations of the two metals (Norris, unpublished). The screening of a range of *Bacillus* species for retention of Co^{2+} and Cd^{2+} at the organism surface has shown two main groups with relatively high (*B. megaterium* KM, *B. globigii* and *B. macerans*) or low (*B. cereus, B. magaterium, B. polymixa* and *B. cereus* var. *mycoides*) cation binding capacity, while a different strain of *B. subtilis* retained intermediate amounts of the cations (Norris & Kelly, 1979). The physicochemical basis for the observed selectivity and tenacity of binding of metals to the *B. subtilis* wall fragments (Beveridge & Murray, 1976) and for the diverse extent of metal binding among the *Bacillus* species is unknown and may be complex. Competition between pairs of metal cations for binding to the *E. coli* surface has indicated an affinity sequence of $Zn^{2+} = Cd^{2+} > Mn^{2+} = Co^{2+} = Ni^{2+} > Mg^{2+} = Ca^{2+}$ (Norris & Kelly, 1979). The same affinity grouping of the five preferred metals has been described with their passive binding by corn mitochondria (Bitell, Koeppe & Miller, 1974). Cd^{2+} and Zn^{2+} again had a higher affinity for binding than that shared by Co^{2+}, Mn^{2+} and Ni^{2+}, while all five cations were apparently bound to the same sites. As with bacteria, the yeast surface can behave as an ion-exchange resin with rapid, reversible binding of cations and a particularly high affinity for uranium ions (Rothstein & Hayes, 1956). This rapid binding of uranium by *S. cerevisiae*, which is also shown by other organisms including *Pseudomonas aeruginosa* and *Penicillium*, is the basis of tests currently investigating the potential of micro-organisms for the recovery of heavy metals from solution (Shumate & Strandberg, 1978). Experimental details are not available, but initial reports indicated that 100 ppm uranium was completely removed from solution in 10 min by *Pseudomonas aeruginosa*. Rothstein & Hayes (1956) noted that the total cation adsorption capacity of the yeast surface was sufficient to bind the equivalent of less than 2% of the total organism cation content. Therefore, it appears that retention of large quantities of metals which have been observed at organism surfaces probably entails entrapment of metals beyond relatively non-specific adsorption to anionic sites and could involve interaction with specific components of organism surface layers or the deposition of insoluble metal forms. Examples of the former of these possibilities includes the binding of mercury to cell walls of *Saccharomyces cerevisiae* (Murray & Kidby, 1975). The walls were capable of binding their own weight of Hg^{2+} to high affinity sites. Although there was some metal retention which could not be correlated with the available wall protein or phosphate binding sites, a protein associated with wall glucan was recognized as being responsible for the binding of a majority of the bound mercury. Similarly, an alkali-soluble, carbohydrate fraction of dried alga, *Hormidium fluitans*, which was harvested from acid mine-water in which it was growing, comprised 1.5% of cell polymers; it bound copper with high affinity as 11.8% of its dry weight (Madgwick & Ralph, 1977). The extracted polymer bound 2.45 mg Cu per mg carbohydrate when added to natural mine water (pH 3.5–5.3 containing 32–719 mg Cu per litre).

(b) *Metal deposition/precipitation at surfaces of organisms* The deposition of insoluble metal compounds or elements at organism surfaces has been observed with various organisms, sometimes after microbial transformation of the metals. Examples of the intracellular deposition of tellurium and selenium were given earlier but there is also evidence for the deposition of elemental selenium at the cell membrane and wall of *E. coli* (Gerrard, Telford & Williams, 1974). Au^{3+} and Fe^{3+} were initially bound at discrete sites of *B. subtilis* wall fragments, but aggregation of hydrated iron oxides and crystallization of elemental gold also occurred (Beveridge & Murray, 1976). The immobilization of large quantities of lead – for example, 490 mg per gram organism dry weight of *Micrococcus luteus* – resulted from aggregation of insoluble lead compounds (Tornabene & Edwards, 1972). More than 99% of the lead was retained at the organism surface by *M. luteus* and *Azotobacter* sp. Fractionation of cells indicated 89.8% and 9.5% of the lead bound to *M. luteus* and 61.5% and 37.6% of the lead bound to *Azotobacter* was retained at the cell membranes and walls respectively. Further study of the interaction of lead with *M. luteus* showed changes in membrane lipids of lead-exposed organisms. Individual lipids did not provide specific stable binding-sites for lead, but it was suggested that natural membrane lipid mixtures might have provided an environment suitable for nucleation of lead through a general complexing phenomenon based on strong affinity of Pb^{2+} to any anionic groups in combination with electronic and steric factors (Tornabene & Peterson, 1975). 'Massive' precipitation of metals at organism surfaces is, however, most often seen in Nature with iron and manganese, both being metals which form some insoluble compounds during microbial transformation.

Micro-organisms have been implicated, although with uncertain roles, in the formation of ocean-floor ferromanganese nodules (see Trudinger, 1976), a potential source of minerals which is of current economic interest. Schütt & Ottow (1977) have described the microflora of the nodules as quantitatively and qualitatively a continuation of the indigenous sea-floor flora. They suggested that the mineral accretion initiates at sites of active bacterial mineralization, such as clays and remnants of bones and teeth, and that it may involve the metabolism of organo-metal complexes. Various bacteria, fungi and algae promote Mn^{2+} oxidation in soil and freshwater, and the sheathed, filamentous bacteria of the *Sphaerotilus–Leptothrix* group and polymorphic bacteria such as *Hyphomicrobium* and *Metallogenium* can become visibly encrusted with manganic oxides. *Hyphomicrobium* is perhaps the predominant organism involved in the deposition of iron and, particularly, manganese in freshwater pipelines throughout the world (Tyler, 1970; Trudinger, 1976) and *Metallogenium* has been implicated in the formation of lacustrine, manganese ore formation (Dubinina, 1973).

The *Sphaerotilus–Leptothrix* group and *Gallionella* are representatives of the 'iron-bacteria' (Cullimore & McCann, 1977). *Gallionella* is probably the only such organism capable of growth using energy from Fe^{2+} oxidation, oxidized iron compounds being deposited particularly in the organism's long, twisted stalks. There is no evidence that the deposition of oxidized iron compounds in the organic sheaths of the *Sphaerotilus–Leptothrix* group is the result of the direct metabolism of iron. The growth and final protein yield of *Sphaerotilus discophorus* was found to be independent of the iron concentration of the medium but this concentration probably determined the amount of iron deposited and this deposition was apparently mediated by constituents of the sheath (Rogers & Anderson, 1976). The concentration of various metals from solution in masses of *Sphaerotilus* in a river at effluent discharge points has been observed and implicated in the inadequate dispersal of effluent metals, with consequent passage of heavy metals through food-chain organisms (see Patrick & Loutit, 1976).

The extracellular polysaccharide matrix produced by another typical sewage organism, *Zoogloea*, has also been shown to contribute to the removal of Cu^{2+}, Co^{2+}, Ni^{2+}, Zn^{2+} and Fe^{3+} from solution (Friedman & Dugan, 1967). The authors suggested the potential application of the organism in waste-water treatment and showed that efficiency of metal removal varied with different strains of bacteria: one strain accumulated 34 mg of copper or 25 mg of cobalt per 100 mg organism dry weight.

Although the accumulation of large quantities of iron and some other metals is usually, therefore, associated with the slime-forming or 'iron-bacteria', many others, including species of *Pseudomonas*, *Mycobacterium*, *Escherichia*, *Corynebacterium* and *Caulobacter*, can become heavily encrusted with iron after adsorption of colloidal ferric iron (Macrae & Edwards, 1972) and some strains of *Pseudomonas* and *Moraxella* cause precipitation of iron after degrading the organic fraction of soluble organic-iron compounds (e.g. ferric gallate) which may occur in some surface waters (Macrae, Edwards & Davis, 1973).

From the foregoing examples, it emerges that it is often difficult to distinguish between the involvement of microbial metabolism (e.g. metal oxidations), specific surface properties of organisms and purely physical factors in removal of metals from solution with micro-organisms. However, organisms, particularly those forming intricate, filamentous networks or a matrix of extracellular material, do provide a basis on which metal cations, precipitates or colloids aggregate. A typical application of such factors is provided by the use of algae to remove metals from streams in the world's largest lead-mining district in Missouri. Large quantities of algae, stimulated by mine and mill effluents, were found to trap heavy-metal colloids, concentrating them thousands of times. The use of shallow lagoons of algae, followed by sedimentation, has greatly reduced undesirable amounts of algae and, with seasonal effectiveness, greatly reduced the levels of heavy metals in the streams (Jennett *et al.*, 1975).

Metal removal from solution via microbial activity

The immobilization of heavy metals from an industrial effluent by their incorporation as organo-metal complexes into sediments has been reported as a satisfactory practical application after encouragement of bacterial growth with plant material (Ilyaletdinov, Enker & Yakubovskii, 1976). When the activity of the aerobic, heterotrophic organisms caused a deficiency of oxygen, copper was efficiently removed from solution as the sulphide after generation of H_2S by sulphate-reducing bacteria (Ilyaletdinov, Enker & Loginova, 1977).

(a) *Precipitation of metal sulphides* The formation of metal sulphides in cultures of dissimilatory sulphate-reducing bacteria has been studied on the laboratory scale since Miller (1950) reported that sulphides of many metals were formed as a result of the activity of *Desulfovibrio desulfuricans*. These studies, and the probable participation of such organisms in the formation of sulphide deposits containing economic quantities of metals, have been reviewed by Trudinger (1976). The potential application of bacterial H_2S generation for metal recovery has also been studied and a process has been proposed for the reduction of high acidity, sulphate and iron concentrations in coal-mine drainage (Tuttle, Dugan & Randles, 1969).

The proposed process included the degradation of wood dust by mixed cellulolytic bacteria to provide the carbon and energy source for the sulphate-reducers. Metabolism of the organisms, with wood dust enriched with 0.1% sodium lactate as the supplied organic material, resulted in the pH increasing from 3.6 to 7 during a ten-day period. Recovery of metals produced by bacterial leaching of a copper sulphide ore has also been demonstrated on a laboratory scale (Tomizuka & Yagisawa, 1978). Maximum rates of metal recovery were 47.5, 18.1 and 25.7 mg 1^{-1} h^{-1} for copper, zinc and iron respectively from a diluted leach solution containing 108, 41 and 54 mg 1^{-1} of copper, zinc and iron respectively, using a continuous culture of sulphate-reducing bacteria at pH 6.0 with sodium lactate as an energy source. A precipitate containing 20%, 6% and 11% copper, zinc and iron respectively was obtained, which represented concentrations 6.9, 38 and 1 times those of the original copper sulphide ore. As demonstrated by Vosjan & Van der Hoek (1972) with removal of mercury and copper in a similar process, growth of the sulphate-reducers and metal recovery requires continuous culture systems in which the metals are precipitated on entering the culture vessel before deleterious concentrations of the toxic metals in solution are reached.

Sulphide production by sulphate-reducing bacteria has also been used for the enrichment of oxidized antimony and lead ores; in this process the recovery of the metals was 5–10% higher than extraction by a standard flotation method using sodium sulphide (Lyalikova, Lyubavina & Solozhenkin, 1977). Prolonged treatment of some minerals with the bacteria and H_2S resulted in some selective depression of flotation characteristics with a slight depression of molybdenite and great depression of chalcopyrite flotation, thus helping to separate a mixed copper–molybdenum concentrate.

As with the precipitations of manganese, iron and lead at organism surfaces which were described earlier, the sulphide precipitations may, at least initially, involve some metal accumulation on the organisms. Precipitated amorphous FeS formed by the action of sulphate-reducing bacteria has been found in direct contact with the cells (Hallberg, 1970), suggesting a possible surface-active process with the organisms, but it must be noted that bacteria also attach to solid metal sulphides and precipitated ferric compounds (McGoran, Duncan & Walden, 1969).

(b) *Precipitation of ferric compounds* Precipitation of ferric compounds on organisms and/or attachment of organisms to the compounds (noted above) could be involved in the removal of iron from acid mine-water in a process designed to utilize the rapid oxidation of ferrous iron by *Thiobacillus ferrooxidans* (Olem & Unz, 1977). Plastic discs on a central shaft were rotated half-immersed in flowing, Fe^{2+}-containing, coal-mine drainage and were colonized by bacteria, assumed to be *T. ferrooxidans*. Bacterial oxidation of the ferrous iron to the less soluble ferric iron greatly reduced the soluble iron levels in the water, with solids accumulating on the discs for five months during continuous operation before there was some displacement and carry-over of solids into the effluent. The authors suggested that the process appeared potentially useful as a first step in the treatment of acid mine-water which could save costs, for example, on any subsequent neutralization.

Miscellaneous microbe–metal reactions

In addition to the main uptake, binding and precipitation processes that remove metals from solution, there are several specific reactions which merit attention in a review of the possible use of organisms or their products in metal recovery. Three examples involving unrelated areas of metabolism serve to indicate the wide range of possibilities.

First, elemental mercury, formed by microbial reduction of mercury ions, and some methylated mercury, selenium and arsenic compounds formed through microbial transformations of the metal compounds (see Konetzka, 1977; Summers & Silver, 1978), may be volatilized from solution, offering some potential for detoxification of limited environments but disadvantageously involving some very toxic compounds and not immediately trapping the metals.

Secondly, the specific requirement of diatoms for silicon has allowed the separation of ^{32}Si isotope from contaminating ^{3}H and ^{60}Co by incorporation of ^{32}Si into diatom 'shells' and its subsequent recovery after ashing (Werner, Pawlitz & Roth, 1975). This silicon requirement also provides the basis of uptake of germanium as germanic acid, an analogue of the silicic acid form in which silicon is accumulated by diatoms (Azam, 1974).

Thirdly, many micro-organisms have high affinity systems for the acquisition of essential iron from their environment in which the metal is otherwise often unavailable for uptake. These systems involve excretion of iron-chelating compounds, generally hydroxamates or phenolates, and subsequent uptake of the ferric chelates by the organisms (see Byers & Arceneaux, 1977). One of these compounds, desferrioxamine produced by *Streptomyces* spp., has found therapeutic use in the treatment of transfusional iron overload (anon., 1978). These compounds have also been considered for the 'recovery' of actinides from animal tissues. Rhodotorulic acid, an iron-chelating compound identified in several yeasts, complexes plutonium *in vivo*, but exhibits no marked superiority to diethylenetriaminepentaacetic acid (the main agent used to stimulate actinide excretion) in removal of plutonium from test animals and does not appear as promising as chemically-synthesized lipophilic derivatives of polyaminocarboxylic acids (R. A. Bulman, personal communication). The production of a molybdenum-chelating peptide compound by *Bacillus thuringiensis* (Ketchum & Owens, 1975) indicates that metals other than iron, or those which replace it, may be specifically complexed by extracellular microbial metabolites.

OUTLOOK

The greater awareness of the effects of metal pollution in general, the prospect (with expanding nuclear power industries) of an increasing environmental burden of radioactive metals in particular, some approaching mineral shortages, the recognition of microbial roles in environmental transformations of metal compounds and the introduction of some metals into food chains should ensure continuing and expanding research into interactions of metals and micro-organisms. The assessment and development of these interactions as applied to metal extraction or recovery processes can be envisaged in two directions.

First, the optimization of natural examples of mineral metabolism. This includes, for example, leaching, metal sulphide precipitation and the potential of 'iron-bacteria' to remove iron and manganese from ground waters.

Secondly, novel processes for the accumulation of metals from solution. Although the maximum usefulness of organisms in metal accumulation has probably not yet been demonstrated, the examples reviewed in this article indicate that the extent and selectivity of uptake of most metals is unlikely to yield microbial processes which can compete economically with physical and chemical treatments such as reverse osmosis and solvent extraction. However, the value of any metal accumulation/recovery process must be judged against the importance (financial or toxicological) of a particular metal. With this in mind, the capacity of organisms to accumulate some valuable and toxic ions, such as uranium, should be investigated further.

Acknowledgements Some of the original work quoted in this article has been supported in part by Imperial Chemical Industries Ltd, the Natural Environment Research Council and the International Atomic Energy Agency.

REFERENCES

AGATE, A. D. & DESHPANDE, H. A. (1977). Leaching of manganese ores using *Arthrobacter* species. In *Conference Bacterial Leaching*, ed. W. Schwartz, pp. 243–50. GBF Monograph No. 4. Weinheim, New York: Verlag Chemie.

AGATE, A. D., KORCZINSKI, M. S. & LUNDGREN, D. G. (1969). Extracellular complex from the culture filtrate of *Ferrobacillus ferrooxidans*. *Canadian Journal of Microbiology*, **15**, 259–64.

AIKING, H. & TEMPEST, D. W. (1977). Rubidium as a probe for function and transport of potassium in the yeast *Candida utilis* NCYC 321, grown in chemostat culture. *Archives of Microbiology*, **115**, 215–21.

ANON. (1978). Desferrioxamine and transfusional iron overload. *Lancet* (i) 479.

ARMSTRONG, W. McD. (1972). Ion transport and related phenomena in yeast and other micro-organisms. In *Transport and Accumulation in Biological Systems*, ed. E. J. Harris, pp. 407–45. London: Butterworth.

ARMSTRONG, W. McD. & ROTHSTEIN, A. (1964). Discrimination between alkali metal cations by yeast. I. Effect of pH on uptake. *Journal of General Physiology*, **48**, 61–71.

AZAM, F. (1974). Silicic-acid uptake in diatoms studied with (^{68}Ge) germanium acid as tracer. *Planta*, **121**, 205–12.

BALASHOVA, V. V., VEDENINA, I. Ya., MARKOSYAN, G. E. & ZAVARZIN, G. A. (1974). The auxotrophic growth of *Leptospirillum ferrooxidans*. *Mikrobiologiya*, **43**, 581–5 (English translation pp. 491–4).

BARBIC, F. F. & KRAJINCANIC, B. V. (1972). The role of thionic-bacteria in leaching of uranium from processed ore. *Mikrobiologiya*, **41**, 346–8.

BARBIC, F. F., BRACILOVIC, D. M., KRAJINCANIC, B. V. & LUCIC, J. C. (1976). Bacterial leaching of waste uranium materials. *Zeitschrift für Allgemeine Mikrobiologie*, **16**, 179–86.

BECK, J. V. (1967). The role of bacteria in copper mining operations. *Biotechnology and Bioengineering*, **9**, 487–97.

BECK, J. V. (1977). Chalcocite oxidation by concentrated cell suspensions of *Thiobacillus ferrooxidans*. In *Conference Bacterial Leaching*, ed. W. Schwartz, pp. 119–28. GBF Monograph No. 4. Weinhem, New York: Verlag Chemie.

BENNETT, J. C. & TRIBUTSCH, H. (1978). Bacterial leaching patterns on pyrite crystal surfaces. *Journal of Bacteriology*, **134**, 310–17.

BERRY, V. K. & MURR, L. E. (1978). Direct observations of bacteria and quantitative studies of their catalytic role in the leaching of low-grade, copper-bearing waste. In *Metallurgical Applications of Bacterial Leaching and Related Microbiological Phenomena*, eds L. E. Murr, A. E. Torma & J. A. Brierley, pp. 108–36. New York: Academic Press.

BERTHELIN, J., BELGY, G. & MAGNE, R. (1977). Some aspects of the mechanisms of solubilization and insolubilization of uranium from granites by heterotrophic micro-organisms. In *Conference Bacterial Leaching*, ed. W. Schwartz, pp. 251–60. GBF Monograph No. 4. Weinheim, New York: Verlag Chemie.

BEVERIDGE, T. J. & MURRAY, R. G. E. (1976). Uptake and retention of metals by cell walls of *Bacillus subtilis*. *Journal of Bacteriology*, **127**, 1502–18.

BITELL, J. E., KOEPPE, D. E. & MILLER, R. J. (1974). Sorption of heavy metal cations by corn mitochondria and the effects on electron and energy transfer reactions. *Physiologia Plantarum*, **30**, 226–30.

BOUTRY, M., FOURY, F. & GOFFEAU, A. (1977). Energy-dependent uptake of calcium by the yeast *Schizosaccharomyces pombe*. *Biochimica et Biophysica Acta*, **464**, 602–12.

BRIERLEY, C. L. (1974). Molybdenite-leaching: use of a high temperature microbe. *Journal of the Less Common Metals*, **36**, 237–47.

BRIERLEY, C. L. (1975). Biogenic extraction of copper and molybdenum at high temperature. *New Mexico Bureau of Mines and Mineral Resources*, Report No. OFR 10–76. Socorro, New Mexico.

BRIERLEY, C. L. (1977). Thermophilic micro-organisms in extraction of metals from ores. *Developments in Industrial Microbiology*, **18**, 273–84.

BRIERLEY, C. L. (1978a). Bacterial Leaching. *CRC Critical Reviews of Microbiology*, **5** (in press).

BRIERLEY, C. L. (1978b). Biogenic extraction of uranium from ores of the Grants region. In *Metallurgical Applications of Bacterial Leaching and Related Microbiological Phenomena*, eds L. E. Murr, A. E. Torma & J. A. Brierley, pp. 345–63. London, New York: Academic Press.

BRIERLEY, C. L. & BRIERLEY, J. A. (1973). A chemoautotrophic and thermophilic micro-organism isolated from an acid hot spring. *Canadian Journal of Microbiology*, **19**, 183–8.

BRIERLEY, C. L. & BRIERLEY, J. A. (1975). Reduction of molybdenum by a thermophilic bacterium. *International Conference on Heavy Metals in the Environment*, pp. C211–C213. *Toronto University Press*.

BRIERLEY, C. L. & MURR, L. E. (1973). Leaching: use of a thermophilic and chemoautotrophic microbe. *Science*, **179**, 488–90.

BRIERLEY, C. L., BRIERLEY, J. A. & MURR, L. E. (1973). Using the SEM in mining research. *Research and Development*, **24**, 24–8.

BRIERLEY, J. A. (1978). Thermophilic iron-oxidizing bacteria found in copper-leaching dumps. *Applied and Environmental Microbiology* (in press).

BRIERLEY, J. A. & BRIERLEY, C. L. (1978). Microbial leaching of copper at ambient and elevated temperatures. In *Metallurgical Applications of Bacterial Leaching and Related Microbiological Phenomena*, eds L. E. Murr, A. E. Torma & J. A. Brierley, pp. 477–90. London, New York: Academic Press.

BRIERLEY, J. A. & LE ROUX, N. W. (1977). A facultative thermophilic thiobacillus-like bacterium: oxidation of iron and pyrite. In *Conference Bacterial Leaching*, ed. W. Schwartz, pp. 55–66. GBF Monograph No. 4. Weinheim, New York: Verlag Chemie.

BRIERLEY, J. A. & LOCKWOOD, S. J. (1977). The occurrence of thermophilic iron-oxidizing bacteria in a copper leaching system. *FEMS Microbiology Letters*, **2**, 163–5.

BRIERLEY, J. A., NORRIS, P. R., KELLY, D. P. & LE ROUX, N. W. (1978). Characteristics of a moderately thermophilic and acidophilic iron-oxidizing *Thiobacillus*. *European Journal of Applied Microbiology and Biotechnology* (in press).

BRISSETTE, C., CHAMPAGNE, J. & JUTRAS, J. R. (1971). Bacterial leaching of cadmium sulphide. *Canadian Mining and Metallurgical Bulletin*, **64**, 85–8.

BROCK, T. D. & GUSTAFSON, J. (1976). Ferric iron reduction by sulfur- and iron-oxidizing bacteria. *Applied and Environmental Microbiology*, **32**, 567–71.

BROCK, T. D., BROCK, K. M., BELLY, R. T. & WEISS, R. C. (1972). *Sulfolobus*: a new genus of sulfur-oxidizing bacteria living at low pH and high temperatures. *Archives of Microbiology*, **84**, 54–68.

BRODA, E. (1972). The uptake of heavy cationic trace elements by micro-organisms. *Annali di Microbiologia ed Enzimologia*, **22**, 93–108.

BRUYNESTEYN, A. & DUNCAN, D. W. (1977). The practical aspects of laboratory leaching studies. In *Conference Bacterial Leaching*, ed. W. Schwartz, pp. 129–37. GBF Monograph No. 4. Weinheim, New York: Verlag Chemie.

BUCHEDER, F. & BRODA, E. (1974). Energy-dependent zinc transport by *Escherichia coli*. *European Journal of Biochemistry*, **45**, 555–9.

BURKIN, A. R. (1971). Progress report on recent advances in extractive metallurgy. A. Hydrometallurgy. *Metals and materials*, **5**, 47–50.

BYERS, B. R. & ARCENEAUX, J. E. L. (1977). Microbial transport and utilization of iron. In *Micro-organisms and Minerals*, ed. E. D. Weinberg, pp. 215–49. New York: Marcell Dekker.

CAPES, C. E., McILHINNEY, A. E., SIRIANNI, A. F. & PUDDINGTON, I. E. (1973). Bacterial oxidation in upgrading pyritic coals. *Canadian Mining and Metallurgical Bulletin*, **66**, 88–91.

CLOUD, P. (1968). Mineral resources from the sea. In *Resources and Man*, pp. 135–55. San Francisco: W. H. Freeman.

CULLIMORE, D. R. & McCANN, A. E. (1977). The identification, cultivation and control of iron bacteria in ground water. In *Aquatic Microbiology*, eds F. A. Skinner & J. M. Shewan, pp. 219–61. London, New York: Academic Press.

DARLAND, G., BROCK, T. D., SAMSONOFF, W. & CONTI, S. F. (1970). A thermophilic, acidophilic mycoplasma isolated from a coal refuse pile. *Science*, **170**, 1416–18.

DE ROSA, M., GAMBACORTA, A. & BU'LOCK, J. D. (1975). Extremely thermophilic acidophilic bacteria convergent with *Sulfolobus acidocaldarius*. *Journal of General Microbiology*, **86**, 156–64.

DERRY, R., GARRETT, K. H., LE ROUX, N. W. & SMITH, S. E. (1976). Bacterially assisted plant process for leaching uranium ores. In *Geology, Mining and Extractive Processing of Uranium*, ed. M. L. Jones, pp. 56–62. London: Institution of Mining and Metallurgy.

DUBININA, G. A. (1973). The significance of microbiological processes in lacustrine manganese ore formation. *Verhandlungen, Internationale Vereinigung für Theoretische und Angewandte Limnologie*, **18**, 1261–72.

DUGAN, P. R. & APEL, W. A. (1978). Microbial desulfurization of coal. In *Metallurgical Applications of Bacterial Leaching and Related Microbiological Phenomena*, eds L. E. Murr, A. E. Torma & J. A. Brierley, pp. 223–50. London, New York: Academic Press.

DUNCAN, D. W. & BRUYNESTEYN, A. (1971). Microbiological leaching of uranium. *New Mexico State Bureau of Mines and Mineral Resources*, Circular 117, pp. 55–61. Socorro, New Mexico.

DUNCAN, D. W. & TRUSSELL, P. C. (1964). Advances in the microbiological leaching of sulphide ores. *Canadian Metallurgical Quarterly*, **3**, 43–55.

DUNCAN, D. W., LANDESMAN, J. & WALDEN, C. C. (1967). Role of *Thiobacillus ferrooxidans* in the oxidation of sulfide minerals. *Canadian Journal of Microbiology*, **13**, 397–403.

DUTRIZAC, J. E. & MACDONALD, R. J. C. (1974). Ferric iron as a leaching medium. *Minerals Science Engineering*, **6**, 59–100.

EBNER, H. G. (1977). Metal extraction from industrial waste with thiobacilli. In *Conference Bacterial Leaching*, ed. W. Schwartz, pp. 217–22. GBF Monograph No. 4. Weinheim, New York: Verlag Chemie.

EBNER, H. G. (1978). Metal recovery and environmental protection by bacterial leaching of inorganic waste materials. In *Metallurgical Applications of Bacterial Leaching and Related Microbiological Phenomena*, eds L. E. Murr, A. E. Torma & J. A. Brierley, pp. 195–206. London, New York: Academic Press.

ECCLESTON, M. & KELLY, D. P. (1978). Oxidation kinetics and chemostat growth of *Thiobacillus ferrooxidans* on tetrathionate and thiosulfate. *Journal of Bacteriology*, **134**, 718–27.

EGOROVA, A. A. & DERYUGINA, Z. P. (1963). The spore-forming thermophilic thiobacterium, *T. thermophilica* (Imschenetskii). *Mikrobiologiya*, **32**, 439–40.

EMOTO, Y. (1929). Über drei neue Arten der Schwefeloxydierenden Bakterien. *Proceedings of the Imperial Academy, Tokyo*, **5**, 148.

FAILLA, M. L. (1977). Zinc: functions and transport in micro-organisms. In *Microorganisms and Minerals*, ed. E. D. Weinberg, pp. 151–214. New York: Marcel Dekker.

FAILLA, M. L., BENEDICT, C. D. & WEINBERG, E. D. (1976). Accumulation and storage of Zn^{2+} by *Candida utilis*. *Journal of General Microbiology*, **94**, 23–36.

FLETCHER, A. W. (1970). Metal mining from low-grade ore by bacterial leaching. *Transactions of the Institution of Mining and Metallurgy*, **79**, C247–52.

FRIEDMANN, B. A. & DUGAN, P. R. (1967). Concentration and accumulation of metallic ions by the bacterium, *Zoogloea*. *Developments in Industrial Microbiology*, **9**, 381–8.

FUHRMANN, G.-F. & ROTHSTEIN, A. (1968). The transport of Zn^{2+}, Co^{2+}, and Ni^{2+} into yeast cells. *Biochimica et Biophysica Acta*, **163**, 325–30.

FUJIWARA, K., IWAMOTO, M., TODA, S. & FUWA, K. (1977). Characteristics of *Escherichia coli* B resistant to cobaltous-iron. *Agricultural and Biological Chemistry*, **41**, 313–22.

GERRARD, T. L., TELFORD, J. N. & WILLIAMS, H. H. (1974). Detection of selenium deposits in *Escherichia coli* by electron microscopy. *Journal of Bacteriology*, **119**, 1057–60.

GMELIN, *Gmelin's Handbuch der anorganischem Chemie* (1955). Kupfer: Teil A1, System Nr60, p. 46. Weinheim: Verlag Chemie.

GOLOVACHEVA, R. S. & KARAVAIKO, G. I. (1977). A new facultative thermophilic *Thiobacillus* isolated from sulphide ore. In *Microbial Growth on C_1-compounds* (Abstracts) pp. 108–9. Pushchino: USSR Academy of Sciences.

GOROLL, D. (1976). Ökologie von *Thiobacillus neapolitanus* und seine mögliche Mitwirkung in Leaching-Process. *Zeitschrift für Allgemeine Mikrobiologie*, **16**, 3–7.

GOW, W. A., MCCREEDY, H. H., RITCEY, G. M., MCNAMARA, V. M., HARRISON, V. F. & LUCAS, B. H. (1971). Bacteria-based processes for the treatment of low-grade uranium ores. In *The Recovery of Uranium*, pp. 195–211. Vienna: International Atomic Energy Agency.

GROUDEV, S. N., GENCHEV, F. N. & GAIDARJIEV, S. S. (1978). Observations on the microflora in an industrial dump leaching copper operation. In *Metallurgical Applications of Bacterial Leaching and Related Microbiological Phenomena*, eds

L. E. Murr, A. E. Torma & J. A. Brierley, pp. 253–74. London, New York: Academic Press.

GUAY, R. & SILVER, M. (1975). *Thiobacillus acidophilus* sp. nov.; isolation and some physiological characteristics. *Canadian Journal of Microbiology*, **21**, 281–8.

GUAY, R., TORMA, A. E. & SILVER, M. (1975). Oxydation de l'ion ferreux et mise en solution de l'uranium d'un minerai par *Thiobacillus ferrooxidans*. *Annales de Microbiologie (Institut Pasteur)*, **126B**, 209–19.

GUAY, R., SILVER, M. & TORMA, A. E. (1976). Microbiological leaching of a low-grade uranium ore by *Thiobacillus ferrooxidans*. *European Journal of Applied Microbiology*, **3**, 157–67.

HALLBERG, R. O. (1970). An apparatus for the continuous cultivation of sulfate-reducing bacteria and its application to geomicrobiological purposes. *Antonie van Leeuwenhoek. Journal of Microbiology and Serology*, **36**, 241–54.

HAROLD, F. M. & ALTENDORF, K. (1974). Cation transport in bacteria: K^+, Na^+, and H^+. In *Current Topics in Membranes and Transport*, eds F. Bronner & A. Kleinzeller, vol. 5, pp. 1–50. London, New York: Academic Press.

HELDWEIN, R., TROMBALLA, H. W. & BRODA, E. (1977). Aufnahme von Cobalt, Blei und Cadmium durch Bäckerhefe. *Zeitschrift für Allgemeine Mikrobiologie*, **17**, 299–308.

HOFFMAN, L. E. & HENDRIX, J. L. (1976). Inhibition of *Thiobacillus ferrooxidans* by soluble silver. *Biotechnology and Bioengineering*, **18**, 1161–5.

ILYALETDINOV, A. N., ENKER, P. B. & YAKUBOVSKII, S. E. (1976). Participation of heterotrophic micro-organisms in removal of heavy metal ions from effluent. *Mikrobiologiya*, **45**, 1092–9 (English translation pp. 932–8).

ILYALETDINOV, A. N., ENKER, P. B. & LOGINOVA, L. V. (1977). Role of sulfate-reducing bacteria in the precipitation of copper. *Mikrobiologiya*, **46**, 113–17 (English translation pp. 92–5).

IMAI, K. (1978). On the mechanism of bacterial leaching. In *Metallurgical Applications of Bacterial Leaching and Related Microbiological Phenomena*, eds L. E. Murr, A. E. Torma & J. A. Brierley, pp. 275–95. London, New York: Academic Press.

JASPER, P. & SILVER, S. (1977). Magnesium transport in micro-organisms. In *Micro-organisms and Minerals*, ed. E. D. Weinberg, pp. 7–47. New York: Marcel Dekker.

JENNETT, J. C., BOLTER, E., GALE, N., TRANTER, W. & HARDIE, M. (1975). The Viburnum Trend, southeast Missouri: the largest lead-mining district in the world – environmental effects and controls. In *International Symposium on Minerals and the Environment, London, June 1974*, pp. 13–26. London: Institution of Mining and Metallurgy.

KARAVAIKO, G. I. & MOSHNIAKOVA, S. A. (1974). Oxidation of sulphide minerals by *Thiobacillus thiooxidans*. *Mikrobiologiya*, **43**, 156–8.

KARAVAIKO, G. I. & PIVOVAROVA, T. A. (1977). Mechanism of oxidation of reduced sulphur compounds by thiobacilli. In *Conference Bacterial Leaching*, ed. W. Schwartz, pp. 37–46. GBF Monograph No. 4. Weinheim, New York: Verlag Chemie.

KARAVAIKO, G. I., SHCHETININA, E. V., PIVOVAROVA, T. A. & MUBARAKOVA, K. YU. (1973). Denitrifying bacteria isolated from deposits of sulphide ores. *Mikrobiologiya*, **42**, 128–35.

KARAVAIKO, G. I., KUZNETSOV, S. I. & GOLOMZIK, A. I. (1977). *The Bacterial Leaching of Metals from Ores*. Stonehouse, Glos.: Technicopy.

KELLY, D. P. (1976). Extraction of metals from ores by bacterial leaching: present status and future prospects. In *Microbial Energy Conversion*, eds H. G. Schlegel & J. Barnea, pp. 329–38. Gottingen: E. Goltze.

KELLY, D. P. (1977). Cheap uranium from bacteria? *Nature, London*, **265**, 406.

KELLY, D. P. (1978). Microbial Ecology. In *The Oil Industry and Microbial Ecosystems*, chapter 2. London: Institute of Petroleum.

KELLY, D. P. & JONES, C. A. (1978). Factors affecting metabolism and ferrous iron oxidation in suspensions and batch cultures of *Thiobacillus ferrooxidans*: relevance to ferric iron leach solution regeneration. In *Metallurgical Applications of Bacterial Leaching and Related Microbiological Phenomena*, eds L. E. Murr, A. E. Torma & J. A. Brierley, pp. 19–44. London, New York: Academic Press.

KELLY, D. P. & TUOVINEN, O. H. (1972). Recommendation that the names *Ferrobacillus ferrooxidans* Leathen and Braley and *Ferrobacillus sulfooxidans* Kinsel be recognized as synonyms of *Thiobacillus ferrooxidans* Temple and Colmer. *International Journal of Systematic Bacteriology*, **22**, 170–2.

KELLY, D. P. & TUOVINEN, O. H. (1975). Metabolism of inorganic sulphur compounds by *Thiobacillus ferrooxidans* and some comparative studies on *Thiobacillus* A2 and *T. neapolitanus*. *Plant and Soil*, **43**, 77–93.

KELLY, D. P., ECCLESTON, M. & JONES, C. A. (1977). Evaluation of continuous chemostat cultivation of *Thiobacillus ferrooxidans* on ferrous iron or tetrathionate. In *Conference Bacterial Leaching*, ed. W. Schwartz, pp. 1–7. GBF Monograph No. 4. Weinheim, New York: Verlag Chemie.

KETCHUM, P. A. & OWENS, M. S. (1975). Production of a molybdenum-coordinating compound by *Bacillus thuringiensis*. *Journal of Bacteriology*, **122**, 412–17.

KHALID, A. M. & RALPH, B. J. (1977). The leaching behaviour of various zinc sulphide minerals with three *Thiobacillus* species. *Conference Bacterial Leaching*, ed. W. Schwartz, pp. 165–73. GBF Monograph No. 4. Weinheim, New York: Verlag Chemie.

KIEL, H. (1977). Laugung von Kupferkarbonat-und Kuppersilikat-Erzen mit heterotrophen Mikro-organismen. In *Conference Bacterial Leaching*, ed. W. Schwartz, pp. 261–70. GBF Monograph No. 4. Weinheim, New York: Verlag Chemie.

KONETZKA, W. A. (1977). Microbiology of metal transformations. In *Microorganisms and Minerals*, ed. E. D. Weinberg, pp. 317–42. New York: Marcel Dekker.

KUZNETSOV, V. I., IVANOV, M. V. & LYALIKOVA, N. N. (1963). *Introduction to Geological Microbiology*. New York: McGraw-Hill.

LE ROUX, N. W. (1969). Mining with microbes. *New Scientist*, **43**, 12–16.

LE ROUX, N. W., NORTH, A. A. & WILSON, J. C. (1973). Bacterial oxidation of pyrite. *10th International Mineral Processing Conference*, paper 45. London: Institution of Mining and Metallurgy.

LE ROUX, N. W., WAKERLEY, D. S. & HUNT, S. D. (1977). Thermophilic *Thiobacillus*-type bacteria from Icelandic thermal areas. *Journal of General Microbiology*, **100**, 197–201.

LE ROUX, N. W., WAKERLEY, D. S. & PERRY, V. F. (1978). Leaching of minerals using bacteria other than thiobacilli. In *Metallurgical Applications of Bacterial Leaching and Related Microbiological Phenomena*, eds L. E. Murr, A. E. Torma, & J. A. Brierley, pp. 167–91. London, New York: Academic Press.

LEWIS, A. J. & MILLER, J. D. A. (1977). Stannous and cuprous ion oxidation by *Thiobacillus ferrooxidans*. *Canadian Journal of Microbiology*, **23**, 319–24.

LIVESEY-GOLDBLATT, E., TUNLEY, T. H. & NAGU, I. F. (1977). Pilot-plant bacterial film oxidation (Bacfox process) of recycled acidified uranium plant ferrous sulphate leach solution. In *Conference Bacterial Leaching*, ed. W. Schwartz, pp. 175–90. GBF Monograph No. 4. New York, Weinheim: Verlag Chemie.

LYALIKOVA, N. N. (1972). Oxidation of trivalent antimony up to higher oxides as a source of energy for the development of a new autotrophic organism, *Stibiobacter*, gen. nov. *Doklady Akademii Nauk SSSR*, **205**, 1228–9.

LYALIKOVA, N. N., LYUBAVINA, L. L. & SOLOZHENKIN, P. M. (1977). Application of sulfate-reducing bacteria for enrichment ores. In *Conference Bacterial Leaching*, ed. W. Schwartz, pp. 93–100. GBF Monograph No. 4. Weinheim, New York: Verlag Chemie.

McELROY, R. O. & BRUYNESTEYN, A. (1978). Continuous biological leaching of chalcopyrite concentrates: demonstration and economic analysis. In *Metallurgical Applications of Bacterial Leaching and Related Microbiological Phenomena*, eds L. E. Murr, A. E. Torma & J. A. Brierley, pp. 441–62. London, New York: Academic Press.

McGORAN, C. J. M., DUNCAN, D. W. & WALDEN, C. C. (1969). Growth of *Thiobacillus ferrooxidans* on various substrates. *Canadian Journal of Microbiology*, **15**, 135–8.

MACKINTOSH, M. E. (1978). Nitrogen fixation by *Thiobacillus ferrooxidans*. *Journal of General Microbiology*, **105**, 215–18.

MACRAE, I. C. & EDWARDS, J. F. (1972). Adsorption of colloidal iron by bacteria. *Applied Microbiology*, **24**, 819–23.

MACRAE, I. C., EDWARDS, J. F. & DAVIS, N. (1973). Utilization of iron gallate and other organic iron complexes by bacteria from water supplies. *Applied Microbiology*, **25**, 991–5.

MADGWICK, J. C. & RALPH, B. J. (1977). The metal-tolerant alga, *Hormidium fluitans* (Gay) Heering from acid mine drainage waters in Northern Australia and Papua–New Guinea. In *Conference Bacterial Leaching*, ed. W. Schwartz, pp. 85–91. GBF Monograph No. 4. Weinheim, New York: Verlag Chemie.

MARKOSYAN, G. E. (1973). A new mixotrophic sulfur bacterium developing in acid media. *Thiobacillus organoparus* sp. n. *Doklady Akademii Nauk SSSR*, **211**, 1205–8 (English translation pp. 318–20).

MARQUIS, R. E., MAYZEL, K. & CARTENSEN, E. L. (1976). Cation exchange in cell walls of Gram-positive bacteria. *Canadian Journal of Microbiology*, **22**, 975–982.

MATZKU, S. & BRODA, E. (1970). Die Zinkaufnahme in das Innere von Chlorella. *Planta*, **92**, 29–40.

MAYLING, A. A. (1969). Bacterial leaching of ores with an alkaline matrix. Canadian Patent (Ottawa) No. 812715.

MILLER, L. P. (1950). Formation of metal sulfides through the activities of sulfate-reducing bacteria. *Contributions. Boyce Thompson Institute for Plant Research*, **15**, 437–65.

MIZOGUCHI, T., SATO, T. & OKABE, T. (1976). New sulfur-oxidizing bacteria capable of growing heterotrophically, *Thiobacillus rubellus*, nov. sp. and *Thiobacillus delicatus* nov. sp. *Journal of Fermentation Technology*, **54**, 181–91.

MURR, L. E. & BRIERLEY, J. A. (1978). The use of large-scale test facilities in studies of the role of micro-organisms in commercial leaching operations. In *Metallurgical Applications of Bacterial Leaching and Related Microbiological Phenomena*, eds L. E. Murr, A. E. Torma & J. A. Brierley, pp. 491–520. London, New York: Academic Press.

MURR, L. E., TORMA, A. E. & BRIERLEY, J. A. (eds) (1978). *Metallurgical Applications of Bacterial Leaching and Related Microbiological Phenomena*. London, New York: Academic Press.

MURRAY, A. D. & KIDBY, D. K. (1975). Sub-cellular location of mercury in yeast grown in the presence of mercuric chloride. *Journal of General Microbiology*, **86**, 66–77.

NEIHOF, R. & ECHOLS, W. H. (1973). Physiochemical studies of microbial cell walls.

I. Comparative electrophoretic behaviour of intact cells and isolated cell walls. *Biochimica et Biophysica Acta*, **318**, 23–32.

NELSON, D. L. & KENNEDY, E. P. (1971). Magnesium transport in *Escherichia coli*. *Journal of Biological Chemistry*, **246**, 3042–9.

NIELSEN, A. M. & BECK, J. V. (1972). Chalcocite oxidation and coupled carbon dioxide fixation by *Thiobacillus ferrooxidans*. *Science*, **175**, 1124–6.

NORRIS, P. R. & KELLY, D. P. (1977). Accumulation of cadmium and cobalt by *Saccharomyces cerevisiae*. *Journal of General Microbiology*, **99**, 317–24.

NORRIS, P. R. & KELLY, D. P. (1978a). Toxic metals in leaching systems. In *Metallurgical Applications of Bacterial Leaching and Related Microbiological Phenomena*, eds L. E. Murr, A. E. Torma & J. A. Brierley, pp. 83–102. London, New York: Academic Press.

NORRIS, P. R. & KELLY, D. P. (1978b). Dissolution of pyrite (FeS_2) by pure and mixed cultures of some acidophilic bacteria. *FEMS Microbiology Letters* (in press).

NORRIS, P. R. & KELLY, D. P. (1979). Accumulation of metals by bacteria and yeasts. *Developments in Industrial Microbiology* (in press).

NORRIS, P. R., MAN, W. K., HUGHES, M. N. & KELLY, D. P. (1976). Toxicity and accumulation of thallium in bacteria and yeast. *Archives of Microbiology*, **110**, 279–86.

OKOROKOV, L. A., LICHKO, L. P., KADOMTSEVA, W. M., KHOLODENKO, V. P., TITOVSKY, V. T. & KULAEV, I. S. (1977). Energy-dependent transport of manganese into yeast cells and distribution of accumulated ions. *European Journal of Biochemistry*, **75**, 373–7.

OLEM, H. & UNZ, R. F. (1977). Acid mine drainage treatment with rotating biological contactors. *Biotechnology and Bioengineering*, **19**, 1475–91.

PARDEE, A. B., PRESTIDGE, L. S., WHIPPLE, M. B. & DREYFUSS, J. (1966). A binding site for sulfate and its relation to sulfate transport into *Salmonella typhimurium*. *Journal of Biological Chemistry*, **241**, 3962–9.

PATON, W. H. N. & BUDD, D. (1972). Zinc uptake in *Neocosmospora vasinfecta*. *Journal of General Microbiology*, **72**, 173–84.

PATRICK, F. M. & LOUTIT, M. (1976). Passage of metals in effluents, through bacteria to higher organisms. *Water Research*, **10**, 333–5.

PINCHES, A. (1975). Bacterial leaching of an arsenic-bearing sulphide concentrate. In *Leaching and Reduction in Hydrometallurgy*, ed. A. R. Burkin, pp. 28–35. London: Institution of Mining and Metallurgy.

PONSFORD, A. (1966). Microbiological activity in relation to coal utilization. *British Coal Utility Research Association Bulletin*, **30**, 41.

RHOADS, D. B., WATERS, F. B. & EPSTEIN, W. (1976). Cation transport in *Escherichia coli*. VIII. Potassium transport mutants. *Journal of General Physiology*, **67**, 325–41.

ROBERTS, K. R. & MARZLUF, G. A. (1971). The specific interaction of chromate with the sulfate permease systems of *Neurospora crassa*. *Archives of Biochemistry and Biophysics*, **142**, 651–9.

ROGERS, S. R. & ANDERSON, J. J. (1976). Measurement of growth and iron deposition in *Sphaerotilus discophorus*. *Journal of Bacteriology*, **126**, 257–63.

ROSS, V. (1957). Geochemistry, crystal structure and minerology of the sulphides. *Economic Geology*, **57**, 755–74.

ROSSI, G. (1978). Potassium recovery through leucite bioleaching: possibilities and limitations. In *Metallurgical Applications of Bacterial Leaching and Related Microbiological Phenomena*, eds L. E. Murr, A. E. Torma & J. A. Brierley, pp. 297–319. London, New York: Academic Press.

ROTHSTEIN, A. & HAYES, A. D. (1956). The relationship of the cell surface to metabolism. XIII. The cation binding properties of the yeast cell surface. *Archives of Biochemistry and Biophysics*, **63**, 87–99.

SAKAGUCHI, H., SILVER, M. & TORMA, A. E. (1976). Microbiological leaching of a chalcopyrite concentrate by *Thiobacillus ferrooxidans*. *Biotechnology and Bioengineering*, **18**, 1091–101.

SAKAGUCHI, H., TORMA, A. E. & SILVER, M. (1976). Microbiological oxidation of synthetic chalcocite and covellite by *Thiobacillus ferrooxidans*. *Applied and Environmental Microbiology*, **31**, 7–10.

SASSON, A. (1975). The earth's invisible garbage men. *The UNESCO Courier*, July 1975, 29–30.

SAYÃO LOBATO, A. (1974*a*). Ocorrencia de bacterias autotroficas – chimioautotroficas – em aguas minerais brasileiras. *Anais do Centro Pesquisas Biologicas*, **4**, 13–16.

SAYÃO LOBATO, A. (1974*b*). *Thiobacillus ferrooxidans* tolerantes a altos teores de Molibdeno sob a forma de MoO₃ existente em minerios de uranio. *Anais do Centro Pesquisas Biologicas*, **4**, 10–12.

SCHÜTT, C. & OTTOW, J. C. G. (1977). Mesophilic and psychrophilic manganese-precipitating bacteria in manganese nodules of the Pacific Ocean. *Zeitschrift für Allgemeine Mikrobiologie*, **17**, 611–16.

SCHWARTZ, W. (ed.) (1977). *Conference Bacterial Leaching*: GBF monograph No. 4. Weinheim, New York: Verlag Chemie.

SHEFFER, H. W. & EVANS, L. G. (1968). Copper leaching practices in the western United States. *U.S. Bureau of Mines, Information Circular* 8341.

SHUMATE, S. E. & STRANDBERG, G. W. (1978). Biological removal of metal ions from aqueous process streams. *Symposium on Biotechnology in Energy Production and Conservation* (May 10–12, 1978), p. 7. Tennessee: Gatlinberg.

SILVER, S. & JASPER, P. (1977). Manganese transport in micro-organisms. In *Micro-organisms and Minerals*, ed. E. D. Weinberg, pp. 105–49. New York: Marcell Dekker.

SILVER, M. & TORMA, A. E. (1974). Oxidation of metal sulfides by *Thiobacillus ferrooxidans* grown on different substrates. *Canadian Journal of Microbiology*, **20**, 141–7.

SILVER, S., JOHNSEINE, P., WHITNEY, E. & CLARK, D. (1972). Manganese-resistant mutants of *Escherichia coli*: physiological and genetic studies. *Journal of Bacteriology*, **110**, 186–95.

SILVERBERG, B. A., WONG, P. T. S. & CHAU, Y. K. (1976). Localization of selenium in bacterial cells using TEM and energy dispersive X-ray analysis. *Archives of Microbiology*, **107**, 1–6.

SILVERMAN, M. P. (1963). Removal of pyritic sulfur from coal by bacterial action. *Fuel*, **42**, 113.

SOLT, J., PASCHINGER, H. & BRODA, E. (1971). Die energieabhängige Aufnahme von Thallium durch *Chlorella*. *Planta*, **101**, 242–50.

SPRINGER, S. E. & HUBER, R. E. (1973). Sulfate and selenate-tolerant strains of *Escherichia coli* K-12. *Archives of Biochemistry and Biophysics*, **156**, 595–603.

STUMM-ZOLLINGER, E. (1972). Die bakterielle Oxydation von Pyrit. *Archiv für Mikrobiologie*, **83**, 110–19.

SUMMERS, A. O. & SILVER, S. (1978). Microbial transformations of metals. In *Annual Review of Microbiology*, **32** (in press).

SZOLNOKI, J. (1977). Utilization of metal bearing industrial waste materials. In *Conference Bacterial Leaching*, ed. W. Schwartz, pp. 223–32. GBF Monograph No. 4. Weinheim, New York: Verlag Chemie.

TABILLION, R. & KALTWASSER, H. (1977). Energieabhängige ⁶³Ni-Aufnahme bei

Alcaligenes eutrophus Stamm HI und H16. *Archives of Microbiology*, **113**, 145–51.

TOMIZUKA, N. & YAGISAWA, M. (1978). Optimum conditions for leaching of uranium and oxidation of lead sulfide with *Thiobacillus ferrooxidans* and recovery of metals from bacterial leaching solution with sulfate-reducing bacteria. In *Metallurgical Applications of Bacterial Leaching and Related Microbiological Phenomena*, eds L. E. Murr, A. E. Torma & J. A. Brierley, pp. 321–44. London, New York: Academic Press.

TOMIZUKA, N., YAGISAWA, M., SOMEYA, J. & TAKAHARA, Y. (1976). Continuous leaching of uranium by *Thiobacillus ferrooxidans*. *Agricultural and Biological Chemistry*, **40**, 1019–25.

TORMA, A. E. (1971). Microbiological oxidation of synthetic cobalt, nickel and zinc sulfides by *Thiobacillus ferrooxidans*. *Revue Canadienne de Biologie*, **30**, 209–16.

TORMA, A. E. (1972). Microbiological extraction of cobalt and nickel from sulfide ores and concentrates. Canadian Patent, 1382357.

TORMA, A. E. (1977). The role of *Thiobacillus ferrooxidans* in hydrometallurgical processes. In *Advances in Biochemical Engineering*, eds T. K. Ghose, A. Fiechter & N. Blakebrough, pp. 1–37. Berlin, Heidelberg, New York: Springer-Verlag.

TORMA, A. E. (1978). Complex lead sulfide concentrate leaching by micro-organisms. In *Metallurgical Applications of Bacterial Leaching and Related Microbiological Phenomena*, eds L. E. Murr, A. E. Torma & J. A. Brierley, pp. 375–87. London, New York: Academic Press.

TORMA, A. E. & GABRA, G. G. (1977). Oxidation of stibnite by *Thiobacillus ferrooxidans*. *Antonie van Leeuwenhoek. Journal of Microbiology and Serology*, **43**, 1–6.

TORMA, A. E. & HABASHI, F. (1972). Oxidation of copper (II) selenide by *Thiobacillus ferrooxidans*. *Canadian Journal of Microbiology*, **18**, 1780–1.

TORMA, A. E. & LEGAULT, G. (1973). Rôle de la surface des minerais sulfurés lors de leur biodégradation par *Thiobacillus ferrooxidans*. *Annales de Microbiologie (Institut Pasteur)*, **124**A, 111–21.

TORMA, A. E. & SUBRAMANIAN, K. N. (1974). Selective bacterial leaching of a lead sulphide concentrate. *International Journal of Mineral Processing*, **1**, 125–34.

TORMA, A. E. & TABI, M. (1973). Mise en solution des métaux de minerais sulfurés par voie bactérienne. *L'Ingenieur*, **294**, 2–8.

TORMA, A. E., WALDEN, C. C. & BRANION, R. M. R. (1970). Microbiological leaching of a zinc sulfide concentrate. *Biotechnology and Bioengineering*, **12**, 501–17.

TORMA, A. E., WALDEN, C. C., DUNCAN, D. W. & BRANNION, R. M. R. (1972). The effect of carbon dioxide and particle surface area on the microbiological leaching of a zinc sulfide concentrate. *Biotechnology and Bioengineering*, **14**, 777–86.

TORMA, A. E., LEGAULT, G., KOUGIOUMOUTZAKIS, D. & OUELLET, R. (1974). Kinetics of bio-oxidation of metal sulphides. *Canadian Journal of Chemical Engineering*, **52**, 515–17.

TORNABENE, T. G. & EDWARDS, H. W. (1972). Microbial uptake of lead. *Science*, **176**, 1334–5.

TORNABENE, T. G. & PETERSON, S. L. (1975). Interaction of lead and bacterial lipids. *Applied Microbiology*, **29**, 680–4.

TRUDINGER, P. A. (1976). Microbiological processes in relation to ore genesis. In *Handbook of Strata-bound and Stratiform Ore Deposits*, ed. K. H. Wolf, pp. 135–90. Amsterdam: Elsevier.

TSUCHIYA, H. M., TRIVEDI, N. C. & SCHULER, M. L. (1974). Microbial mutualism in ore leaching. *Biotechnology and Bioengineering*, **16**, 991–5.

TUCKER, F. L., THOMAS, J. W., APPLEMAN, M. D., GOODMAN, S. H. & DONAHUE, J. (1966). X-ray diffraction studies on metal deposition in Group D streptococci. *Journal of Bacteriology*, **92**, 1311–14.

TUOVINEN, O. H. & KELLY, D. P. (1972). Biology of *Thiobacillus ferrooxidans* in relation to the microbiological leaching of sulphide ores. *Zeitschrift für Allgemeine Mikrobiologie*, **12**, 311–46.

TUOVINEN, O. H. & KELLY, D. P. (1973). Studies on the growth of *Thiobacillus ferrooxidans*. I. Use of membrane filters and ferrous iron agar to determine viable numbers, and comparison with $^{14}CO_2$ fixation and iron oxidation as measures of growth. *Archiv für Mikrobiologie*, **88**, 285–98.

TUOVINEN, O. H. & KELLY, D. P. (1974a). Use of micro-organisms for the recovery of metals. *International Metallurgical Reviews*, **19**, 21–31.

TUOVINEN, O. H. & KELLY, D. P. (1974b). Studies on the growth of *Thiobacillus ferrooxidans*. II. Toxicity of uranium to growing cultures and tolerance conferred by mutation, other metal cations and EDTA. *Archives of Microbiology*, **95**, 153–64.

TUOVINEN, O. H. & KELLY, D. P. (1974c). Studies on the growth of *Thiobacillus ferrooxidans*. III. Influence of uranium, other metal ions and 2,4-dinitrophenol on ferrous iron oxidation and carbon dioxide fixation by cell suspensions. *Archives of Microbiology*, **95**, 165–80.

TUOVINEN, O. H. & KELLY, D. P. (1974d). Studies on the growth of *Thiobacillus ferrooxidans*. IV. Influence of monovalent cations on ferrous iron oxidation and uranium toxicity in growing cultures. *Archives of Microbiology*, **98**, 167–74.

TUOVINEN, O. H. & KELLY, D. P. (1974e). Studies on the growth of *Thiobacillus ferrooxidans*. V. Factors affecting growth in liquid and development of colonies on solid media containing inorganic sulphur compounds. *Archives of Microbiology*, **98**, 351–64.

TUOVINEN, O. H., NIEMELA, S. I. & GYLLENBERG, H. G. (1971). Tolerance of *Thiobacillus ferrooxidans* to some metals. *Antonie van Leeuwenhoek. Journal of Microbiology and Serology*, **37**, 489–96.

TUOVINEN, O. H., KELLY, D. P., DOW, C. S. & ECCLESTON, M. (1978). Metabolic transitions in cultures of acidophilic thiobacilli. In *Metallurgical Applications of Bacterial Leaching and Related Microbiological Phenomena*, eds L. E. Murr, A. E. Torma & J. A. Brierley, pp. 61–81. London, New York: Academic Press.

TUTTLE, J. H., DUGAN, P. R. & RANDLES, C. I. (1969). Microbial sulfate reduction and its potential utility as an acid mine water pollution abatement procedure. *Applied Microbiology*, **17**, 297–302.

TYLER, P. A. (1970). Hyphomicrobia and the oxidation of manganese in aquatic ecosystems. *Antonie van Leeuwenhoek. Journal of Microbiology and Serology*, **36**, 567–78.

TYNECKA, Z., ZAJAC, J. & GOS, Z. (1975). Plasmid-dependent impermeability barrier to cadmium ions in *Staphylococcus aureus*. *Acta Microbiologica Polonica* Ser. A, **7**, 11–20.

VENKATESWERLU, G. & SASTRY, K. S. (1970). The mechanism of uptake of cobalt ions by *Neurospora crassa*. *Biochemical Journal*, **118**, 497–503.

VOSJAN, J. H. & VAN DER HOEK, G. J. (1972). A continuous culture of *Desulphovibrio* on a medium containing mercury and copper ions. *Netherlands Journal of Sea Research*, **5**, 440–4.

WEBB, M. (1970). Interrelationships between the utilization of magnesium and the uptake of other bivalent cations by bacteria. *Biochimica et Biophysica Acta*, **222**, 428–39.

WEINBERG, E. D. (ed.) (1977). *Micro-organisms and Minerals*. New York: Marcel Dekker.

WELLS, A. F. (1975). *Structural Inorganic Chemistry*, 4th Edition, pp. 907–8. Oxford: Clarendon Press.

WERNER, D., PAWLITZ, H. D. & ROTH, R. (1975). The separation of ^{32}S from contaminating ^{3}H and ^{60}Co by incorporation into diatoms. *Zeitschrift für Naturforschung*, **30** (C), 423–4.

WHITTENBURY, R. & KELLY, D. P. (1977). Autotrophy: a conceptual phoenix. In *Microbial Energetics*, eds B. A. Haddock & W. A. Hamilton, pp. 121–49. *27th Symposium of the Society for General Microbiology*. Cambridge University Press.

WILLECKE, K., GRIES, E.-M. & OEHR, P. (1973). Coupled transport of citrate and magnesium in *Bacillus subtilis*. *Journal of Biological Chemistry*, **248**, 807–14.

WILLIAMS, R. A. D. & HOARE, D. S. (1972). Physiology of a new facultatively autotrophic thermophilic Thiobacillus. *Journal of General Microbiology*, **70**, 555–66.

WYCKOFF, R. W. G. (1963). *Crystal Structures*, vol. 1. New York: Interscience.

WYCKOFF, R. W. G. (1970). Two kinds of chalcopyrites demonstrated by bacterial oxidation. *Bulletin de la Societé Française de Minéralogie et de Cristallographie*, **93**, 120–2.

ZAJIC, J. E. (1969). *Microbial Biogeochemistry*. New York, London: Academic Press.

ZAJIC, J. E., JACK, T. R. & SULLIVAN, E. A. (1977). Chemical and microbially-assisted leaching of Athabasca oil sands coke. In *Conference Bacterial Leaching*, ed. W. Schwartz, pp. 233–42. GBF Monograph No. 4. Weinheim, New York: Verlag Chemie.

ZAVARZIN, G. A. (1966). Iron bacteria on the volcanoes of Kunashir lake. *Trudy Moskva ob-va Ispeet Prirody*, **24**, 217.

INDUSTRIAL ALCOHOL

JOHN D. BU'LOCK

Weizmann Microbial Chemistry Laboratory, Department of Chemistry, The University, Manchester M13 9PL, UK

The object of this review is to describe some new developments in industrial alcohol production and perhaps to indicate some desirable directions for new research. First, however, it will be useful to outline the context in which the current revival of interest in this, the oldest of fermentations (Genesis **9**, 21), is occurring.

THE ECONOMIC CONTEXT

The production of industrial alcohol by fermentation draws substantially upon the accumulated expertise of the brewer and the distiller, but it must be considered quite separately from the production of potable alcohols since neither the process objectives nor the economic constraints are the same. The brewer, the wine maker and the distiller produce more or less diluted alcohol mixed with sundry minor ingredients; their concern is to do so consistently and to standards determined by organoleptic qualities. The size of their market is governed by consumer preferences and advertising success and their direct production costs are overshadowed by the costs of packaging, distribution, marketing and (above all) the excise structure. In contrast, the producer of industrial alcohol has a chemically-graded product. His market may be very large but it is determined by national and even international decisions and, even when there is an element of subsidy, his selling price is determined with reference to the external scale of world petrochemical costs.

Petrochemical ethanol is made by the hydration of ethylene, the decline of fermentation alcohol dates from the large-scale production of ethylene (for polymer manufacture) from the 1940s. Within twenty years of the advent of large-scale petroleum cracking, industrial production of fermentation alcohol fell below potable alcohol production in most industrial countries, so that by 1970 an authoritative account (Harrison & Graham, 1970) placed the topic somewhere between industrial archaeology and cultural anthropology: 'In the technologically more advanced countries, ethanol...is now made by chemical means.... However in regions where suitable cheap materials are available, spirit for industrial use is still made.'

Since then, however, while the world price of crude oil has increased some four-fold, the prices of 'suitable cheap materials' have risen far less on average, despite wide short-term fluctuations, and moreover they differ considerably from place to place (Hepner, 1978). At the same time, oil-importing nations are anxious to reduce their import costs and may even subsidize home-produced alternatives. Since ethanol can be used directly as a motor fuel, and can also be converted quite readily into ethylene and other process intermediates, its production from indigenous and renewable resources is highly attractive. The resultant pattern of activity is one which varies considerably from one nation to another according to the local interplay of political, social and economic factors.

In countries as different as India and Cuba, alcohol production from cane sugar reduces economic dependence, strengthens the industrial base, provides employment and creates a buffer market for the agricultural product. Meanwhile in North America an effective surplus of cereal starch is converted into glucose and increasingly into glucose–fructose syrups; abroad, this forces the sucrose producers to seek new outlets, while domestically, in an economy surprisingly exposed to the effects of petroleum prices, conversion of the starch-glucose into ethanol has renewed attractions. In countries where a similar agricultural starch or sugar surplus is conceivable, such as Australia and South Africa, serious feasibility studies for large-scale agro-industrial projects have been carried out (McCann & Saddler, 1976). Worldwide, research attention is being given to the problem of utilizing cheaper but less tractable carbohydrate sources, particularly primary and waste ligno-cellulose, with the option of alcohol fermentation as a final step. Countries which depend upon technology export (which includes Britain, at least in principle) are recognizing the opportunities these developments provide for their own initiatives; perhaps, not least among the causes of all this activity, we might also acknowledge the existence of a small army of biotechnologists frustrated in their desire to feed the world on single-cell protein, SCP, and anxious to find plausible employment.

No account of the present situation of industrial fermentations, however sketchy, would be complete without reference to the Brazilian alcohol programme. This was adopted in 1975 with the specific aims of reducing petroleum imports, providing a home base for chemical industry, creating a guaranteed market for existing sugar cane acreage and providing for substantial new agro-industrial development (Jackman, 1976). The 1975 plan was for fermentation production of 3–4 × 10^6 m³ of ethanol by the early 1980s, and 12–16 × 10^6 m³ by 1990, initially from the existing sugar cane crop but eventually from a more balanced production of sugar and starch crops, particularly manioc (cassava). These figures should be compared with the total annual US production, which in 1970–1975 was around 1.5 × 10^6 m³, mainly from ethylene (Miller, 1975). The agricultural development envisaged is considerable, perhaps 10^6 ha for 1980 and 5 × 10^6 ha for 1990 (Goldemberg, 1978), and the total investment is commensurate. Current development is only slightly behind schedule, with 0.7 × 10^6 m³ ethanol produced in 1977 and about twice that in 1978 (Hammond, 1978), and the first manioc-based plant, rated at 60 m³ per day, has worked since January 1978. An important lesson for UK industry is the considerable involvement of overseas companies, not only in the alcohol programme proper but also in developing ancillary industries such as enzyme supply and the production of ethylene, acetic acid, etc.

So far, innovation in the Brazilian programme has been understandably restricted to marginal improvements in essentially traditional processes (Jackman, 1976). More widely, however, there is an apparent concensus that development programmes of similar or larger scale will be running in many other regions by the end of the century and it is in this context that the development of new technologies for the alcohol fermentation is opportune and even urgent.

Traditionally, the successive steps in alcohol production – substrate preparation, fermentation and product recovery – are treated independently. Much current thinking emphasizes the interdependence of these steps but they will provide convenient headings for our analysis.

SUBSTRATES

Any substrate which can be fermented to alcohol can also, at least in principle, be used as human or animal food, either directly or by conversion into a single-cell protein product. Hence these substrates have independent minimum values which play a major role in determining the economics of alcohol production and hence in limiting the applicable technology. Whereas little more than half of the cost of petrochemical alcohol is due to raw-material costs, the proportion for alcohol from (for example) molasses is between 75% and 85% (Hepner, 1978). Until the balance of world prices changes even more radically than it has done in the last seven years, economies of scale in the fermentation route are less important than economies in substrate provision and utilization. Small plants using local raw materials may be more economic than one large plant dependent upon substrates brought in over a larger radius and with intermediate processing.

Because production of fermentation alcohol is so closely linked to agriculture, it is also important to provide for adjustments between different products from the primary crop. For example, in an important Australian study on cassava utilization it was found desirable to strike a balance between production of refined starch and starch hydrolysate, and between SCP and alcohol fermentations of the hydrolysate (Mc-Cann & Saddler, 1976). This need for flexibility is important in proposals for various kinds of 'integrated' crop processing, all of which have the aim of reducing individual product costs by increasing the total recovery of a wider range of products. Such processes may well determine the precise nature of future fermentation substrates, particularly where fermentation routes provide a way of adjusting the relatively fixed proportions in which different crop fractions are produced.

The precise mode of substrate preparation has a considerable effect on other process details. For example, where alcohol is to be produced from cereals, several different approaches are possible. Non-starch components of the grain may be separated in milling, after saccharification, before distillation or carried right through to the stillage. In the traditional production of potable alcohols from grain, all of these are exemplified but the procedures actually adopted are determined mainly by effects on product qualities. In a well-designed industrial alcohol process, the most appropriate procedures will instead be determined by process costs and the totality of product values and to some extent this is already done, for example in the early separation of corn oil from maize. However the presence or absence of other non-starch components is important in the mainstream process itself. The need to control breakdown of cereal protein during saccharification determines the degree of purity desirable in the crude amylases normally used and the temperature–time profile of the saccharification step; this in turn influences the kinetic balance between yeast growth and fermentation (Harrison & Graham, 1970) and has a major effect on the formation of fusel oils, which complicate distillation, and can have a toxic effect in some types of fermentation system. It also influences the nutritional value of by-products to be recovered either before or after distillation. Any large-scale process will need to be much more carefully optimized in these respects than most current European practice.

On the other hand the techniques of enzymic saccharification are now very efficiently developed and, indeed, are more useful for industrial alcohol production than they are for potable alcohol. In particular, complete hydrolysis to glucose, now standard, is clearly more useful than the traditional malting procedures. In place of a mixture of mono- and oligo-saccharides, which are metabolized sequentially and at different rates (Rainbow, 1970) and are particularly undesirable in a continuous process, enzymic hydrolysis allows optimum kinetics and correspondingly higher productivity.

A rather different example of interactions between improved crop utilization and fermentation process design is provided by sugar cane (Lipinsky, 1978). Traditionally, the cane is harvested and processed to make crystalline sucrose; the molasses used for fermentation is a by-product and only about half its solids content is fermentable. The extracted cane residues (bagasse) constitute a very low value waste and only part of it can be used as fuel for the extraction process. The cane leaf, which itself could provide much of the fuel requirement, is burnt off in the field. In the newly-developed 'Canadian Separator Equipment' process (Rajic & Atchison, 1978) much more valuable by-products are obtained – some suited for microbiological processing – and the sugar is extracted in forms more suitable for a flexible balance between sucrose production and fermentation. Less drastic modifications of traditional sugar milling are already practised in conjunction with alcohol production in Brazil (Jackman, 1976). For cassava, similarly integrated whole-crop utilization is again feasible and will almost certainly become essential as its industrial use increases (McCann & Saddler, 1976).

The concentration of sugar obtained at the end of substrate preparation is also important. Where intermediate transportation is avoided, the output sugar concentration should ideally equal the optimum input concentration for the fermentation system and any departures from this ideal suggest that heat and/or water is being wastefully consumed. Flexibility in this respect is variable. With sugar cane, the limits of sugar concentration depend on the precise nature of the milling and extracting system. With cereal carbohydrate, the upper limit is determined by starch viscosity prior to saccharification, whereas with potatoes it is determined by the natural water content of the tubers. In the case of process wastes, of which soft-wood pulping liquor and whey ultrafiltrate are the most important examples, the initial sugar concentration available is outside the processor's control and it becomes a matter of detailed thermal balancing whether to effect concentration of the liquor prior to fermentation (Abrams, 1975).

FERMENTATION

Physiological aspects

For the large-scale production of a low-value, growth-linked metabolite such as ethanol, productivity considerations make continuous fermentation an imperative. The problems which attend continuous brewing of potable alcohols (Hough & Button, 1972) can be avoided since process design is not required to make a compromise between productivity and organoleptic qualities. Moreover, the relevant basic physiology and biochemistry of the yeast fermentation is adequately understood and fully documented (Rose & Harrison, 1971), and here only the most significant special features need to be pointed out.

Substrate inhibition

In continuous fermentations, the steady-state sugar concentration will be minimal, whatever its concentration in the feed, so that true substrate inhibition can be ignored. Inhibition by other components of the feedstock may be more serious but, to the extent that some of these can arise from mistreatment of the substrate (e.g. furfural and hydroxymethyl-furfural), it is avoidable.

Diauxic effects

Differential utilization of glucose, fructose and sucrose in cane sugar has little effect on process kinetics and in enzymic starch hydrolysates the effect of oligosaccharides can be made negligible. Full utilization of the raffinose content of beet sugar requires selection of a melibiase-producing yeast either for the fermentation itself or for pre-treatment. For the fermentation of whey ultrafiltrate with a lactose-utilizing yeast like *Saccharomyces fragilis*, it is better to leave the lactose intact and thus avoid the glucose/galactose diauxie (O'Leary *et al.*, 1977).

Growth versus fermentation

The nutritional effects which control both the upper and the lower limits of growth of *Saccharomyces cerevisiae* in the anaerobic system are well understood in relation to traditional batch processes for potable alcohols (Harrison & Graham, 1970). In continuous systems, with substantial yeast recycle (see later), the higher cell density requires correspondingly higher maintenance which can be balanced empirically against the substrate input if allowance is made for cell mortality in the recycle system (Cysewski & Wilke, 1977). However, more detailed studies are required to understand this process completely (Goma & Strehaiano, 1978).

Temperature

For most yeasts, the optimum temperature for the anaerobic fermentation is some 10 °C higher than that for growth (Stokes, 1971), so in a continuous process some compromise is necessary. In traditional batch processes, the temperature programme is rather simply contrived, making some use of the heat of fermentation (Harrison & Graham, 1970), but for a continuous high-productivity system more effective controls are needed to ensure the least unfavourable energy balance. In general the heat of fermentation will be sufficient to make cooling systems necessary in an atmospheric-pressure system but quite insufficient to sustain isothermal evaporation in a reduced-pressure system. Selection of more thermotolerant yeasts is feasible and would assist both types of process but thermotolerant yeasts generally have higher maintenance requirements and, in any case, operating temperatures much over 50 °C are unlikely to be feasible. For truly high-temperature fermentations the investigation of suitable prokaryotes, i.e. thermophilic anaerobic bacteria, will almost certainly be required (see later).

Oxygen

The yeast fermentation is not truly anaerobic since growing yeasts require a small but finite oxygen supply to produce unsaturated fatty acids and sterols essential for membrane function. In batch or homogeneously mixed continuous fermentations, sufficient oxygen can usually be supplied if the incoming liquor is saturated with air, but at higher cell densities (in continuous systems with recycle) pre-saturation, even with neat oxygen, may be insufficient and a supply to the fermenter itself is required. The oxygen tension necessary is in the 0.1 % air-saturation range (Cowland & Maule, 1966; Cysewski & Wilke, 1976) and thus below the range of simple electrodes, so that the supply cannot be regulated by direct reference to on-line data. At reduced pressure the rate of oxygen transfer is proportionately decreased and the need can only be met by using oxygen, not air (Cysewski & Wilke, 1977).

Product toxicity

The well-known inhibition of the yeast fermentation by ethanol has an auto-regulatory role in many traditional fermentations, but in industrial alcohol production it is wholly disadvantageous. Strain selection can be helpful (Rose, 1976), but the alcohol susceptibility of yeasts is increased at higher temperatures and it is influenced by the previous history of the fermentation (Kodama, 1970). In particular, when ethanol is produced very rapidly, diffusion out of the cells becomes rate-limiting and the internal concentration may be strongly inhibitory. Nagodawithana & Steinkraus (1976) compared 'rapid' and 'slow' fermentations (using the same cell density and medium but at 30 °C and 15 °C respectively) and found a four-fold difference in the internal ethanol levels, with up to 2×10^{11} molecules per cell (over 3 M) in the worst case. Effects on both viability and fermentation rate are already apparent at external ethanol levels as low as 6% (w/v) and this justifies interest in systems for removing excess ethanol as it is formed (see later). The higher alcohols of 'fusel oil' are even more toxic: Ingram (1955) states that 0.6% (w/v) of pentanol or 1.5% (w/v) of isobutanol is completely inhibitory. Accumulation of these alcohols to such levels only occurs in reduced-pressure fermentations (see later).

Fermentation systems

On the basis of the physiological requirements, and from considerations of process economy, it is possible to state the ideal characteristics of a fermentation system for industrial alcohol.

The process should run continuously on a single-sugar substrate with controlled levels of other nutrients, including oxygen, sufficient to maintain the cell density. The steady-state sugar concentration will be minimal but the input sugar concentration should be as high as possible. The cell population should be as high as possible in the fermenter and as low as possible in the effluent. The steady-state ethanol concentration in the fermenter should be 5% (w/v) or less and the higher alcohol level below 0.5% (w/v).

Such requirements are not met by any traditional fermentation and indeed some are mutually contradictory. In practical terms, we will also require that the plant shall be cheap to build and simple to operate – ideally, self-regulating. The combined challenge is formidable but in recent years several developments towards new systems have occurred and will repay examination.

Continuous fermentations with recycle

The yeast yield in anaerobiosis is low, around 1–2% (which is why traditional batch fermentations are controlled by inoculum size and the initial aerobic stage). For this reason the productivity of a homogeneous continuous fermentation is rather limited. For example, a recently-described Danish plant (Rosen, 1978) runs on molasses equivalent to a

12% (w/v) sugar input. There are two successive fermenter stages in which the yeast concentration is c. 1% (w/v), and the combined residence time is 16 h or more. Thus ethanol productivity is about $4 \text{ g l}^{-1} \text{h}^{-1}$ or less. More productive systems require a higher cell population, which must be maintained either by selectively retaining cells within the fermenter or by returning separated cells from the effluent to the fermenter. Cell retention and cell return are two different means of obtaining the same kinetic effect and are conveniently both termed 'cell recycle' mechanisms.

Cell return is most easily introduced into a conventional system since all that is needed is a return line from the final yeast separators, although, under the rather poor hygiene which often prevails in the post-fermentation stages, an intermediate 'cleaning' stage in the return line is often necessary. Both centrifugation and sedimentation can be adapted to produce yeast for recycle purposes and the performance of the separator stage can be a limiting factor. For example, Cysewski & Wilke (1977), using a laboratory-scale system to ferment a 10% (w/v) glucose feed, were able to build up a cell content of 5% (w/v) by returning yeast from a simple settler. This worked well at dilution rates up to 0.75 h^{-1}, but at faster throughputs the loading on the settler was too high and the system washed out. Even so, the ethanol productivity with recycle was $28 \text{ g l}^{-1} \text{h}^{-1}$, compared with $7 \text{ g l}^{-1} \text{h}^{-1}$ in homogeneous operation. With a more effective recycle of strongly flocculent yeast to raise the yeast density to (say) 10% (w/v), productivities as high as $50 \text{ g l}^{-1} \text{h}^{-1}$ should be attainable, provided the dense suspension can be adequately transferred, mixed, etc. Approaches to such a system have indeed been described (Engelbart & Dellweg, 1976). In an Alfa-Laval patent, Ehnstron (1976) described the use of a specialized centrifugal separator to obtain a return stream with about 15% (w/v) yeast. In the system described, this was mixed with up to 0.2 volumes of wort and pumped under pressure into a long tubular reactor, but other applications for such a separator can be visualized.

The simplest systems for cell retention are simple tower fermenters using strongly flocculent yeasts (Hough & Button, 1972). Here, the substrate rises from the base of the tower through a bed of sedimented yeast. As it rises, the combined effects of CO_2 production and (usually) deflocculation of the yeast (in response to the substrate sugar level) break up the bed and disperse the yeast. Nearer the top of the tower, as the sugar is fermented out, the yeast reflocculates and, as the liquor is degassed at the top of the tower, returns to the middle section. Operation is thus part-way between a mixed system and a plug flow. Such towers have been used, with mixed success, for brewing beer at residence times of the liquor down to 4 h. Their weakness is that if flocculation fails because of a change in some process-variable the whole system becomes uncontrolled. A significant proportion of the yeast is under-utilized because of poor mixing. This type of fermenter does not seem to have been used for industrial alcohol production but, with improvement, it offers a rather simple route to quite high productivities.

Fermentation–Distillation

If the cell density can be adequately increased by recycle, alcohol toxicity becomes the limiting factor when more concentrated substrates are used. A multistage system can be used to give an approximation to plug flow (in which inhibition only affects the later stages in the system) which is also given by the interrupted tubular reactor described in the Alfa-Laval system (Ehnstron, 1976). Removal of alcohol during the fermentation should give even greater benefits.

A laboratory-scale system for removing ethanol (mixed with CO_2 and water vapour) by carrying out the fermentation under reduced pressure was first described by Ramalingham & Finn (1977). Their fermentation was carried out at a nominal 30 °C and 32 mmHg pressure though it seems that controls were not sufficiently well-balanced to give steady distillation. With an 18% (w/v) sugar feed, pre-saturation with oxygen was not enough to maintain cell density and a sterol and fatty acid supplement was employed instead. Stable running at a residence time of 11 h was then attained and glucose utilization was complete. With a similarly-supplemented 50% (w/v) glucose feed and with batch-wise additions of yeast centrifuged from the effluent, residence times of 19 h for 98% utilization of the glucose were reached; maximum ethanol productivity was around 12 g l^{-1} h^{-1}. This work showed that the reduced pressure *per se* had no adverse effects and that alcohol removal allowed a faster fermentation and also made it possible to use a feedstock of minimal water content.

Independently, Cysewski & Wilke (1977) described a similar system in rather more detail. The fermenter was supplied with oxygen and coupled to a simple cooled settler from which a metered cell recycle was returned. With the fermenter at 35 °C, they made direct comparisons between runs at atmospheric pressure and at 50 mmHg, with and without cell recycle. At atmospheric pressure, they used a 10% (w/v) glucose feed and by maintaining the oxygen requirement they obtained ethanol productivities of 7 g l^{-1} h^{-1} without recycle and 28 g l^{-1} h^{-1} with recycle (see above). At 50 mmHg pressure, it was possible to use a

33% (w/v) glucose feed but it was necessary to bleed out a proportion of the liquor to avoid accumulations of non-volatile inhibitors; at the optimum 'bleed : boil' ratio, the steady-state ethanol was around $40 \text{ g l}^{-1} \text{ h}^{-1}$. By taking the excess liquor from the settler supernatant and running with cell recycle the productivity was raised to over $80 \text{ g l}^{-1} \text{ h}^{-1}$. This was attained at a very high cell density (12% (w/v)) and mixing at this density was undoubtedly facilitated by the combined effects at reduced pressure of boiling, CO_2 evolution and the O_2 stream.

The very considerable productivity advantages of systems of this kind are not obtained without cost. The complexity added by the vacuum system and the requisite heating and cooling controls is considerable. Moreover, the thermodynamic costs are also disadvantageous (Abrams, 1975). Unnecessary work is done in bringing the evolved CO_2 back to atmospheric pressure, though this could probably be avoided by minor changes to the system. More seriously, even though a relatively concentrated alcohol can be condensed from the vapour (16% (w/v) in the work of Cysewski & Wilke), the vapour pressure at which it is removed is largely contributed by the fermentation water which has an even higher latent heat of evaporation. The effect is exacerbated by the whole process being designed to maintain the minimum level of alcohol in the fermenting liquor and it would be a useful improvement if the system could be controlled more precisely, so that the alcohol level were maintained at the highest level compatible with a good rate of fermentation, possibly about 5%. An alternative would be to remove the alcohol by entrainment distillation, i.e. in a vapour stream provided by a third component of high vapour pressure and low latent heat. Such an approach using hexane as the carrier has been outlined (Finn & Boyajian, 1976) but without practical details; the search for a suitable carrier liquid combining the necessary physical properties (water-immiscibility, low boiling-point, and low latent heat) with biological inertness and large-scale plant safety is likely to be difficult, though the approach is theoretically sound (Abrams, 1975).

The thermodynamic disadvantages would be less significant if the plant were sited so as to make use of relatively low-grade heat from another installation and they would also be greatly reduced if a fermentation could be run at 50 °C, or higher, with a suitable organism (see below).

Fixed cells

A quite different approach to the problem of increasing productivity is to use a plug-flow system in which alcohol inhibition only affects the last phases of the fermentation. This can be done, as already noted, by using a concentrated yeast liquor in a long tubular reactor but it is necessary to interrupt such a reactor rather frequently to allow the CO_2 to escape (Ehnstron, 1976). Alternatively, the substrate can be made to flow past stationary yeast cells and, with the advent of a range of methods for obtaining moderately viable 'fixed' cells (Abbott, 1977), several groups have reported on the use of column reactors containing immobilized yeast. For example, Griffith & Compare (1976) used yeast absorbed in gelatin, coated on a glass packing and cross-linked with glutaraldehyde. In prolonged running, they could produce a column effluent with 14–15% (w/v) alcohol at residence times of 2–8 h. Such productivities, between 20 and 80 g l^{-1} h^{-1} of ethanol, are comparable to those in the most efficient of conventional fermentations. Other immobilization methods and support materials have been reported (Kennedy, Barker & Humphreys, 1976; Kierstan & Bucke, 1977; Gaddy & Sitton, 1978; Mavituna & Sinclair, 1978). It appears surprisingly easy to maintain yeast viability in such systems, though no really detailed study has yet been published, while their capacity to produce relatively concentrated ethanol at short residence times is particularly promising, especially where fairly clean substrates free from suspended solids are available.

It will be difficult to scale up suitable industrial reactors from the column systems devised in most laboratories because of the need to provide for CO_2 release. A column producing 50 g l^{-1} h^{-1} of ethanol releases 0.4 l CO_2 min^{-1} for each litre of its liquid capacity and all of this must pass out through the top of the column. Therefore, either a large array of rather short columns must be used or some other arrangement devised in which the advantage of plug flow kinetics is retained.

PRODUCT RECOVERY

Conventionally, fermentation alcohol is recovered from spent liquor by one or more stages of distillation. Solids (including surplus yeast) may be separated previously, leaving only soluble residues in the stillage, or they may be carried through the first still to give stillage residues of higher solids content. This has a considerable effect both on still design and on final treatment methods.

The energy requirement for distillation has been much discussed. It is very sensitive to the proportion of alcohol in the liquor to be distilled; the heat required to distil unit weight of alcohol from a 6% (w/v) liquor is some 1.6 times greater than from a 12% (w/v) liquor. The most detailed recent study (Abrams, 1975) emphasizes that this is true whatever the pressure at which distillation occurs and adds the requirement that both the heating and the cooling (condenser) temperatures must be practical. Only in entrainment distillation (see above) would the heat requirement be substantially changed; on the other hand the actual cost of the heat requirement depends very much on local circumstances (including the on-site availability of waste heat from other processes).

It may therefore be found that a reduction in fermenter productivity entailed by running to higher alcohol levels is more than compensated by a reduction in total heat requirement. Considering the newer systems described above, reduced-pressure systems, evaporating alcohol from a low steady-state concentration, are at a disadvantage compared with the fixed-cell systems. However, the heat requirements for stillage treatment may be equally important and have often been left out of consideration. A conventional molasses distillery produces approximately one tonne of BOD (see Slater & Somerville, this book, p. 230) for every tonne of alcohol, and the 'population equivalent' of a 100 m³ per day molasses distillery is a city of 1.7 million people (Jackman, 1977). For such cases, Jackman has argued very cogently that the most feasible treatment system is evaporation of the stillage to a stabilized residue (Bass, 1975) for animal feed (or even fertilizer) use. In such a case the total heat required for the entire process is determined by the water content of the substrate and we have seen that the distillation–fermentation systems are particularly advantageous in their ability to use concentrated substrates, since the process water throughout is minimal.

SOME CONCLUSIONS FOR THE YEAST FERMENTATION

Combining traditional practice with some of the newer data we can reach some provisional conclusions regarding future directions in fermentation alcohol production. Systems must be specifically adapted to specific combinations of local conditions – type of substrate, local technical capability, local heat costs, local environmental requirements and such economy of scale as these combined factors permit. The range of systems now becoming available will allow this adaptation to be much more precise than has hitherto been possible. We should not expect there to be a unique solution for all times and all places, but to make correct decisions we shall need a fuller body of working experience with some of the newer systems, particularly on pilot-plant scale.

OTHER ORGANISMS, OTHER PRODUCTS

Classically, anaerobic fermentations with organisms other than yeasts have been run on a very large scale in the past and, in the face of competition from petrochemicals, their decline was even more spectacular than that of the industrial alcohol fermentation. Until very recently there has been virtually no new work on these fermentations so that to review them was mainly an exercise in fermentation history (Bu'Lock, 1975; Hastings, 1978).

The potentially useful fermentations are with a variety of *Clostridium* spp. which will ferment a wide range of carbohydrates to organic acids (principally acetic, propionic, butyric and lactic), hydrogen and CO_2 or, with a lower hydrogen evolution, to mixtures of ethanol, acetone, isopropanol and n-butanol. The products are nearly always a mixture and the significant proportion of hydrogen evolved clearly represents a waste of reducing-power. The classic acetone–butanol fermentation was particularly wasteful, since the acid products which accumulated in the early part of the fermentation were never completely reduced in the later stages. Despite this, the original process can still be economically operated provided the high fuel requirement can be met cheaply (Lurie, 1975). The basic problems of the butanol processes are, indeed, similar to some of those already considered for the yeast alcohol fermentation, namely product toxicity and the heat needed for product separation. In the case of butanol these problems become even more acute, however.

From an analysis of the problems posed by these bacterial fermentations, and of possible directions for their improvement (Bu'Lock, 1978), the need to develop methods for isothermal product removal concurrent with the fermentation appears paramount. This is also, as we have seen, one route towards improvements in the alcohol fermentation; in discussing this approach it was pointed out that the most economical way of implementing it depended upon the development of organisms capable of carrying out the fermentation at substantially higher temperatures and that such organisms would almost certainly be found amongst the bacteria rather than the yeasts.

The first steps towards finding such organisms are now being reported and they are relevant both to ethanol production and to the revival of other anaerobic fermentations of industrial interest. In one such study, Wiegel & Ljungdahl (1978) have screened soil isolates for clostridia with temperature optima for growth around 68–70 °C, and obtained isolates capable of quite rapid fermentations of a variety of substrates, including glucose, sucrose, starch and cellobiose, to mixtures in which either acetic acid, hydrogen and CO_2 or ethanol and CO_2 were the predominant components. Another thermophilic clostridium capable of carrying out a direct fermentation of suitably pre-treated cellulose to a mixture of ethanol, acetic acid, hydrogen and CO_2 has also been investigated (Weimer & Zeikus, 1977). We may expect a spate of such reports in the near future and industrial interest is certain to be rekindled but it will be some time before the technical aspects of the use of such organisms can be adequately explored. Yeast may have been 'immobilized' but it is not yet 'spent'.

REFERENCES

ABBOTT, B. J. (1977). Immobilized cells. *Annual Reports on Fermentation Processes*, **1**, 205–33.

ABRAMS, H. J. (1975). Distillation, and the reduction of product separation costs in fermentation processes. In *Octagon Papers Two: Large-Scale Fermentations for Organic Solvents*, eds J. D. Bu'Lock & A. J. Powell, pp. 49–67. Manchester: University Department of Extra-Mural Studies.

BASS, H. H. (1975). Process for the preparation of fertilizers and animal feed from molasses fermentation residues. Australian Patent 73867/74 (December 4, 1975).

BU'LOCK, J. D. (1975). Acetone–butanol, butanediol and other fermentations. In *Octagon Papers Two: Large-Scale Fermentations for Organic Solvents*, eds J. D. Bu'Lock & A. J. Powell, pp. 5–19. Manchester: University Department of Extra-Mural Studies.

BU'LOCK, J. D. (1978). The acetone-butanol fermentation. In *Utilisation Industrielle du Carbone d'Origine Végétale par Voie Microbienne*. Toulouse: INRA.

COWLAND, T. W. & MAULE, D. R. (1966). Effects of aeration on the growth and metabolism of *Saccharomyces cerevisiae* in continuous culture. *Journal of the Institute of Brewing*, **72**, 480–8.

CYSEWSKI, G. R. & WILKE, C. R. (1976). Utilization of cellulosic materials through enzymatic hydrolysis. I. Fermentation of hydrolyzate to ethanol and single-cell protein. *Biotechnology and Bioengineering*, **18**, 1297–313.

CYSEWSKI, G. R. & WILKE, C. R. (1977). Rapid ethanol fermentations using vacuum and cell recycle. *Biotechnology and Bioengineering*, **19**, 1125–43.

EHNSTRON, L. K. J. (1976). Method for carrying out continuous fermentation. British Patent 1 422 076 (January 21, 1976).

ENGELBART, W. & DELLWEG, H. (1976). Continuous alcoholic fermentation by circulating agglomerated yeast. In *Abstracts Fifth International Fermentation Symposium*, ed. H. Dellweg, p. 48. Berlin: Westkreuz.

FINN, R. K. & BOYAJIAN, R. A. (1976). Preliminary economic evaluation of the low-temperature distillation of alcohol during fermentation. In *Abstracts Fifth International Fermentation Symposium*, ed. H. Dellweg, p. 376. Berlin: Westkreuz.

GADDY, J. L. & SITTON, O. C. (1978). Immobilized cell systems for production of ethanol and methane. Paper presented at First European Congress of Biotechnology, Interlaken, 1978.

GOLDEMBERG, J. (1978). Brazil: Energy options and current outlook. *Science*, **200**, 158–64.

GOMA, G. & STREHAIANO, P. (1978). Kinetics of ethanol production by fermentation. Paper presented at First European Congress of Biotechnology, Interlaken, 1978.

GRIFFITH, W. L. & COMPARE, A. L. (1976). A method for coating fermentation tower packing so as to facilitate micro-organism attachment. *Developments in Industrial Microbiology*, 17, 241–6.

HAMMOND, A. L. (1978). Energy: Elements of a Latin American strategy. *Science*, 200, 753–4.

HARRISON, J. S. & GRAHAM, J. C. J. (1970). Yeasts in distillery practice. In *The Yeasts*, vol. 3, eds A. H. Rose & J. S. Harrison, pp. 283–348. London, New York: Academic Press.

HASTINGS, J. J. H. (1978). Acetone–butanol fermentation. In *Economic Microbiology*, vol. 2. *Primary Products of Metabolism*, ed. A. H. Rose, pp. 31–45. London, New York: Academic Press.

HEPNER, L. (1978). The feasibility of basic chemicals for fermentation processes. *Engineering and Process Economics*, 3, 17–23.

HOUGH, J. S. & BUTTON, A. M. (1972). Continuous brewing. *Progress in Industrial Microbiology*, 11, 89–132.

INGRAM, M. (1955). *Introduction to the Biology of Yeasts*. London: Pitman.

JACKMAN, E. A. (1976). Brazil's National Alcohol Programme. *Process Biochemistry*, 11(5), 19–30.

JACKMAN, E. A. (1977). Distillery effluent treatment in the Brazilian National Alcohol Programme. *The Chemical Engineer*, 1977, 239–42.

KENNEDY, J. F., BARKER, S. A. & HUMPHREYS, J. D. (1976). Microbial cells living immobilized on metal hydroxides. *Nature, London*, 261, 242–4.

KIERSTAN, M. & BUCKE, C. (1977). The immobilization of microbial cells, subcellular organelles and enzymes in calcium alginate gels. *Biotechnology and Bioengineering*, 19, 387–97.

KODAMA, K. (1970). Saké yeast. In *The Yeasts*, vol. 3, eds A. H. Rose & J. S. Harrison, pp. 225–82. London, New York: Academic Press.

LIPINSKY, E. S. (1978). Fuel from biomass: integration with food and materials systems. *Science*, 199, 644–51.

LURIE, J. (1975). Discussion contributions. In *Octagon Papers Two: Large-Scale Fermentations for Organic Solvents*, ed. J. D. Bu'Lock & A. J. Powell, pp. 18–19, 70–6. Manchester: University Department of Extra-Mural Studies.

MCCANN, D. J. & SADDLER, H. D. W. (1976). Photobiological energy conversion in Australia. *Search*, 7, 17–23; see also *A Cassava-based Agro-industrial Complex – a New Industry for Australia?* Sydney: Energy Research Centre, University of Sydney.

MAVITUNA, F. & SINCLAIR, C. G. (1978). The efficiency of gel-entrapped microorganisms as biocatalysts. Paper presented at First European Congress of Biotechnology, Interlaken 1978.

MILLER, D. L. (1975). Ethanol production and potential. *Biotechnology and Bioengineering Symposia*, 5, 345–52.

NAGODAWITHANA, T. W. & STEINKRAUS, K. H. (1976). Influence of the rate of ethanol production and accumulation on the viability of *Saccharomyces cerevisiae* in 'rapid fermentation'. *Applied and Environmental Microbiology*, 31, 158–62.

O'LEARY, V. S., GREEN, R., SULLIVAN, G. C. & HOLSINGER, V. H. (1977). Alcohol production by selected yeast strains in lactase-hydrolyzed acid whey. *Biotechnology and Bioengineering*, 19, 1019–35.

RAINBOW, C. (1970). Brewers yeasts. In *The Yeasts*, vol. 3, eds A. H. Rose & J. S. Harrison, pp. 147–224. London, New York: Academic Press.

RAJIC, F. & ATCHISON, J. (1978): unpublished, cited in Lipinsky (1978).

RAMALINGHAM, A. & FINN, R. K. (1977). The Vacuferm process: a new approach to fermentation alcohol. *Biotechnology and Bioengineering*, **19**, 583–9.

ROSE, A. H. & HARRISON, J. S. (eds) (1971). *The Yeasts*, vol. 2. London, New York: Academic Press.

ROSE, D. (1976). Yeasts for molasses alcohol. *Process Biochemistry*, **11**(2), 10–12, 36.

ROSEN, K. (1978). Continuous production of alcohol. *Process Biochemistry*, **13**(5), 25–6.

STOKES, J. L. (1971). Influence of temperature on the growth and metabolism of yeasts. In *The Yeasts*, vol. 2, eds A. H. Rose & J. S. Harrison, pp. 119–34. London, New York: Academic Press.

WEIMER, P. J. & ZEIKUS, J. G. (1977). Fermentation of cellulose and cellobiose by *Clostridium thermocellum* in the absence and presence of *Methanobacterium thermoautotrophicum*. *Applied and Environmental Microbiology*, **33**, 289–97.

WIEGEL, J. & LJUNGDAHL, L. G. (1978). Extreme thermophilic Clostridia isolated from soil. *American Society of Microbiology: Abstracts of Annual Meeting*, p. 93.

MATHEMATICAL MODELLING OF FERMENTATION PROCESSES: SCOPE AND LIMITATIONS

NICOLAAS W. F. KOSSEN

Biotechnology Group, Department of Chemical Engineering,
Delft University of Technology, Jaffalaan 9, Delft,
The Netherlands

List of symbols

Symbol	Specification	Units
a	Specific surface	m^{-1}
A	Surface area	m^2
C	Concentration	$g\ l^{-1}$ ($g\ m^{-3}$)
D	Dilution rate	s^{-1} (h^{-1})
$\mathbb{D}$	Diffusion coefficient	$m^2\ s^{-1}$
E	Arbitrary extensive property	
G	Gibbs free energy	$J\ mole^{-1}$
H	Equilibrium constant	$g\ l^{-1}\ atm^{-1}$
j	Flux	$g\ m^{-2}\ s^{-1}$
J	Flow	$g\ s^{-1}$
k_1	Mass-transfer coefficient	$m\ s^{-1}$
K_{eq}	Equilibrium constant	
K_s	Monod saturation constant	$g\ l^{-1}$
K_w	Mass conductivity	$l\ s^{-1}$
m_s	Maintenance factor	s^{-1}
M	Equilibrium constant	
N	Number of organisms	
p	Pressure	atm
Pe	Péclèt number	
q	Amount of heat	J (Joule)
r	Production rate (mass)	$g\ l^{-1}\ h^{-1}$
r'	Production rate (numbers)	$l^{-1}\ h^{-1}$
R	Gas constant	$J\ mole^{-1}\ °K^{-1}$
t	Time	s (h)
T	Temperature	$°K$
U	Internal energy	$J\ mole^{-1}$
v	Velocity	$m\ s^{-1}$
V	Volume	l

Symbol	Specification	Units
w	Work	J mole^{-1}
x	Distance	m
Y	Yield	
α	Growth-rate parameter	
β	Conversion fraction	
λ	Death rate	s^{-1} (h^{-1})
μ	Specific growth rate	s^{-1} (h^{-1})
μ_{max}	Maximum specific growth rate	s^{-1} (h^{-1})
ϕ	Volumetric flow	l s^{-1}

Subscripts

b	Due to bulk flow
d	Due to diffusion
E	Extensive property
g	Gas
i	At the interface
l	Liquid
M	Micro-organisms
N	Numbers
OM	Oxygen to micro-organisms
s	Substrate
SM	Substrate to micro-organisms

1. INTRODUCTION

This chapter is aimed primarily at those microbiologists not familiar with mathematical modelling and who would like to know more about the basic principles of this subject. It is not meant as a survey of all the possible mathematical techniques that can be used in fermentation modelling. The author, therefore, has tried to keep the mathematics as simple as possible.

Before starting a discussion about mathematical models a few words have to be said about models in general. A model is a simplified image of a part of reality. Manipulation of the model gives information about the behaviour of that particular part of reality, at least about those aspects that the model stands for. This is also the main reason to use models: they give us a grasp on reality. Finally a model must always be formulated in such a way that it can be refuted.

One can distinguish different kinds of models: verbal (qualitative) models; scale models (e.g. a model of a DNA molecule); mathematical (quantitative) models etc. A mathematical model consists of one or more equations which give us insight in the quantitative behaviour of a part of reality. It is, like any model mentioned above, always a simplification and concerned with a restricted number of aspects of reality. Mathematical models can be subdivided into different classes (see section 2).

An example of a verbal model is the following statement: 'the growth rate of *Escherichia coli* on glucose increases with increasing substrate concentration until a certain maximum is reached'. This model tells us how a part of reality (*E. coli*) behaves with regard to the aspect 'growth'. It can also be refuted (which would be the case if growth rate decreased with increasing glucose concentration or if no maximum is reached). A possible mathematical model for this phenomenon is the following equation:

$$\mu = \mu_{max} \frac{C_s}{K_s + C_s} \tag{1}$$

This equation also implies that the specific growth rate (μ) increases with increasing substrate concentration (C_s) to a certain level (μ_{max}), as the verbal model does. But it tells more. If μ_{max} and K_s are known, and if the model is right, it gives us the exact form of the dependence of growth rate on substrate concentration. This offers a much more rigorous opportunity to refute the model. This advantage of mathematical models, together with other advantages (and disadvantages) is the subject of Section 3. With regard to fermentation processes, different kinds of mathematical models can be grouped together. This is the subject of Section 4 which is the kernel of this paper. Sections 5 and 6 deal with model-building and limitations of mathematical modelling respectively.

In every natural science there is a tendency away from verbal toward mathematical modelling. Microbiology is by no means an exception to this rule. The book by Dean & Hinshelwood (1966) is an old but still very good example of mathematical modelling in microbiology. A vast and rapidly increasing amount of modern literature has been devoted to this subject. A number of books have appeared which deal with the tools of mathematical modelling, mathematics, with special emphasis on (micro)biological problems (Grossman & Turner, 1974; Hadeler, 1974; Rubinow, 1975). It is probably useful to make the remark that verbal models are not necessarily inferior to mathematical ones. Often no quantitative model is needed. If, for example, a verbal model predicts an increase in growth rate with increasing substrate concentration, while the experiment shows a decrease, then it is useless to translate the verbal model into a mathematical one. Furthermore, verbal models always precede mathematical models and usually are more suitable for communication.

Finally, it is important to mention an essential distinction between models: descriptive or 'black box' models on one hand and predictive or 'grey box' or explanatory models on the other hand. A descriptive model only tells us *how* the system behaves and *not why*. It can be either verbal or mathematical. The mathematical descriptive models are usually the outcome of curve-fitting of experimental results, a popular but hazardous activity if one uses the results for extrapolation. Extrapolation of a descriptive model is always dangerous because if one does not know why the system behaves as it does under the experimental circumstances used one has no guarantee at all that it will behave in a similar way under other circumstances. Predictive models contain the (most important) mechanisms that govern the system's behaviour (they are therefore also called mechanistic models). If these mechanisms are understood, it can be predicted how the system behaves under circumstances different from the ones used during the experiments. Of course one has to stay within the limits of validity of the mechanisms. More details are given in a publication by Roels & Kossen (1978).

2. CLASSES OF MATHEMATICAL MODELS

As mentioned in the preceding section, there exist a number of different classes of mathematical models. For a particular problem it is very important to choose the right class of model. This choice also depends on the type of answers one hopes to get from the model, e.g. if one is only interested in time- or place-averaged values the model usually is much simpler than when one is interested in instantaneous or local values.

The problem of choice will be illustrated with an example of kinetics of microbial growth. In general, growth is described with a limited set of simple equations:

$$\mu = \mu_{\max} \frac{C_s}{K_s + C_s} \tag{1}$$

$$r_M = \mu \, C_M \tag{2}$$

r_M is the increase in dry mass (expressed in g l^{-1} h^{-1} or equivalent units).

For a batch culture, $r_M = dC_M/dt$, i.e. r_M is the slope of the curve that represents biomass as a function of time. For a continuous culture $r_M = D\,C_M$ (D is dilution rate). Equations of this kind implicitly use the assumption that growth can be expressed as the increase of biomass as such. Nothing is said about the structure (or composition or quality) of biomass. We therefore use the term *unstructured model* for this kind of mathematical model. This kind of model can only describe growth phenomena under conditions where the composition does not change, so-called balanced growth. If the growth is non-balanced, as is the case in a great number (probably the majority) of non-laboratory situations, then a model has to be used that can describe the change in structure of the micro-organism. We call these *structured models*. Furthermore, the term biomass is used as a quantity averaged (distributed) throughout the culture fluid. We use the term *distributed model* for this kind of approach. If the cellular nature of microbial life is taken into account, one uses a *segregated model* – for example

$$r_M{}' = \mu_N N_M \tag{3}$$

where N_M is the number of cells (per unit volume). There are many situations where the cell mass changes without a similar change in cell numbers (e.g. the onset of a batch culture with non-adapted cells). Under those circumstances equations (2) and (3) give quite different results.

If equations (1) and (2) are used for the calculation of biomass as a function of time, then the situation is completely determined, given the values of μ_{max} and K_s (and C_{M0} and C_{s0}, the concentrations of biomass and substrate at the beginning). There is no room for uncertainty in this model. Models with no room for uncertainty are called *deterministic models*.

If one takes uncertainty into account by appointing a probability for an event to take place one has a *stochastic model*. Stochastic models are often necessary for sterilization processes (Fredrickson, 1966) and for growth of yeast on hydrocarbon droplets (Verkooijen, 1977). A survey of stochastic models is given by Goel & Richter-Dyn (1974). Another hidden assumption of model equations (1) and (2) is that events happen continuously in time (*continuous-time models*). This is a good approximation if many micro-organisms are present, growing in a non-synchronous way. If we have a synchronous culture and/or a small number of micro-organisms, then events do not happen continuously in time; we then have to use a *discrete-time model* such as

$$N_{t+\Delta t} = N_t + \alpha\,N_t \tag{4}$$

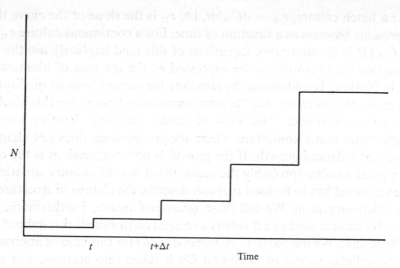

Fig. 1. Growth of a synchronous culture as described by a discrete-time model.

This equation results in a growth curve like the one shown in Fig. 1. Finally, the model equations (1) and (2) are generally used in situations without concentration gradients, i.e. in well stirred fermenters. Physically this means that the time for homogenization of the fermenter contents is small compared with other 'time constants' such as the time to consume the substrate added and the doubling time of the micro-organisms. This also is a situation that only exists at laboratory scale. In non-laboratory cultures one often has considerable gradients in substrate concentration and temperature and sometimes also in biomass concentration. Gradients particularly occur in continuous and fed batch cultures. Usually the average substrate concentration is then very low. This means that in the more or less stagnant regions of the fermenter the substrate concentration is essentially zero. Near the substrate entrance the concentration is relatively high. This effect increases with increasing scale and viscosity.

If there are no gradients one can use a *lumped-parameter model*. In case of gradients one has to use a *distributed-parameter model*. This last expression is a most unfortunate one, because we have used the term 'distributed' already in a different meaning. It would be better to use the term *gradient model* (as opposed to *non-gradient* or lumped-parameter model).

The mathematical models mentioned above are grouped together as contrasting pairs in Table 1.

From the fact that equations (1) and (2) belong to the group of unstructured, distributed, deterministic, continuous-time models, one comes to the conclusion that any mathematical model belongs to a number of classes of models. A survey of the above-mentioned classes of models is given by Fredrickson, Megee & Tsuchiya (1970) and, more recently by Roels & Kossen (1978).

Table 1. *Contrasting pairs of groups of mathematical models*

Unstructured	Structured
Distributed	Segregated
Deterministic	Stochastic
Continuous-time	Discrete-time
Lumped-parameter (non-gradient)	Distributed-parameter (gradient)

The importance of the right choice of model can be illustrated with experimental results obtained by van der Beld (1976). A batch of activated sludge suspension is starved for a given time. Then suddenly an amount of substrate is added (skim milk). The TOC (total organic carbon, see Slater & Somerville, this book pp. 230–1) and the oxygen consumption rate found are given in Fig. 2. The substrate is consumed

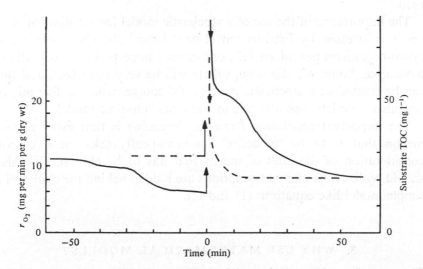

Fig. 2. Total organic carbon (TOC, broken line) and rate of oxygen consumption (r_{O_2}, solid line) in a suspension of activated sludge. The batch was starved to time 0, then substrate was added. Data of van der Beld (1976).

almost instantaneously and oxygen consumption during this period is extremely high. Similar results have been obtained in our laboratory with glucose and nitrate. From these results it could be concluded that when using equations (1) and (2) and the equation

$$r_s = \frac{1}{Y_{SM}} r_M + m_s C_M \qquad (5)$$

extremely high growth rates were obtained ($\mu = 10$ h^{-1}).

As van der Beld shows, the phenomena in Fig. 2 can be explained by assuming that the substrate is first stored in a buffer in the micro-organisms. This process takes place within minutes. Then, slowly, this stored substrate is used for synthesis of cell components. The cell is thus assumed to consist of two compartments: a buffer and the rest of the cell, so van der Beld uses a two-compartment model in this situation. A two-compartment model is the simplest form of a structured model, it can be very effective and can explain a lot of growth phenomena under unsteady-state conditions. This has already been pointed out by Williams (1967). These models have to be used with care, as shown by Fredrickson (1976) and in one of the forementioned publications (Roels & Kossen (1978)). The conclusion from this example is that, in order to explain the behaviour described by Fig. 2, a structured model has to be used.

The importance of the use of a stochastic model for sterilization processes is stressed by Fredrickson (1966). Especially when only a few micro-organisms per ml are left, one cannot hope that they will all behave alike. Some will die soon, others will be very persistent, and this can be treated as a stochastic process. Of course, when so few micro-organisms are left, one also has to use a discrete-time model.

The important conclusion from the foregoing is that every phenomenon that is to be 'modelled' mathematically asks for a careful consideration of the kind of model that has to be chosen. One also should realize how many assumptions are hidden behind the choice of a simple model like equations (1) and (2).

3. WHY USE MATHEMATICAL MODELS?

The use of mathematical models is by no means a necessity for all scientists. Generations of microbiologists lived without them in a period in which the development of microbiology was impressive. Nevertheless there is, as said already, a tendency to use more and more mathematical models in every natural science. Four reasons for this tendency are discussed below.

(i) *The construction of a mathematical model asks for a precise formulation of the problem under investigation*

Questions that have to be answered follow immediately from the kinds of mathematical models summarized in Section 2:

(*a*) Can we assume balanced growth? If not, what are the essential compartments in the micro-organism that have to be distinguished?

(*b*) Do we have to take into account the cellular nature of micro-organisms? If so, do we have to distinguish between different states of the cells (composition, morphology, development stage of an organism like, for example, *Hyphomicrobium*)?

(*c*) Can we describe the situation as a deterministic process or do we have insufficient information, so that the system appears to be of a stochastic nature?

(*d*) Do the events take place evenly distributed in time?

(*e*) Can we assume the culture to be well mixed, or do gradients occur?

(ii) *Mathematical models open the possibility for rigorous testing of a theory with the aid of advanced statistical techniques*

This very important point has been mentioned already in the introduction. Verbal models only give an expected direction of a change (e.g. increase or decrease), mathematical models not only give the expected direction of change, but also the magnitude of change.

(iii) *Mathematical models can lead to a more efficient experimentation*

When the model has been constructed, the parameters and variables in the model can be varied one by one, and calculations can be made of how sensitive the outcome of the model (for example biomass production) is for these variations. This procedure is called *sensitivity analysis*. Those quantities that have a minor influence on the outcome do not have to be investigated thoroughly.

(iv) *Mechanistic mathematical models, when tested carefully, give the opportunity for predictive quantitative statements about the system behaviour*

This opens the way for process design and control in fermentation industry and biological wastewater treatment.

Of course there are also drawbacks to mathematical modelling. Some of them are listed below:

Figural fetishism

People are in general more impressed by figures than by facts. Extensive mathematical modelling, usually followed by a massive computer outcome, often leads to the conclusion that 'this must be true'. It is probably worthwhile to mention here a statement by May (1973): 'Be it added that some other massive computer studies could benefit most from the installation of an on-line incinerator.'

The attitude of believing in figures has been enforced by the fact that few people can really look through the model and the computer work. The majority do not want to admit their lack of capability at this point, and surrender. This shows a second possible disadvantage:

Complexity

Quite a few authors use very complex models that only a small in-crowd can understand. The models contain lots of parameters that should be determined independently, but often are used to 'fit' the model with the experiments. We will come back to this point in section 6. Some other authors start their publications with a set of impressive partial differential equations with three or four independent variables (time and three lateral dimensions), preferably written in vector notation. After three pages it all boils down to a simple equation, often already well known. The majority of the harmless readers have already given up before this stage, however. Complexity leads to a following disadvantage:

High energy of activation of the reader

The more complex the model, the higher the energy barrier the reader has to overcome to struggle through it.

Straight-line syndrome

Many of the more simple mathematical models lead to a graphical representation in the form of a straight line (the Lineweaver–Burk plot, the Eadie–Hofstee plot and the like). These plots already have the disadvantage that the experimental points cluster in a relatively small area. Often, however, objective observation of the experimental points suggests a non-straight curve, but because the model asks for a straight line such a line is drawn. Many publications suffer from this habit. A good publication that analyses this habit is due to Rowe (1963), albeit a publication dealing with an example from chemical engineering.

On the whole, mathematical modelling can have great advantages. The computer has taken over the dull calculations. Computer languages are more and more 'user orientated', for example CSMP (Continuous Systems Modelling Program) from IBM, a language of which the essentials can be learned within one day. Mathematical models, however, should always be a result of, and not a substitute for, creative thinking. Finally it is very important to choose 'the difficult path between unrealistic oversimplicity and unwieldly and untestable complexity' (Watt, 1975).

4. MODELS IN FERMENTATION

Introduction

This section deals with an alternative subdivision of models, with special emphasis on fermentation processes. It is not in conflict with the subdivision given in Section 2 but just another way of grouping things together, in some respects in a more functional way.

As an introduction it should be mentioned that microbial life is essentially chemical in nature. In a living cell hundreds of chemical reactions take place usually in a well balanced way. The speed of chemical reaction (or the reaction rate) is described with a group of mathematical models called *kinetic equations*. This rate is a function of the concentration of the reactants. These reactants have to be transported to the area when the reaction takes place; this transport is described with a second group of models, *transport equations*. Finally, no reaction and no transport is really complete and they both tend to an equilibrium. This equilibrium is described with a third group of models, *thermodynamic equations*. These equations can be put together in models that simultaneously describe kinetics and transport; this group of models forms *balance equations*. They can either describe the situation at one point of the system (microbalance) or the situation in the system as a whole (macrobalance). More subtle subdivisions are possible (Himmelblau & Bischoff, 1968). An example of a macrobalance is the substrate balance for a well mixed continuous fermenter (Fig. 3) with constant volume:

$$V \frac{dC_s}{dt} \quad = \quad J_{s_{in}} \quad - \quad J_{s_{out}} \quad - \quad Y_{SM}^{-1} \, C_M \, \mu_{max} \frac{C_s}{K_s + C_s} \qquad (6)$$

Accumulation   Transport in   Transport out   Rate of reaction

In this balance we see that something disappears (substrate). There are also balances for conservative properties like elements, total mass or total energy; these properties cannot form or disappear (in absence of nuclear reactions). Their balance equations do not contain a kinetic

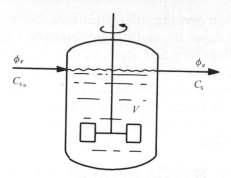

Fig. 3. A well mixed continuous fermenter.

term. The groups of models that can be distinguished are summarized in Table 2.

Table 2. *Models for fermentation*

Kinetic equations
Transport equations
Thermodynamic equations
Balances: micro ⎱ for conservative or
        macro ⎰ non-conservative properties

Fundamentally, any of these groups of models can be stochastic or deterministic, structured or unstructured etc. All the models of Table 2 will now be dealt with successively.

Kinetic models

For kinetic models, the distinction between unstructured and structured models is essential. In section 2 it was mentioned that unstructured kinetic models are valid only when growth is balanced. For batch growth this means that growth must be in the exponential phase (Fredrickson *et al.*, 1970). For continuous cultures, growth is balanced only under stationary conditions. The subject will be treated only briefly here, because an extensive literature survey on kinetic models has been published recently (Roels & Kossen, 1978).

Unstructured kinetic models

The best-known unstructured kinetic model is due to Monod (1950)

$$r_{\mathrm{M}} = \mu \, C_{\mathrm{M}} = \mu_{\max} \frac{C_{\mathrm{s}}}{K_s + C_{\mathrm{s}}} \, C_{\mathrm{M}} \tag{7}$$

This equation is a combination of equations (1) and (2). Generally, equation (7) is used in combination with some linear law for substrate consumption (like equation (5)). This combination describes a lot of experimental results reasonably well, but so do a lot of other equations. As Yang & Humphrey (1975) showed, experimental results for phenol-limited growth could be described equally well (from a statistical point of view) by five different growth equations. Boyle & Berthouex (1974) also stress the point that the Monod equation is by no means the only suitable one. Furthermore, they make clear that closeness of fit is not a sufficient condition for model acceptance, and that a critical consideration of the parameters, obtained with statistical methods, is always necessary. We will come back to this problem in Section 6.

Many variations are possible, either on the Monod equation or on one of the other equations mentioned. One can take into account effects like growth on more than one substrate as Tsao & Hanson (1975) do, or effects like inhibition (Yang & Humphrey, 1975).

Structured kinetic models

As said before, unstructured models can only be used if growth is balanced. This means that the physiological state of the organisms should remain unchanged. This physiological state is expressed in terms of concentrations of proteins, RNA, DNA and the like. Usually it is assumed that these concentrations are sufficient to describe this state. However, as Fredrickson (1976) said, 'it does not allow one to account for the geometrical structure of cells, nor for the dependence of diffusional processes upon such structures'. It is like describing a house in terms of percentages of brick, cement, wood and plastic instead of the number and size of the rooms, insulation, solidity etc. Nevertheless the percentages mentioned above give more information than the simple statement: 'it's a house'. Concerning micro-organisms, even simple structured models (two-compartment models) are already very effective in describing unbalanced growth phenomena (Roels & Kossen, 1978).

The complexity of structured models can be expressed as the number of compartments in the cell that are treated separately in the model. The minimum of compartments is two, an example is the model of Williams (1967) (a one-compartment model is by definition unstructured); the maximum is reached when every component of the cell (every enzyme, intermediate product, tRNA, mRNA etc.) is dealt with separately. The models used by Garfinkel are at this end of complexity (Garfinkel, 1969).

A three-compartment model for the nutrient dynamics of phytoplankton growing in a nitrate-limited environment has been published by Grenney, Bella & Curl (1973). A model of intermediate complexity has been presented by van Dedem & Moo-Young (1975).

The building of structured models consists of the setting up of a material balance for every compartment. This balance contains terms for accumulation, transport and rate of reaction, just like equation (5). Vector notation is often used as an easy shorthand writing method (Roels & Kossen, 1978). The use of computers to solve the numerous differential equations is inevitable. A number of very useful remarks about the formulation of structured growth models has been given by Fredrickson (1976). His point that a change in concentration of an intracellular component can be caused by growth of the biomaterial alone is often neglected by structured-model builders.

If one wants to use structured models to explain experimental phenomena, and not as *l'art pour l'art*, models with a small number of compartments (two or three) are quite sufficient. Thus the model of Williams gives a semi-quantitative explanation of phenomena observed experimentally, such as lag phase, changing cell composition and changing cell size during batch growth, although the model suffers from some inconsistencies, as mentioned by Fredrickson (1976). Generally speaking the use of simple unstructured models can be very useful, but this is still *terra incognita* in applied microbiology. Concluding this section it has to be mentioned that the publication of Roels & Kossen (1978) contains a number of illustrative examples of the use of unstructured and structured kinetic models.

Transport equations

Growth of micro-organisms is only possible because of transportation of material and energy. These transport processes take place from the surroundings of the cell to the cell membrane, through the membrane, from the membrane to the internal parts of the cell and vice versa. All the heat transport and part of the mass transport is passive, which means that transport goes downhill (in the direction of decreasing concentration or temperature). Partly the mass transport is active, which means uphill transport and the need for an energy source. It is not possible within the scope of this publication even to give a survey of all the possible different transport mechanisms; the reader is referred to Konings (1976) for an extensive literature review. Mathematical models for these processes lie in the domain of the thermodynamics of irreversible processes for which the books of Katchalsky & Curran (1974) and Prigogine (1967) provide a good introduction. The theories of these authors are valid only for small deviations from equilibrium, where flow can be considered to be proportional to the gradient. Biological systems as a whole are usually far from equilibrium; this subject is much more complicated and is dealt with by Glansdorf & Prigogine (1974). For parts of micro-organisms (a membrane system) the close-to-equilibrium concept can often be used, however.

A qualitative idea of 'how things work' is given by Morowitz (1968) (see Fig. 4). Two chambers filled with gas are separated by a membrane with pores of small size compared with the mean free path of the

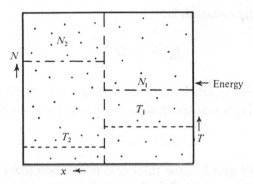

Fig. 4. Transport against a gradient, due to energy flow.

molecules. Heat is transported through the system from right to left ($T_1 > T_2$). Molecules on the right-hand side thus have a higher kinetic energy than on the left-hand side and so have a higher velocity (and collision frequency). This causes a transportation of molecules from the right to the left. A steady state (not an equilibrium!) is reached when the total number of collisions per time on the left- and right-hand side of the membrane are equal. One can easily show that under those circumstances

$$\frac{N_1}{N_2} = \sqrt{\frac{T_2}{T_1}} \qquad (8)$$

This example shows that flow of energy through a system can lead to some kind of structure, and transport against a concentration gradient. A membrane is present here but it is not necessary. Without a membrane, however, the average concentration difference is smaller with the same energy flow. This means the efficiency of the process is lower. When the energy flow stops, disorder results (molecules are evenly distributed over the two chambers). Transport of molecules from the left to the right is then passive (downhill).

From this simple example to the complicated situation in micro-organisms, there is still a long way to go. Progress has been made, however, on many aspects of active transport systems. Mathematical models are numerous in this area but outside the scope of this paper.

The situation for passive transport is much easier. Two mechanisms are of importance: bulk flow and diffusion. Bulk flow is given by:

$$j_b = v \times C \quad \text{(moles per unit area per time)} \qquad (9a)$$

(v is velocity)
For diffusion the equation is:

$$j_d = - \mathbb{D} \frac{dC}{dx} \qquad (9b)$$

($\mathbb{D}$ is the diffusion coefficient, dC/dx is the concentration gradient)
For boundary layer transport, equation ($9b$) is written as:

$$j_d = k_1 \Delta C \qquad (9c)$$

(where $k_1 = \mathbb{D}/\delta$ and δ is the thickness of the boundary layer)
The relative importance of both mechanisms is given by the Péclèt number (Pe) which can be considered as the ratio between j_b and j_d

$$\text{Pe} = \frac{v \times x}{\mathbb{D}} \qquad (10)$$

Pe $= 1$ means that j_b and j_d are roughly of equal importance.
Two examples will illustrate the use of the Péclèt number: for the flow in a tube towards a continuous fermenter the Péclèt number is ($v \approx 10^{-3}$ m s^{-1}, $\mathbb{D} \approx 10^{-9}$ m^2 s^{-1}, $x \approx 0.5$ m) Pe $= 5 \times 10^5$.
This means that bulk transport is far more important than diffusion. To decide whether or not flow of protoplasm inside micro-organisms ($d = 5 \times 10^{-6}$ m) gives a real contribution to transport (compared with molecular diffusion) (with $\mathbb{D} = 10^{-9}$ m^2 s^{-1}) one can calculate the necessary velocity to result in Pe $= 1$:

$$\text{Pe} = 1 = \frac{v \times x}{\mathbb{D}} = \frac{v \times 5 \times 10^{-6}}{10^{-9}}$$

So $v = 2 \times 10^{-4}$ m s^{-1}.
This should mean that every second the protoplasm flows 40 times ($2 \times 10^{-4}/5 \times 10^{-6}$) around in a micro-organism, not a very probable situation. So transport is dominated by diffusion. It is very important

to realize that both kinetics and transport are rate processes. When these processes are in series, which they usually are, then the slower one of the two determines the overall rate. Quite a few publications about kinetics are of limited value because this point has not been recognized. We will come back to this point in the subsection on balances.

Thermodynamics

Classical thermodynamics gives information about conditions for equilibrium, and energy transformations.

Conditions for equilibrium The conditions for equilibrium correspond at constant E and V with a maximum in entropy ($\Delta S = 0$) or under conditions of constant temperature and pressure, with a minimum in the Gibbs free energy ($\Delta G = 0$). Because micro-organisms are usually in an environment of constant temperature and pressure, and since G can be expressed as a function of concentration, G is a very convenient quantity.

For a (chemical) reaction of the form $aA + bB \leftrightarrows cC + dD$, the change in the Gibbs free energy can be written as

$$\Delta G = \Delta G^0 + RT \ln \underbrace{\left(\frac{C_C^c \cdot C_D^d}{C_A^a \cdot C_B^b} \right)}_{K_{eq}} \tag{11}$$

ΔG^0 is the standard free energy change (gain or loss of free energy if one mole of reactant is converted to one mole of product under standard conditions: 25 °C, pH 7 and one-molar concentrations; see Lehninger (1973)). For equilibrium conditions $\Delta G = 0$, so that

$$K_{eq} = e^{-\Delta G^0/RT} \tag{12}$$

Values for ΔG^0 for various reactions can be found in literature (see Lehninger, 1973; Klotz, 1967 and Christensen & Cellarius, 1972). These books are also very suitable for further study of thermodynamics of biological systems.

For simple situations like dissolving oxygen in water, equation (12) reduces to

$$C = H \cdot p \tag{13}$$

where H is still a function of temperature, C is concentration of oxygen in water and p is the partial oxygen pressure.

Equations (11 to 13) are only valid in the low concentration range. In the higher ranges one should use activities instead of concentrations.

Energy transformations Energy transformations are the basis of life. Calculations of energy transformations permit us to give statements about the amount of heat produced when, for example, glucose is oxidized to carbon dioxide and water, or about the maximum possible conversion-efficiencies in micro-organisms. The principle of all energy-transformation calculations is that energy is a conservative property: it can neither be created nor destroyed. This is the verbal form of the first law of thermodynamics. The mathematical form is:

$$\Delta U = q + w \tag{14}$$

where ΔU is the change in internal energy of a system due to the amount of heat added (q) and the amount of work done on the system (w). The second law of thermodynamics formulates restrictions for the possible transformations: in practice it is always possible to convert work quantitatively into heat but the reverse is not always possible.

It is beyond the scope of this publication to go in detail about energy transformations. Of all the numerous publications in this field the one of Roels (1979) is mentioned here in particular, because it gives an excellent survey of the possibilities and limitations of the application of thermodynamics in microbiology.

Balances

Introduction

One can distinguish between micro- and macrobalances for both conservative and non-conservative properties, as mentioned on pp. 337–8. In macrobalances, gradients are neglected because they belong to the class of lumped-parameter models. Equation (6) is an example of such a balance. Macrobalances do not permit the calculation of local concentrations. Microbalances describe the situation in a part of a system that is not homogeneous; an example is the substrate microbalance for a situation of stationary diffusion and consumption of substrate in a layer of micro-organisms:

$$\mathbb{D}\frac{d^2C_s}{dx^2} - r_s = 0 \tag{15}$$

When the rate of substrate consumption (r_s), the diffusion coefficient

(D) and the boundary conditions are known, equation (15) permits the calculation of local substrate concentration (C_s) as a function of distance (x).

In the next two subsections, balances for conservative and non-conservative properties will be dealt with. For conservative properties only macrobalances will be used, for non-conservative properties micro-balances will be used also.

Balances for conservative properties

The balance equation for conservative properties has no production terms. The general form is:

$$\text{accumulation} = \text{inflow} - \text{outflow}$$

Or in mathematical terms:

$$\frac{dE}{dt} = J_{in} - J_{out} \qquad (16)$$

E is an extensive property, J is a flow ($J = j \times A$, where A is the area) and for the steady state, $J_{in} = J_{out}$.

A most important application of a simple balance as mentioned above is the conservation of atomic species. For a batch culture the inflow and outflow of the fermenter are not zero. There is a continuous flow of oxygen and carbon dioxide, and eventually of other volatile products. In the fermenter, carbon and nitrogen go from the liquid phase to the micro-organisms; the micro-organisms excrete carbon dioxide. The subject can be dealt with in different ways. One can split up substrate consumption into maintenance processes, precursor synthesis, biomass synthesis and product formation (Roels & Kossen, 1978), or take the first three processes together. In that case:

$$a\,C_x\,H_y\,O_z + b\,O_2 + c\,NH_3 \rightarrow d\,C_\alpha\,H_\beta\,O_\gamma\,N_\varepsilon + e\,H_2O + f\,CO_2 \qquad (17)$$
carbon/energy source   biomass

A similar equation can be written for product formation. If one defines a fraction of substrate converted to product then balances for carbon, hydrogen, oxygen and nitrogen can be written down. With the measurement of only a few quantities (for example O_2, CO_2 and NH_3) all the other quantities can be calculated (including biomass and yield) if the biomass elemental composition is known. This technique has been presented in detail by Cooney, Wang & Wang (1977). In a second publication (Wang, Cooney & Wang, 1977) they used this technique for the analysis of baker's yeast fermentation. Roels & Kossen (1978) show how element balances can be used to calculate maintenance coefficients and yield factors for every substrate component if the maintenance coefficient or yield factor for one component only is known.

No attempt will be made to present here all the lengthy equations which are present in the above-mentioned publications. It must be clear, however, how powerful the use of simple element balances can be. In the forementioned publication of Roels (1979) element balances have been put on a more general level and the possibilities of the use of heat balances have also been mentioned.

Balances for non-conservative properties

The general form of a balance for a non-conservative property is:

$$\text{accumulation} = \text{inflow} - \text{outflow} + \text{production}$$

or in mathematical terms:

$$\frac{dE}{dt} = J_{in} - J_{out} + V r_E \tag{18}$$

Examples of the application of these balances are numerous. Only a few will be presented here: the solution of equation (15) (a distributed microbalance where $dE/dt = J_{out} = 0$); the growth of micro-organisms in a shake flask (a series of macrobalances where growth is distributed and $J_{out} = 0$); and the growth of *Hyphomicrobium* sp. in batch culture (a segregated macrobalance where $J_{in} = J_{out} = 0$).

Diffusion and consumption of substrate in a layer of micro-organisms

Equation (15) has to be solved for the boundary conditions at $x = x_0$ $\rightarrow C_s = 0$ and $D \, dC_s/dx = 0$ (x_0 is the distance where all the substrate has been consumed). The solution of equation (15) is now

$$C_s = \frac{r_s}{2 D} (x_0 - x)^2 \tag{19a}$$

r_s is supposed to be constant, which is true for a substrate like oxygen down to very low concentrations. If the concentration at $x = 0$ is C_{si} then x_0 can be found:

$$x_0 = \sqrt{\frac{2 D C_{si}}{r_s}} \tag{19b}$$

Elimination of x_0 from equation (19a) with equation (19b) results in

$$C_s = \frac{r_s}{2 \, \mathbb{D}} \left(\sqrt{\frac{2 \, \mathbb{D} \, C_{si}}{r_s}} - x \right)^2 \qquad (19c)$$

If we assume $r_s = 3.4 \times 10^{-6}$ kg m^{-3} s^{-1}, $\mathbb{D} = 10^{-9}$ m^2 s^{-1} and C_{si} $= 5 \times 10^{-3}$ kg m^{-3}, then the values of C_s (x) given in Table 3 are obtained.

Table 3.

x (mm)	C_s (g m^{-3})
0	5.00
0.5	2.51
1.0	0.87
1.5	0.08
1.71	0

In Fig. 5 C_s (x) is presented as a graph. Oxygen depletion occurs at a depth of 1.71 mm.

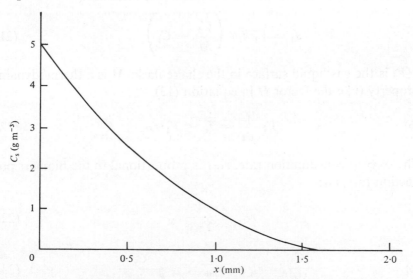

Fig. 5. Oxygen concentration profile in a layer of micro-organisms.

A result as presented in Fig. 5 can never be obtained with a macro-balance. A macrobalance for this system would be

$$j_{x=0} = r_s \, x_0 \qquad (20)$$

If $j_{x=0}$ and r_s are known, this permits the calculation of x_0, but nothing can be said about the shape of the concentration profile.

Growth of micro-organisms in a shake flask This model, from a publication of Van Suijdam, Kossen & Joha (1978), gives the opportunity to calculate when oxygen supply in shake flasks becomes a limiting factor. It consists of four balances:

(a) Oxygen balance for the gas phase;
(b) Oxygen balance for the liquid phase;
(c) Biomass balance for the liquid phase;
(d) Substrate balance for the liquid phase.

Similar balances can be used for an industrial fermenter.

(a)
$$V_g \frac{dC_g}{dt} = J_g - J_1 \tag{21a}$$

J_g is the oxygen diffusion through the cotton plug:

$$J_g = K_w (C_o - C_g) \tag{21b}$$

J_1 is the oxygen transfer across the gas/liquid interface:

$$J_1 = V_1 k_1 a \left(\frac{C_g}{M} - C_1 \right) \tag{21c}$$

$a \cdot V_1$ is the gas/liquid surface in the shake flask, M is a thermodynamic property (like the factor H in equation (13).

(b)
$$V_1 \frac{dC_1}{dt} = J_1 - V_1 \cdot r_{O_2} \tag{22a}$$

The oxygen consumption rate, r_{O_2}, is proportional to the biomass production rate, r_M:

$$r_{O_2} = \frac{1}{Y_{OM}} r_M \tag{22b}$$

(c)
$$\frac{dC_M}{dt} = r_M = \mu_{max} C_M \frac{C_s}{K_s + C_s} \tag{23a}$$

If the oxygen concentration in the liquid has become zero, growth is limited by oxygen supply:

$$\frac{dC_M}{dt} = Y_{OM} \frac{J_1}{V_1} \tag{23b}$$

(d)
$$C_M - C_{M_o} = Y_{SM} (C_{s_0} - C_s) \tag{24}$$

The following assumptions are implicit in equations (21) to (24):

gas phase is mixed homogeneously;
liquid phase is mixed homogeneously;
no lag phase;
negligible substrate consumption for maintenance.

The oxygen concentration has been calculated from this equation using a simple computer programme. Values for K_w and $k_l a$ have been obtained by experiments. The results are presented in Fig. 6; as one can see the mathematical model and the experimental results almost coincide.

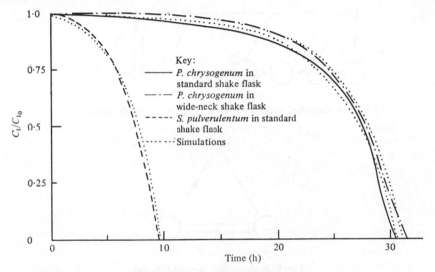

Fig. 6. Oxygen concentration as a funtion of time in shake-flask cultures. From Van Suijdam *et al.* (1978).

The equations above permit a rapid calculation of the moment when oxygen is growth-rate limiting. This is a typical example of a situation where transport and conversion processes are in series. The slower one of the two determines the overall rate (see pp. 340–3).

Growth of Hyphomicrobium *sp.* The growth of *Hyphomicrobium* sp. in a batch culture is very interesting from the point of view of mathematical modelling, because a segregated model has to be used. This is caused by the fact that this micro-organism grows in a five-step, *m*-birth cycle (see Fig. 7). Because of the fact that it is very difficult to distinguish experimentally between forms 1 and 2 and also between forms 4 and 5, a three-step, *m*-birth cycle can be used (Fig. 8). For this problem a

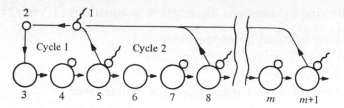

Fig. 7. The birth-cycle of *Hyphomicrobium* sp.

mathematical model has been set up* for the situation when $m \to \infty$. In that case only three different forms have to be distinguished (Fig. 9).

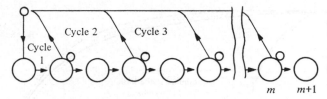

Fig. 8. A simplified version of the birth-cycle of *Hyphomicrobium* sp.

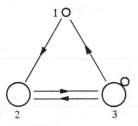

Fig. 9. A three-step eternal life-cycle.

The balances for the three different forms of the micro-organism are:

$$\frac{dN_1}{dt} = \mu_3 N_3 - \mu_1 N_1 - \lambda N_1 \tag{25a}$$

$$\frac{dN_2}{dt} = \mu_1 N_1 - \mu_2 N_2 + \mu_3 N_3 - \lambda N_2 \tag{25b}$$

$$\frac{dN_3}{dt} = \mu_2 N_2 - \mu_3 N_3 - \lambda N_3 \tag{25c}$$

where λ is the death rate.

* This model has been developed by Dr Ir. C. van Leeuwen of our Laboratory.

The substrate balance is given in a simplified version:

$$\frac{dC_s}{dt} = (\beta\lambda - \mu_1)\frac{N_1}{Y} + (\beta\lambda - \mu_2)\frac{N_2}{Y} + (\beta\lambda - \mu_3)\frac{N_3}{Y} \qquad (26)$$

where β is the fraction of the dead micro-organisms which can be used as substrate by the others.

Furthermore Monod kinetics are assumed:

$$\mu_i = \mu_{\max_i}\frac{C_s}{K_{s_i} + C_s} \qquad (27)$$

With these equations, graphs like those in Fig. 10 have been obtained.

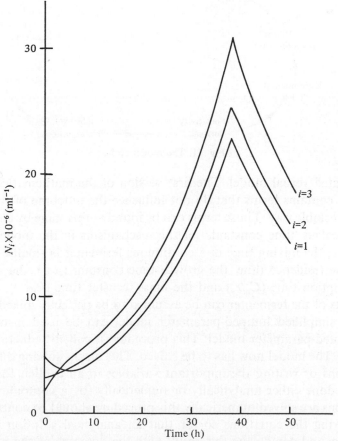

Fig. 10. Model curves for the growth of *Hyphomicrobium* sp. in batch culture.

The examples given in this section are an illustration of some kinds of models presented in the preceding sections. They show the uniformity in mathematical modelling: setting up of equations for transport, kinetics or thermodynamics, or of combinations of these equations in balances. In the next section some recommendations will be given for the construction and refutation of mathematical models.

5. SETTING UP AND REFUTATION OF MATHEMATICAL MODELS

The construction of a mathematical theory is a cyclic event as presented in Fig. 11. A mathematical model is always set up on the basis of a well

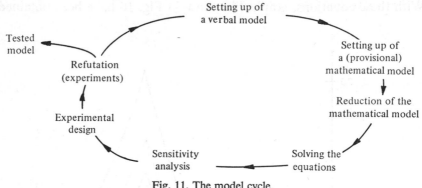

Fig. 11. The model cycle.

formulated verbal model. The first version of the mathematical model usually contains terms that do not influence the outcome of the model in a detectable way. Those terms can be found at this stage by comparing the so-called 'time constants' of the mechanisms in the model. If, for example, the mixing time in a continuous fermenter is small compared with the residence time, the growth time constant (μ^{-1}), the substrate consumption time (C_s/r_s) and the mass transfer time (k_1a^{-1}), then the contents of the fermenter can be assumed to be perfectly mixed. In that case a simplified lumped-parameter model can be used instead of a distributed-parameter model. This procedure is called 'reduction of the model'. The model now has to be solved. This means solving differential equations or writing the important variables in an explicit form. This can be done either analytically or numerically (by a computer). If the equations are solved numerically this procedure should be controlled by simplifying the equations so far that an analytical solution becomes possible, and comparing this result with the outcome of the numerical computations for this simplified situation. Then a sensitivity analysis is possible. Before 'putting the model in jeopardy' by experimentation, one has to know which variables have the greatest influence on the outcome of the model. These influential variables can be found by varying all the variables in the mathematical model one after another. It should be realized that the sensitivity of a given variable can be a function of the values of the other variables (see Kleynen, 1975).

The next stages are the experimental design and the refutation of the model. Literature on this point is overwhelming. It includes experimental design, statistical analysis and the like. Not only is the refutation of the model as a whole important, but also the determination of parameters for the model (confidence intervals). This question has been put forward by De Kwaadsteniet, Jager & Stouthamer (1976). Experimental design for kinetic experiments has been presented in two publications by Johnson & Berthouex (1975a, b) and by Williamson & McCarty (1975). There is an excellent publication on parameter estimation and model identification by Beckey & Yamashiro (1976). Especially in the field of systems engineering the number of techniques developed on model identification is enormous. One can only hope that some of these very useful techniques will penetrate into bioscience; until now the application of them in this field is disappointing.

The first version of a mathematical model can often be refuted. One has to refine the first version, usually going through the verbal stage again.

6. LIMITATIONS OF MATHEMATICAL MODELLING

At present there are a number of limitations to mathematical modelling. Sometimes the limits are being set up by *computational capacity*. Stochastic models now and then suffer from this problem, especially when so-called Monte Carlo Techniques are being used (Verkooijen, 1977). Quite a different kind of limit is set by what can be called the *discipline gap*. There are very elegant verbal models in microbiology, furthermore the mathematical arsenal to translate these models into mathematical ones is impressive, finally the techniques available in systems engineering to estimate the parameters or to get an idea of the kind of (black box) model from experimental results are numerous and often very elegant. The problem is 'who tells whom'. The author knows examples of fruitful collaboration, but there are ever-increasing misunderstandings. As in the contact between different nations, interpreters are needed. They should be people who excel in one area but have enough knowledge of the other areas to be able to communicate. The problem here is that sometimes the mathematical or systems-engineering techniques that can be of value for a microbiologist are not of 'standing' for the mathematician and systems engineer respectively. It is the author's opinion, however, that one should avoid at all costs the production of people who know a little bit of everything. A good overview in one area is essential to be of real help.

Finally, limitations of mathematical modelling are due to *refutational problems*. It is a well known fact that any model, once enough of its parameters are adapted, can describe any physical situation. This was stated clearly by Wei in his entertaining publication about the 'Least-Square Fitting of an Elephant' (1975). There is nothing against many-parameter models, but one should always be able to determine the parameters independently from the experiments that have been performed to test the model as a whole. If for example a structured model contains lots of mechanisms like enzyme- and ATP-production, inhibition, repression, turnover and the like, then the model can only be refuted if these mechanisms can be tested separately (by measurement of the enzymes *in vivo*, ATP concentrations etc.). If one uses these mechanisms only to have enough parameters at hand to adapt the model to the experiments then this can hardly be considered as a refutation. Another disadvantage of a complicated model is that probably no-one apart from the inventor can understand it. Refutational problems also come about because it is often impossible to get experimental results which are sufficiently accurate. This has been mentioned by Yang & Humphrey (1975) and by Boyle & Berthouex (1974). Also, closeness of fit is not a sufficient condition for acceptance of the model. Another necessary condition is that the parameters used should have physically acceptable magnitudes.

These limits to mathematical modelling are not absolute. They move in a positive direction with increasing computational facilities, communication between disciplines and refinement of experimental techniques.

7. CONCLUSIONS

(1) Every problem asks for a careful choice of the kind of mathematical model.

(2) Mathematical models give the opportunity for more precise formulation and refutation than verbal models. They can also be used for design and control purposes. However, there are also some disadvantages in mathematical modelling.

(3) The most important models in microbiology are those in the field of kinetics, transport phenomena, thermodynamics and combined models (balances).

(4) For the setting up and refutation of mathematical models a rigorous scheme can be followed.

(5) Limits to mathematical modelling are usually set by communication and refutation problems, though computational capacities are sometimes limiting.

(6) There is room for improvement of mathematical modelling and systems engineering in microbiology, both qualitatively and quantitatively. Better contact between microbiologists, mathematicians and systems engineers is desirable and necessary.

REFERENCES

BECKEY, G. A. & YAMASHIRO, S. M. (1976). Parameter estimation in mathematical models of biological systems. *Advances in Biomedical Engineering*, **6**, 1–43.

BOYLE, W. C. & BERTHOUEX, P. M. (1974). Biological wastewater treatment model building fits and misfits. *Biotechnology and Bioengineering*, **16**, 1139–59.

CHRISTENSEN, H. N. & CELLARIUS, R. A. (1972). *Introduction to Bioenergetics: Thermodynamics for the Biologist*. Philadelphia: Saunders.

COONEY, C. L., WANG, H. Y. & WANG, D. I. C. (1977). Computer aided material balancing for prediction of fermentation parameters. *Biotechnology and Bioengineering*, **19**, 55–67.

DEAN, A. C. R. & HINSHELWOOD, C. N. (1966). *Growth, Function and Regulation in Bacterial Cells*. Oxford University Press.

DE KWAADSTENIET, J. W., JAGER, J. C. & STOUTHAMER, A. H. (1976). A quantitative description of heterotrophic growth in micro-organisms. *Journal of Theoretical Biology*, **57**, 103–20.

FREDRICKSON, A. G. (1966). Stochastic models for sterilization. *Biotechnology and Bioengineering*, **8**, 167–82.

FREDRICKSON, A. G. (1976). Formulation of structured growth models. *Biotechnology and Bioengineering*, **18**, 1481–6.

FREDRICKSON, A. G., MEGEE, R. D. & TSUCHIYA, H. M. (1970). Mathematical models for fermentation processes. *Advances in Applied Microbiology*, **13**, 419–65.

GARFINKEL, D. (1969). Simulation of glycolytic systems. In *Concepts and Models of Biomathematics: Simulation Techniques and Methods*, ed. F. Heinmets. New York: Marcel Dekker.

GLANSDORF, P. & PRIGOGINE, I. (1974). *Thermodynamic Theory of Structure, Stability and Fluctuations*. London, New York: Wiley Interscience.

GOEL, N. S. & RICHTER-DYN, N. (1974). *Stochastic Models in Biology*. London, New York: Academic Press.

GRENNEY, W. J., BELLA, D. A. & CURL, H. C. (1973). A mathematical model of the nutrient dynamics of phytoplankton in a nitrate-limited environment. *Biotechnology and Bioengineering*, **15**, 331–58.

GROSSMAN, S. I. & TURNER, J. E. (1974). *Mathematics for the Biological Sciences*. London, New York: MacMillan.

HADELER, K. P. (1974). *Mathematik für Biologen*. Springer-Verlag, Berlin.

HIMMELBLAU, D. M. & BISCHOFF, K. B. (1968). *Process Analysis and Simulation, Deterministic Systems*. New York: Wiley Interscience.

JOHNSON, D. B. & BERTHOUEX, P. M. (1975a). Efficient biokinetic experimental design. *Biotechnology and Bioengineering*, **17**, 557–70.

JOHNSON, D. B. & BERTHOUEX, P. M. (1975*b*). Using multiple response data to estimate biokinetic parameters. *Biotechnology and Bioengineering*, **17**, 571–583.

KATCHALSKY, A. & CURRAN, P. F. (1974). *Nonequilibrium Thermodynamics in Biophysics.* Cambridge, Massachusetts: Harvard University Press.

KLEYNEN, J. P. C. (1975). A comment on Blanning's metamodel for sensitivity analysis: the regression metamodel in simulation. *Interfaces*, **5** (May), No. 3, 21–5.

KLOTZ, I. M. (1967). *Energy Changes in Biochemical Reactions.* London, New York: Academic Press. (Also in a German edition published by Georg Thieme Verlag, Stuttgart.)

KONINGS, W. N. (1976). Active solute transport in bacterial membrane vesicles. *Advances in Microbial Physiology*, **15**, 175–252.

LEHNINGER, A. L. (1973). *Bioenergetics.* Menlo Park, California: Benjamin.

MAY, R. M. (1973). *Stability and Complexity in Model Ecosystems.* Princeton: Princeton University Press.

MONOD, J. (1950). La technique de culture continue. Théorie et application. *Annales de l'Institut Pasteur*, **79**, 390–410.

MOROWITZ, H. J. (1968). *Energy Flow in Biology.* London, New York: Academic Press.

PRIGOGINE, I. (1967). *Introduction to Thermodynamics of Irreversible Processes.* New York: Wiley Interscience.

ROELS, J. A. & KOSSEN, N. W. F. (1978). On the modelling of microbial metabolism. *Progress in Industrial Microbiology*, **14**, 95–203.

ROELS, J. A. (1979). The application of macroscopic principles in bioenergetics. Submitted to *Journal of Theoretical Biology*.

ROWE, P. N. (1963). The correlation of engineering data. *The Chemical Engineer* (March), CE 69–CE 76.

RUBINOW, S. I. (1975). *Introduction to Mathematical Biology.* New York: Wiley Interscience.

TSAO, G. T. & HANSON, T. P. (1975). Extended Monod equation for batch cultures with multiple exponential phases. *Biotechnology and Bioengineering*, **17**, 1591–8.

VAN DEDEM, G. & MOO-YOUNG, M. (1975). A model for diauxic growth. *Biotechnology and Bioengineering*, **17**, 1301–12.

VAN DER BELD, H. (1976). A model for the dynamic control of activated sludge plants. *Systems and Models in Air and Water Pollution.* Congress of the Institute of Measurement and Control, London, 22–4 September, 1976. London: Chameleon Press.

VAN SUIJDAM, J. C., KOSSEN, N. W. F. & JOHA, A. C. (1978). A model for the oxygen transfer in a shake flask. *Biotechnology and Bioengineering* (in press).

VERKOOIJEN, A. H. M. (1977). A stochastic model for the growth of yeasts on hydrocarbons. Ph.D. thesis, Eindhoven Technical University.

WANG, H. Y., COONEY, C. L. & WANG, D. I. C. (1977). Computer aided baker's yeast fermentation. *Biotechnology and Bioengineering*, **19**, 69–86.

WATT, K. E. F. (1975). Critique and comparison of biome ecosystem modelling. In *Systems Analysis and Simulation in Ecology*, vol. III, ed. B. C. Patten, pp. 139–52. New York, London: Academic Press.

WEI, J. (1975). Least square fitting of an elephant. *Chemtech* (February), 128–129.

WILLIAMS, F. M. (1967). A model for cell growth dynamics. *Journal of Theoretical Biology*, **15**, 190–207.

WILLIAMSON, K. J. & MCCARTY, P. L. (1975). Rapid measurement of Monod half-velocity coefficients for bacterial kinetics. *Biotechnology and Bioengineering*, **17**, 915–24.

YANG, R. D. & HUMPHREY, A. E. (1975). Dynamic and steady state studies of phenol biodegradation in pure and mixed cultures. *Biotechnology and Bioengineering*, **17**, 1211–35.

MICROBIAL GENERATION AND INTER-CONVERSION OF ENERGY SOURCES

I. JOHN HIGGINS* AND H. ALLEN O. HILL†

*Biological Laboratory, University of Kent,
Canterbury, Kent CT2 7NJ, UK*

*† Inorganic Chemistry Laboratory, University of Oxford,
Oxford OX1 3QR, UK*

INTRODUCTION

Potential for microbiological contributions to future energy policy

No-one doubts that, if an energy crisis does not yet exist, it is only a matter of time before it will. Its origins are obvious; solutions, if they exist, less so.

We are concerned with the contribution that microbes and their constituents might make to a solution of this problem. So far, the only significant uses have been, firstly the generation of methane gas from waste materials such as sewage and domestic refuse; most usually the gas is generated as a by-product of a waste-treatment process. Secondly the conversion of various energy-sources such as light, methane, methanol, petroleum fractions and carbohydrate waste into protein or other valuable biochemicals, such as organic acids. The main purpose of this article is to discuss other as yet largely unexplored possibilities for using micro-organisms in future energy technology. It is also appropriate to discuss potential for biotransformation of simple organic chemicals, since many of these enjoy dual roles as both energy-sources and starting materials or intermediates for chemical syntheses. These chemical fuels deserve our special consideration because they are the only forms in which energy can be stored economically for periods greater than a few days, they form the basis of the modern chemical industry, we depend upon them for transportation, and they are primarily derived from oil, the fossil fuel which will probably be exhausted within the next fifty years. In this context, the significance of microbes is that, as with all organisms, they contain molecular assemblies capable of catalysing reactions with efficiencies, under moderate conditions, that synthetic systems are rarely able to match. We seek to make use of these properties either by employing the intact organism or components thereof. Many problems already identified in the search for solutions to the energy problem are in fact concerned with catalysis, or rather, the lack of effective catalysts. We can, therefore, attempt to use biological systems as catalysts, or incorporate lessons learnt from biological systems into synthetic systems; we should not forget that the latter have attractive features, such as thermal stability and the ease with which they lend themselves to industrial manufacture.

The 'energy crisis' has catalysed several important international meetings which have considered the role biology may play in solving these problems, and three particularly important proceedings have recently been published (Buvet, Allen & Massue, 1977; Schlegel & Barnea, 1977; Gysi, 1978). In the following sections we discuss the most promising areas of microbiology which may lead to important contributions to the energy technology of the twenty-first century. We have not included single-cell protein as this topic has recently been thoroughly reviewed (Tannenbaum, 1977; see also Mateles, this volume, pp. 29–52).

PHOTOBIOLOGY

Quantitatively, solar energy is clearly the most important energy source (Table 1) and micro-organisms or photosynthetic systems derived from

Table 1. *World availability of non-nuclear energy*

Type	Amount (10^{19} cal a^{-1})
Sunlight	17 000
Coal, oil, gas	80
Water	20
Tidal	0.09
Geothermal	0.008

them may be used in various ways to trap solar energy. Much has been written recently about the use of solar energy (e.g. Bockris, 1976) and considerable imagination, increasing endeavour and even a certain mysticism have been brought to the problem. Use of solar energy in low-grade heating systems is well advanced in some localities; the generation of potential differences in certain materials (the photovoltaic effect) is a more attractive method of collection, since the efficiencies are quite high (12–25%). Development is presently limited by the high capital cost and lack of suitable storage methods. Little work has yet been done on the photogeneration of current (the photogalvanic effect).

The efficiency of photosynthesis is usually quoted as $0.5-3\%$. This is true for the overall process but is misleading, since the efficiency of photon capture is much higher, and if the energy could be diverted into other processes the overall efficiency of an isolated unit might be comparable to photovoltaic devices. Our understanding of the process of photosynthesis has advanced to such a stage that these more exciting possibilities begin to entice. It has been known for many years that hydrogen can be produced by algae (Gaffron & Rubin, 1942) and purple bacteria (Gest & Kamen, 1949). Since that time a large number of photosynthetic micro-organisms have been shown to generate hydrogen in the light and the subject has been reviewed (Pfennig, 1967; Gest 1972; Kessler, 1973; Kondratieva & Gogotov, 1976; Kondratieva, 1977; Mitsui, 1978).

The photosynthetic electron-transport system can be regarded as a solar cell which generates a potential difference of approximately 1.2 V. Normally these high-potential electrons can be used via ferredoxin to reduce NADP to NADPH for biosynthetic purposes, but in the presence of a hydrogenase (Mortenson & Chen, 1974) they can be converted into molecular hydrogen:

$$2e + 2H^+ \leftrightharpoons H_2$$

Under anaerobic conditions, organic and inorganic wastes can be used for hydrogen generation by the purple bacteria. These micro-organisms need organic acids or sulphur compounds as electron sources for hydrogen production. For example, *Rhodospirillum rubrum* can photometabolize acetate, fumarate, malate and succinate to carbon dioxide and hydrogen, whilst some species use thiosulphate as electron donor. Amongst photosynthetic micro-organisms, the purple bacteria are the best hydrogen-producers with rates up to 25 ml per g dry wt per h. However, if photosynthetic micro-organisms are to make a major energy contribution in the form of hydrogen, we must turn to those which use water as electron source: the algae and the blue-green bacteria. This light-dependent generation of hydrogen from water is called biophotolysis and the system is summarized in Fig. 1.

As mentioned above, spontaneous hydrogen evolution by phototrophs is a common phenomenon and usually its physiological role is one of regulation, enabling organisms to dissipate excess reducing power. Since photochemically-generated reducing power is mainly used for CO_2 fixation, to obtain efficient whole-organism hydrogen generating systems it may be necessary to interfere with the normal process of photosynthesis in order to stimulate hydrogen production. Although attempts are being made to do this, an alternative approach is to construct artificial systems capable of biophotolysis (Fig. 1), using, for

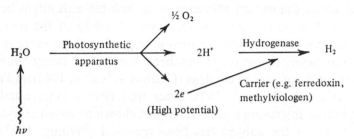

Fig. 1. A general scheme for cell-free biophotolysis.

example, isolated chloroplasts in conjunction with purified bacterial hydrogenase (Benemann *et al.*, 1973; Rao, Rosa & Hall, 1976). However at the time of writing only rather low rates of hydrogen production have been obtained with whole organisms or cell-free systems. For example, the best rates obtained using spinach chloroplasts, *Spirulina* ferredoxin and *Desulfovibrio* hydrogenase were 40 μmole hydrogen per mg of chlorophyll per h, linear for 3 h (Reeves *et al.*, 1977). The main problems are lack of stability and inhibition of hydrogenase by oxygen. It is possible to solve the latter difficulty by separating the oxygen-generating system from the hydrogen-generating one with connection via a redox intermediary, or by using an oxygen-scavenging system. Recently, a procedure that may lead to more stable preparations has been developed. This involves microencapsulation of chloroplasts or photosystem 1 particles with *Chromatium* hydrogenase in a semi-permeable membrane (Kitajima & Butler, 1976). The diversion of the photoelectrons to effect other useful reductions has been considered (Fig. 2). It may also be possible to use photoelectrons derived from water to reduce oxygen to hydrogen peroxide (Losada, 1978), thus:

$$H_2O + O_2 \xrightarrow{h\nu} H_2O_2 + \tfrac{1}{2}O_2$$

The use of algal ponds, perhaps in association with bacteria, for the production of protein (usually as an animal foodstuff) is straightforward. The process represents an example of energy capture, conversion and storage. The last feature is not an insignificant component of the energy problem. The algae, if used in subsequent fermentations, could be converted to more convenient energy stores e.g. methane or

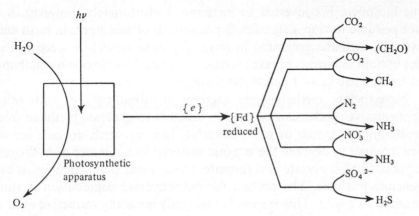

Fig. 2. Possibilities for using photoactivated electrons via ferredoxin for useful reduction reactions.

methanol. Obviously there are significant energy losses in such systems but, in certain localities, especially when used for waste treatment, they may prove attractive (Hall, 1978).

Economic considerations make construction of commercial cell-free systems rather unlikely in the near future, and some of the inherent difficulties have been discussed in detail elsewhere (Benemann, 1977; Estigneev, 1977). At present, whole-organism systems seem a better proposition, the most promising ones being offered by the heterocystous blue-green bacteria. The nitrogenase of these organisms generates hydrogen in the absence of molecular nitrogen (Benemann & Weare, 1974; Weissman & Benemann, 1977). Using cultures starved of nitrogen, which encourages heterocyst formation, roughly stoichiometric hydrogen and oxygen evolution has been maintained for several weeks, with a light-conversion efficiency of about 0.5% (Weissman & Benemann, 1977). It has been suggested that the purple membrane of *Halobacterium halobium* (Henderson, 1977; Schreckenbach, 1977) could be used to produce hydrogen and oxygen (Skulachev, 1976) but this remains to be demonstrated.

METHANOGENESIS

Methane, the major component of many natural gas deposits, is a valuable energy source, widely used for industrial and domestic purposes. Although most methane used comes from natural deposits, it is continually produced in vast amounts, together with carbon dioxide, as the ultimate carbon end-product of all microbial anaerobic degradation processes. Thus a considerable proportion of all carbon cycled through the biosphere is converted to methane. Unfortunately, however, it is not possible to trap a significant proportion of this since, in most ecosystems, methane generated in anaerobic regions such as deep soil or the bottom sediments of lakes is oxidized to carbon dioxide by methane-utilizing bacteria as it percolates through aerobic microflora.

Nevertheless, methanogenic bacteria in mixed cultures with other fermentative species can be used in reactors to generate methane from waste organic matter or plant material. The non-methanogenic species are required to degrade the organic material to a mixture of hydrogen, carbon dioxide, acetate and formate, from which the methanogens can generate methane. The process for carbohydrate degradation is summarized in Fig. 3. This system is thermodynamically attractive and, in

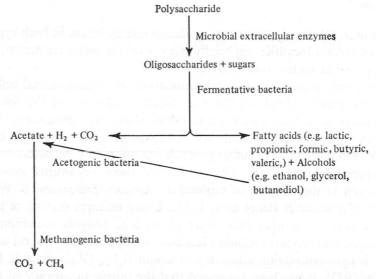

Fig. 3. Anaerobic degradation of carbohydrate by a free-living microbial consortium.

the case of carbohydrate, conversion to methane results in a loss of only about 10% of the calorific value of the starting material. This loss is primarily due to the formation of biomass. Because of the different physiological requirements of the non-methanogenic species and the methanogens in the consortium, it has been suggested that optimization of the process requires physical separation of the two microbial groups and a mechanism for delivering the products of initial biodegradation to the methanogens. This has prompted the recent development of two-phase anaerobic digestion processes (Ghosh & Klass, 1978).

Because of the technical difficulties associated with studying methanogens in the laboratory, knowledge of their physiology and biochemistry is slow to accumulate. The mechanism of methane generation is still uncertain. When CO_2 is the ultimate precursor, eight electrons from four hydrogen molecules catalyse the reduction, the overall reaction being

$$4H_2 + HCO_3^- + H^+ \rightarrow CH_4 + 3H_2O \ (\Delta G'_0 = -32.4 \text{ kcal})$$

It was originally thought that a methylcobamide is an intermediate in methanogenesis (Blaylock & Stadtman, 1963) but there is now some doubt of it as there is no indication of its involvement in hydrogen-grown organisms (Wolfe & Higgins, 1978). There is, however, strong evidence that Coenzyme M (2-mercaptoethane sulphonic acid) is involved as a methyl carrier in the form of 2-(methylthio)ethane sulphonic acid (Gunsalus et al., 1976). It is the only substrate for the methylreductase of methanogenic bacteria to be found to date. The nature of this enzyme has only recently been elucidated (Gunsalus, 1977). It is composed of three components: A is a heat-sensitive protein, B is a heat-stable but oxygen-sensitive factor with a low molecular weight (about 1000) and C is a protein of high molecular weight exhibiting hydrogenase activity. The reaction catalysed is:

$$H_2 + CH_3\text{–S–CoM} \xrightarrow[\text{ATP, Mg}^{2+}]{\text{A, B, C}} CH_4 + HS\text{–CoM}$$

Little is known about the mechanisms of ATP synthesis or carbon assimilation in these bacteria. The subject has recently been reviewed (Mah et al., 1977; Zeikus, 1977; Wolfe & Higgins, 1978).

The use of anaerobic digesters in sewage-treatment processes has been practised for many years, and the current state of the art has been discussed by Hobson, Bousfield & Summers (1974). In modern plants, the methane formed is used, for example, to generate electricity. Many have suggested that methane could be generated on a large scale from anaerobic digesters using household and industrial refuse and 'low-grade' plant material as feedstocks. The latter is not as plentiful or suitable as is often assumed. Wood is not degraded rapidly by micro-organisms and is a useful fuel when burnt directly, as well as being of value as a building material.

The main potential for methane production lies in the degradation of waste material, where the costs of building and operating the system are largely offset by the need to dispose of the waste. Dilute organic matter in water, for example sewage, is the most suitable source and there seems little point in diluting material which is dry and can be used directly as fuel for burning. Although only about 1% of the world's current energy requirements could be met by anaerobic digestion of liquid wastes, this would nevertheless clearly be worth harnessing. There seems to be considerable potential for small-scale 'home' digesters (Hungate, 1977) to be fed on a mixture of sewage, refuse, weeds, etc.

POTENTIAL FOR ELECTRICITY GENERATION USING MICRO-ORGANISMS

Electricity may, of course, be generated using microbiologically produced hydrogen or methane to drive conventional mechanical generators, but this is relatively inefficient. There are three main types of more direct and potentially more efficient processes that may be developed. The first involves using intact micro-organisms to generate electricity in a fuel cell as a result of the metabolism of a substrate; such a cell could also be constructed using enzymes derived from micro-organisms. Secondly, it may prove possible to use immobilized, microbiologically derived hydrogenase coupled to electrode systems to catalyse the efficient interconversion of hydrogen and electricity. The third type of process involves using whole phototrophs, or photoactive systems derived from them, to generate electricity directly from sunlight. This would be a primary source of electricity of great quantitative potential. However, the question arises whether biological photocells based on micro-organism or higher-plant systems can in practice ever compete with conventional electrochemical systems.

The direct conversion of light to electricity is attractive on thermodynamic grounds, since it is theoretically more efficient than using heat as an intermediate energy form. In most biological photochemical membranes high-potential electrons are generated by photoactivation. It should be possible, therefore, to develop biological photocells in which this effect is used to generate current and voltage in an external circuit. A simple photovoltaic device has already been demonstrated that, using thylakoid preparations from higher green plants, converts light into electricity with up to 1.5% efficiency (Allen, 1977).

The bacteriorhodopsin-containing purple membrane of *Halobacterium halobium*, mentioned above under hydrogen generation, generates a proton gradient in response to light *in vivo*, and this can be used to generate ATP via a proton-translocating ATPase (Mitchell, 1976). It has been pointed out that bacteriorhodopsin could theoretically serve as the basis of a photocell (Skulachev, 1976) although such a system remains to be developed.

Biological fuel cells

The direct conversion of the free energy of a chemical reaction into electricity is widely acknowledged to be an attractive proposition, but the prevailing attitude to the development of fuel cells which exploit this principle is, sadly, one of procrastination. A simple hydrogen–oxygen fuel cell is depicted in Fig. 4. Oxidation occurs at the anode and reduction at the cathode, the two electrode compartments being separated by a membrane. As the hydrogen gas passes over the anode surface, it is electrochemically oxidized to hydrogen ions which enter the electrolyte and migrate towards the cathode, where oxygen is reduced to water. Electrons released at the anode flow through the external circuit. (For detailed discussions of fuel cells see Bockris & Srinivasan, 1969; Bockris & Nagy, 1974.)

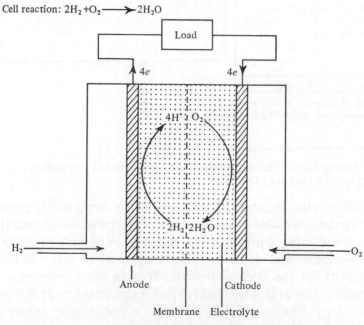

Cell reaction: $2H_2 + O_2 \longrightarrow 2H_2O$

Fig. 4. A simple fuel cell.

A comparison of 'cold' (fuel cell) and 'hot' (heat engine) means of combustion (Fig. 5) highlights the inherent defects of the latter. Theoretically, any reaction which involves oxidation and reduction can be used in a fuel cell and the advantages of such a system are considerable. Since the reactions in a fuel cell involve charge transfer, and not heat transfer, they are not subject to the limitations of the Carnot cycle.

Consequently, the theoretical maximum efficiency, as conventionally defined, of fuel cells is often 100% and for certain reactions, e.g. $2C + O_2 \rightarrow 2CO$, can even exceed it. Obviously such maximum efficiencies are not realized but 50–70% is not uncommon.

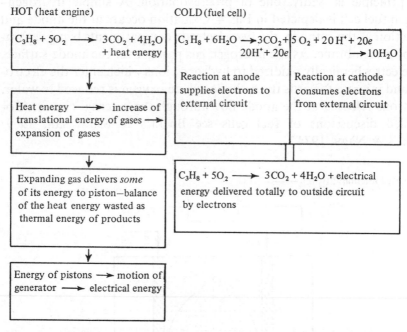

HOT (heat engine)   COLD (fuel cell)

$C_3H_8 + 5O_2 \longrightarrow 3CO_2 + 4H_2O$ + heat energy

$C_3H_8 + 6H_2O \rightarrow 3CO_2 + 20H^+ + 20e$ | $5O_2 + 20H^+ + 20e \rightarrow 10H_2O$

Reaction at anode supplies electrons to external circuit | Reaction at cathode consumes electrons from external circuit

Heat energy $\longrightarrow$ increase of translational energy of gases $\longrightarrow$ expansion of gases

Expanding gas delivers *some* of its energy to piston—balance of the heat energy wasted as thermal energy of products

$C_3H_8 + 5O_2 \longrightarrow 3CO_2 + 4H_2O$ + electrical energy delivered totally to outside circuit by electrons

Energy of pistons $\longrightarrow$ motion of generator $\longrightarrow$ electrical energy

Fig. 5. A comparison of the essential steps in the production of energy from propane by a heat engine with those for a fuel cell.

If fuel cells are so attractive why aren't they more widely employed? Their limitations reside in the slowness of electron transfer reactions at electrodes. Consequently most fuel cells rely on the use of platinum or other precious-metal electrodes, since they combine this role with that of catalysts for the electron transfer. It is in the development of replacement catalysts that we might expect a significant contribution from microbiology. The so-called biofuel cells, which employ whole microorganisms to effect the relevant oxidation, do not appear attractive, since it is difficult to envisage efficient and effective electron transfer between whole organisms and electrodes. The opportunities for direct microbiological intervention in the electrochemical energy conversion seem most promising via enzymes isolated from them. The hydrogen/oxygen fuel cell is the most developed and the use of hydrogenases and cytochrome oxidases from different sources as catalysts for electron transfer directly to the electrodes would warrant investigation. At present, hydrocarbons are still attractive energy sources and fuel cells based on them have been developed. Our understanding of organisms which effect alkane oxidation, whilst still inadequate, is sufficiently advanced to allow isolation of enzymes or enzyme complexes which catalyse these oxidations. An attractive idea, which might be commercially feasible, would be to couple fuel cell technology to the use of alkanes as starting materials for chemical synthesis. For example, Fig. 6 depicts a hypothetical fuel cell in which enzymes are incorporated into the electrodes. At the anode, a methanol dehydrogenase would oxidize methanol via formaldehyde to formate, yielding four electrons for the external circuit (possibly via a carrier) and four protons which would diffuse to the cathode in which methane mono-oxygenase is incorporated. This enzyme would reduce oxygen to water and oxidize methane to methanol, consuming protons and electrons from the circuit. Thus two useful chemical products would be generated from methane, in addition to electrical power.

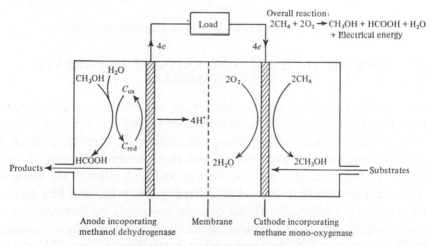

Fig. 6. A hypothetical enzymic fuel cell. C is a hydrogen carrier.

The most promising application of microbiological techniques concerns their *indirect* use in supplying the electroactive material. Many schemes are possible. The simplest involve the conversion of waste organic material to a suitable fuel such as hydrogen, methanol or ethanol. Given the high development of the hydrogen/oxygen fuel cell, the conversion of waste material to molecular hydrogen would seem the most immediately promising route. If biophotolysis as described above becomes feasible, then it would seem sensible to combine it with fuel cell technology. In essence such a system constitutes a regenerative fuel cell (Fig. 7). The attraction of these schemes lies in the coupling of a useful enzymically catalysed chemical reaction to an efficient electron carrier.

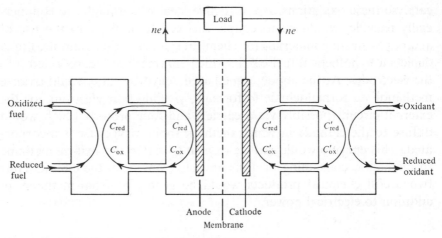

Fig. 7. A regenerative fuel cell. C and C' are electron carriers.

THE USE OF MICRO-ORGANISMS TO GENERATE FUEL CHEMICALS

Amongst the hundred organic chemicals of greatest quantitative importance, there are only six for which there is a microbiological process using existing biotechnology. These are acetic acid, acetone, n-butanol, ethanol, glycerol and isopropanol. All these compounds can be produced by fermentation, but only acetic acid and ethanol are made in significant quantities in this way. Generally speaking, in a free market the production of simple, stable organic chemicals by fermentation is not competitive with chemical methods. However, since chemical methods are based on petroleum oil as a starting product, the relative economic situation could change if there were to be further dramatic increases in the price of oil. There are also three major local factors that can render microbiological processes viable under current conditions. First, there may be the opportunity of coupling the processing of waste materials or agricultural surplusses to the synthetic processes: secondly, there may be local production of relatively cheap fermentable plant material, e.g. molasses; and thirdly, local economic or political restrictions may favour fermentation processes. For example, in Italy ethanol is produced entirely by fermentation, since there is a prohibitively high tax on synthetic ethanol. Some countries such as Brazil, have encouraged fermentative alcohol production from plant material by price-fixing because of lack of local oil resources.

It is possible that some aerobic micro-organisms might prove valuable for synthesizing fuel chemicals through the exploitation of co-metabolism (Horvath, 1972) or non-growth metabolism. In the former case, many species have been shown to catalyse biotransformations of substrate analogues to products which they cannot further metabolize whilst they utilize the substrate for growth. In the latter case it is possible, using inhibitors or trapping agents or by genetic manipulation, to collect an intermediate in a catabolic sequence. These processes have been regarded as having potential for the manufacture of more exotic or fine chemicals such as steroid and terpenoid derivatives (Abbott & Gledhill, 1971).

In recent years, however, an extensive knowledge of the physiology and biochemistry of methane-utilizing bacteria has been acquired (for a recent review, see Wolfe & Higgins, 1978). It is clear that these bacteria can effect a wide range of oxidative biotransformations of hydrocarbons (and other compounds) to products (mainly alcohols, aldehydes, acids) which accumulate and are not further metabolized. This is largely due to the extremely wide substrate specificity of some methane mono-oxygenase enzymes (Colby, Stirling & Dalton, 1977; Tonge, Harrison & Higgins, 1977; F. S. Sariaslani & I. J. Higgins, unpublished observations). Perhaps the most obvious biotransformation that could be catalysed by these bacteria is the production of methanol from methane, and this has been discussed recently (Foo & Hedén, 1977; Foo, 1978). They could also be used to synthesize formaldehyde or formate from methane and ethanol, acetaldehyde and acetate from ethane. Together with other hydrocarbon-oxidizing species, they may prove of value for specific hydroxylations of higher hydrocarbons. Introduction of such processes will depend upon successful application of immobilization techniques, and probably the development of high-pressure microbiological reactors.

There is also potential for using micro-organisms to catalyse interconversions involving hydrogen, for example some methanogenic strains will convert hydrogen to methane (i):

$$4H_2 + CO_2 \rightarrow CH_4 + 2H_2O \tag{i}$$

However, it may also be possible to effect reactions (ii) and (iii):

$$3H_2 + CO_2 \rightarrow CH_3OH + H_2O \tag{ii}$$
$$2H_2 + CO \rightarrow CH_3OH \tag{iii}$$

Other possibilities are summarized in the following equations:

$$3H_2 + N_2 \rightarrow 2NH_3 \qquad \text{(iv)}$$
$$H_2 + O_2 \rightarrow H_2O_2 \qquad \text{(v)}$$
$$H_2 + CH_4 + O_2 \rightarrow CH_3OH + H_2O \qquad \text{(vi)}$$
$$H_2 + RH + O_2 \rightarrow ROH + H_2O \qquad \text{(vii)}$$

Obviously, it may be more feasible to effect such biotransformations not by using the whole organism but enzymes isolated from it.

Intense effort has been made to effect (iv) and, though far from competitive with the present industrial process, the current progress is promising. A key component of all three systems is the enzyme hydrogenase and, though it has a reputation for fragility, hydrogenases isolated from different organisms vary greatly in their thermal stability and even their reactivity towards molecular oxygen, some being more selective in their substrates. The isolation of a really stable hydrogenase might make feasible many processes at present requiring high-temperature catalysts. Most hydrogenases isolated to date have little thermal stability; those from thermophilic organisms would repay investigation, especially if they proved capable of immobilization.

Several microbial enzymes that catalyse pertinent hydrocarbon oxidations have been studied. Cytochrome-P450 systems are present in a number of micro-organisms and such an n-alkane mono-oxygenase was first reported in a *Corynebacterium* (Cardini & Jurtshuk, 1968). Another alkane mono-oxygenase has been obtained from a strain of *Pseudomonas putida* and contains the non-haem iron protein, rubredoxin (Peterson *et al.*, 1967). Both are complex enzyme systems oxidizing some n-alkanes to the corresponding primary alcohols; they are discussed in detail elsewhere (Higgins & Gilbert, 1978). More recently, two different enzyme systems that oxidize methane to methanol have been purified from methane-utilizing bacteria (Tonge *et al.*, 1977; Colby & Dalton, 1978). As mentioned above, both these preparations also catalyse a wide variety of other oxidations. All four enzymes catalyse the following general reaction:

$$R-H + O_2 + XH_2 \rightarrow R-OH + H_2O + X$$

In three cases X is obligatorily NAD or NADP, whilst in the case of the *Methylosinus trichosporium* system, NAD may be replaced by ascorbate. This requirement for expensive and unstable reducing agents has militated against such enzymes being considered seriously for catalysing commercial chemical interconversions, except possibly for some of the more exotic fine chemicals of the flavouring and perfumery industries.

However, a potential solution to this problem has recently appeared in the form of electroenzymology (I. J. Higgins & H. A. O. Hill, patents pending). In this process, enzymes such as these mono-oxygenases, which require reductants to function, have them supplied electrochemically. Where the product is of high value and the enzyme mechanism requires NADH binding, it may even prove economic to regenerate NADH electrochemically. This technique is heavily reliant on the selection and growth of organisms capable of interesting biotransformations from which the appropriate enzymes can be isolated.

Whatever the solutions to the energy problem, one prediction is possible: electricity will still be the most convenient form in which energy is 'transported'. In addition to electroenzymology there are other ways in which biological systems can 'use' electricity. Already, it has been employed as the energy source for growth of the autotrophs *Hydrogenomonas eutropha* (Schlegel & Lafferty, 1965) and *Thiobacillus ferrooxidans* (D. Herbert, H. A. O. Hill & D. R. Turner, unpublished work).

USE OF MICRO-ORGANISMS IN OIL PRODUCTION

Hydrocarbons are common components of micro-organisms, although the types present and both absolute and relative quantities are very variable (Albro & Ditmer, 1970). It is conceivable that, by suitable strain and/or genetic manipulation, processes for producing oil derived from micro-organisms may be developed (Tornabene, 1977). Indeed, it has already been shown that hydrocarbon production can be dramatically increased by genetic transformation procedures (Morrison, Tornabene & Kloos, 1971; Kloos, Tornabene & Schleiffer, 1974). Such processes are clearly long-term possibilities.

The use of micro-organisms to aid in oil-recovery processes seems a much likelier proposition in the near future (see Sutherland & Ellwood, this volume, pp. 107–50). The problems of oil recovery will become increasingly pressing in the next decade, since approximately two-thirds of current known reserves are not recoverable at present on technological and/or economic grounds. Currently, water-based displacement techniques are used to drive out oil, but they are inefficient for the more viscous oils because of the low viscosity of water. There is considerable potential for using microbiological water-soluble polymers in more effective displacing agents (for example, xanthan gums and poly-β-hydroxybutyrate have been considered). Surfactants can also assist in secondary recovery, and it may be possible to produce these microbiologically, *in situ*. High-viscosity polymers from micro-organisms might also be used as plugging agents to prevent the displacing fluid by-passing the oil through fractures in the rockbed.

The suggestion that micro-organisms could be used to decrease oil viscosity by selective degradation of components seems an unlikely possibility, since the lighter fractions are generally the more biodegradable (Higgins & Gilbert, 1978) and in many reservoirs the microbial biomass would cause blockage of the pores in the rock, many of which have diameters of about 30 μm. However, there is potential for using the degradative capacity of micro-organisms to assist in oil recovery from the extensive shale-oil deposits in the world. Here there are both mineral and oil components, the latter a highly viscous tar, difficult and expensive to recover by conventional techniques. For example, *Thiobacilli* have been used to generate sulphuric acid *in situ* which rapidly and effectively dissolves the carbonate component of the shale (Findley, Appleman & Yen, 1974). It may also be possible to use micro-organisms to partially degrade the complex, insoluble, organic components, keragen and bitumen (Findley *et al.*, 1976).

There is currently considerable research effort, especially in industrial laboratories, in the general area of oil biotechnology, and some microbiologically-based processes seem likely to emerge within the next few years.

CONCLUDING REMARKS

There is considerable potential for exploiting micro-organisms in a number of ways to make very significant contributions to future energy technology. However, with the notable exception of methane generation, most possibilities are far from practical realities. There are many problems waiting to be solved, especially those concerning stability of microbial systems, enzyme preparations and electrode and fuel cell technology. Most possibilities discussed in this brief overview not only challenge the microbiologist but also the biochemist, the electrochemist and the engineer. Future work in this area will surely benefit from an interdisciplinary approach.

REFERENCES

ABBOTT, J. B. & GLEDHILL, W. E. (1971). The extracellular accumulation of metabolic products by hydrocarbon-degrading micro-organisms. *Advances in Applied Microbiology*, **14**, 249–88.

ALBRO, P. W. & DITMER, J. G. (1970). Bacterial hydrocarbons: occurrence, structure and metabolism. *Lipids*, **5**, 320–5.

ALLEN, M. J. (1977). Direct conversion of radiant into chemical energy using plant systems. In *Living Systems as Energy Converters*, eds R. Buvet, M. J. Allen & J.-P. Massue, pp. 271–4. Amsterdam: North-Holland.

BENEMANN, J. (1977). Hydrogen and methane production through microbial photosynthesis. In *Living Systems as Energy Converters*, eds R. Buvet, M. J. Allen & J.-P. Massue, pp. 275–84. Amsterdam: North-Holland.

BENEMANN, J. R. & WEARE, N. M. (1974). Hydrogen evolution by nitrogen-fixing *Anabaena cylindrica*. *Science*, **184**, 174–5.

BENEMANN, J. R., BERENSON, J. A., KAPLAN, N. O. & KAMEN, M. D. (1973). Hydrogen evolution by a chloroplast–ferredoxin–hydrogenase system. *Proceedings of the National Academy of Sciences, USA*, **70**, 2317–20.

BLAYLOCK, B. A. & STADTMAN, T. C. (1963). Biosynthesis of methane from the methyl moiety of methylcobalamin. *Biochemical and Biophysical Research Communications*, **11**, 34–43.

BOCKRIS, J. O'M. (1976). *Energy: The Solar–Hydrogen Alternative*. London: The Architectural Press.

BOCKRIS, J. O'M. & NAGY, Z. (1974). *Electrochemistry for Ecologists*. New York: Plenum Press.

BOCKRIS, J. O'M. & SRINIVASAN, S. (1969). *Fuel Cells, Their Electro-Chemistry*. New York: McGraw Hill.

BUVET, R., ALLEN, M. J. & MASSUE, J.-P. (eds) (1977). *Living Systems as Energy Converters*. Amsterdam: North-Holland.

CARDINI, G. & JURTSHUK, P. (1968). Cytochrome P-450 involvement in the oxidation of n-octane by cell-free extracts of *Corynebacterium* sp. strain 7EIC. *Journal of Biological Chemistry*, **243**, 6070–2.

COLBY, J. & DALTON, H. (1978). Resolution of the methane monooxygenase of *Methylococcus capsulatus* (Bath) into three components. Purification and properties of component C, a flavoprotein. *Biochemical Journal*, **171**, 461–8.

COLBY, J., STIRLING, D. I. & DALTON, H. (1977). The soluble methane monooxygenase of *Methylococcus capsulatus* (Bath): its ability to oxygenate n-alkanes, ethers and alicyclic, aromatic and heterocyclic compounds. *Biochemical Journal*, **165**, 395–402.

ESTIGNEEV, V. B. (1977). Utilization of the photosynthetic apparatus of green plants and algae for the production of gaseous hydrogen. In *Living Systems as Energy Converters*, eds R. Buvet, M. J. Allen & J.-P. Massue, pp. 275–84. Amsterdam: North-Holland.

FINDLEY, J. E., APPLEMAN, M. D. & YEN, T. F. (1974). Degradation of oil shale by sulphur-oxidizing bacteria. *Applied Microbiology*, **28**, 460–4.

FINDLEY, J. E., APPLEMAN, M. D. & YEN, T. F. (1976). Microbial degradation of oil shale. In *Science and Technology of Oil Shale*, ed. T. F. Yen, pp. 175–81. Ann Arbor: Ann Arbor Science.

FOO, E. L. (1978). Microbial production of methanol. *Process Biochemistry*, **13**, 23–8.

FOO, E. L. & HEDÉN. (1977). Is a biocatalytic production of methanol a practical proposition? In *Microbial Energy Conversion*, eds H. G. Schlegel & J. Barnea, pp. 267–80. Oxford: Pergamon Press.

GAFFRON, H. & RUBIN, J. (1942). Fermentative and photochemical production of hydrogen in algae. *Journal of General Physiology*, **26**, 219–40.

GEST, H. (1972). Energy conservation and generation of reducing power in bacterial photosynthesis. In *Advances in Microbial Physiology*, eds A. H. Rose & D. W. Tempest, vol. 7, pp. 243–83. London, New York: Academic Press.

GEST, H. & KAMEN, M. D. (1949). Photoreduction of molecular hydrogen by *Rhodospirillum rubrum*, *Science*, **109**, 558–9.

GHOSH, S. & KLASS, D. L. (1978). Two-phase anaerobic digestion. *Process Biochemistry*, **13**, 15–24.

GUNSALUS, R. P. (1977). Methylreductase of *Methanobacterium thermoautotrophicum*. Ph.D thesis, University of Illinois, Urbana, USA.

GUNSALUS, R. P., EIRICH, D., ROMESSER, J., BALCH, W., SHAPIRO, S. & WOLFE, R. S. (1976). Methyltransfer and methane formation. In *Microbial Production and Utilization of Gases*, eds H. G. Schlegel, G. Gottschalk & N. Pfennig, pp. 191–9. Göttingen: Goltze.

GYSI, C. (ed.) (1978). *Bioenergy: Energy from Living Systems*. Zurich: Gottlieb Duttweiler Institute.

HALL, D. O. (1978). Solar energy conversion through biology – is it a practical energy source? In *Bioenergy: Energy from Living Systems*, ed. C. Gysi, pp. 26–68. Zurich: Gottlieb Duttweiler Institute.

HENDERSON, R. (1977). The purple membrane from *Halobacterium halobium*. *Annual Review of Biophysics and Bioengineering*, **6**, 87–109.

HIGGINS, I. J. & GILBERT, P. D. (1978). Biodegradation of hydrocarbons. In *The Oil Industry and Microbial Ecosystems*, eds K. Chater & H. J. Somerville, pp. 80–117. London: Heyden and Son.

HOBSON, P. N., BOUSFIELD, S. & SUMMERS, R. (1974). Anaerobic digestion of organic matter. *Critical Reviews of Environmental Control*, **4**, 131–9.

HORVATH, R. S. (1972). Microbial co-metabolism and the degradation of organic compounds in nature. *Bacteriological Reviews*, **36**, 146–55.

HUNGATE, R. E. (1977). Suitability of methanogenic substrate, health hazards and terrestrial conservation of plant nutrients. In *Microbial Energy Conservation*, eds H. G. Schlegel & J. Barnea, pp. 339–46. Oxford: Pergamon Press.

KESSLER, E. (1973). Hydrogenase, photoreduction and anaerobic growth. In *Algae: Physiology and Biochemistry*, ed. W. P. Stewart, vol. 10, pp. 456–73. Oxford: Blackwell Scientific Publications.

KITAJIMA, M. & BUTLER, W. L. (1976). Microencapsulation of chloroplast particles. *Plant Physiology*, **57**, 746–50.

KLOOS, W. E., TORNABENE, T. G. & SCHLEIFFER, K. H. (1974). Isolation and characterization of micrococci from human skin, including two new species of *Micrococcus lylae* and *Micrococcus kristinae*. *International Journal of Systematic Bacteriology*, **24**, 79–101.

KONDRATIEVA, E. N. (1977). Phototrophic micro-organisms as sources of hydrogen and hydrogenase formation. In *Microbial Energy Conversion*, eds H. G. Schlegel & J. Barnea, pp. 205–16. Oxford: Pergamon Press.

KONDRATIEVA, E. N. & GOGOTOV, I. N. (1976). Micro-organisms – hydrogen producers. (In Russian.) *Izvestija Akademii Nauk, SSSR*, seria biol., 69–85.

LOSADA, M. (1978). Photoproduction of ammonia and hydrogen peroxide. In *Bioenergy: Energy from Living Systems*, ed. C. Gysi, pp. 147–83. Zurich: Gottlieb Duttweiler Institute.

MAH, R. A., WARD, D. M., BARESI, L. & GLASS, T. L. (1977). Biogenesis of methane. *Annual Reviews of Microbiology*, **31**, 309–41.

MITCHELL, P. (1976). Vectorial chemistry and the molecular mechanics of chemiosmosis: power transmission by proticity. *Biochemical Society Transactions*, **4**, 399–430.

MITSUI, A. (1978). Hydrogen from bacteria and algae. In *Bioenergy: Energy from Living Systems*, ed. C. Gysi, pp. 205–36. Zurich: Gottlieb Duttweiler Institute.

MORRISON, S. J., TORNABENE, T. & KLOOS, W. E. (1971). Neutral lipids in the study of relationships of members of the family *Micrococcoceae*. *Journal of Bacteriology*, **108**, 353–8.

MORTENSON, L. E. & CHEN, J. S. (1974). Hydrogenase. In *Microbial Iron Metabo-*

lism, ed. J. B. Neilands, vol. 1, pp. 231–82. London, New York: Academic Press.

PETERSON, J. A., KUSUNOSE, M., KUSUNOSE, E. & COON, M. J. (1967). Enzymatic ϖ-oxidation. II. Function of rubredoxin as the electron carrier in ϖ-hydroxylation. *Journal of Biological Chemistry*, **242**, 4334–40.

PFENNIG, N. (1967). Photosynthetic bacteria. *Annual Reviews of Microbiology*, **21**, 285–324.

RAO, K. K., ROSA, L. & HALL, D. A. (1976). Prolonged production of hydrogen gas by a chloroplast biocatalytic system. *Biochemical and Biophysical Research Communications*, **68**, 21–8.

REEVES, S. G., RAO, K. K., ROSA, L. & HALL, D. O. (1977). Biocatalytic production of hydrogen. In *Microbial Energy Conversion*, eds H. G. Schlegel & J. Barnea, pp. 235–43. Oxford: Pergamon Press.

SCHLEGEL, H. G. & BARNEA, J. (eds) (1977). *Microbial Energy Conversion*. Oxford: Pergamon Press.

SCHLEGEL, H. G. & LAFFERTY, R. M. (1965). Growth of 'Knallgas' bacteria (*Hydrogenomonas*) using direct electrolysis of the culture medium. *Nature, London*, **205**, 308–9.

SCHRECKENBACH, TH. (1977). Light-energy conversion by the purple membrane from *Halobacterium halobium*. In *Microbial Energy Conversion*, eds H. G. Schlegel & J. Barnea, pp. 245–66. Oxford: Pergamon Press.

SKULACHEV, V. P. (1976). Conversion of light energy into electric energy by bacteriorhodopsin. *FEBS Letters*, **64**, 23–5.

TANNENBAUM, S. R. (1977). Single cell protein. In *Food Proteins*, eds J. R. Whitaker & S. R. Tannenbaum, pp. 315–30. Westport, Connecticut: Avi.

TONGE, G. M., HARRISON, D. E. F. & HIGGINS, I. J. (1977). Purification and properties of the methane mono-oxygenase enzyme system from *Methylosinus trichosporium* OB 3b. *Biochemical Journal*, **161**, 333–44.

TORNABENE, T. G. (1977). Microbial formation of hydrocarbons. In *Microbial Energy Conversion*, eds H. G. Schlegel & J. Barnea, pp. 281–99. Oxford: Pergamon Press.

WEISSMAN, J. & BENEMANN, J. R. (1977). Hydrogen production by nitrogen-starved cultures of *Anabaena cylindrica*. *Applied and Environmental Microbiology*, **33**, 123–31.

WOLFE, R. S. & HIGGINS, I. J. (1978). The microbial production and utilization of methane – a study in contrasts. In *MTP International Review of Biochemistry*, series II, *Microbial Biochemistry*, ed. J. R. Quayle, Lancaster: MTP Press (in press).

ZEIKUS, J. G. (1977). The biology of methanogenic bacteria. *Bacteriological Reviews*, **41**, 514–41.

GENETIC MANIPULATIONS FOR INDUSTRIAL PROCESSES

*KEITH T. ATHERTON, †DAVID BYROM AND *EDWARD C. DART

* Corporate Research Laboratory,
ICI Limited, Runcorn Heath WA7 4QE, Cheshire, UK

† Agricultural Division,
ICI Limited, Billingham TS23 1LB, Cleveland, UK

INTRODUCTION

In his Presidential address to the Second International symposium on the Genetics of Industrial Micro-organisms, Pontecorvo (1976) drew attention to the enormous gap between basic knowledge and applications in genetics and gene-transfer processes, and accused the fermentation industry of paying scant attention to the techniques of modern genetics.

Although in most cases there are quite justifiable reasons why industrial strain-improvement programmes have been limited to mutation and selection, the situation is ripe for change. With the explosive growth and development of techniques for the in-vitro and in-vivo transfer of genes between a growing number of micro-organisms it is our belief that industrial commitment to microbial genetics is likely to expand rapidly.

The aim of this paper is to consider from the industrial standpoint the current state of the art in 'modern' genetics and project how this is likely to extend from *Escherichia coli*, its plasmids and phages, to organisms of more industrial interest.

Strain improvement by mutation

The usual genetic manipulation required in an industrial process is yield improvement, whether this is for single-cell protein, antibiotic, amino acid, enzyme or ethanol production. The most common way of obtaining the improvement after physiological variables such as temperature, pH, oxygen tension etc., have been optimized is by the time-honoured process of mutation.

Though essentially random, the process is helped tremendously if a strong selection pressure for the product, and some knowledge of the relevant biochemistry, is available. For example, Nakayama (1972) has successfully used resistance to amino acid analogues to identify high-yielding amino acid producing strains of *Corynebacterium glutamicum*. Amino acid analogues inhibit growth by competing with normal cell metabolites for an essential binding site. Selection for resistance reveals some mutants which do not respond to normal metabolite feedback inhibition and hence 'overproduce' the appropriate amino acid.

In many cases direct selection is not possible, and the process becomes more of a random procedure. Recognizing this, attempts have been made to minimize the effort required. In one, a statistical treatment has been developed which enables the user to calculate how many flasks to grow up and how many replicates to use to stand the best chance of finding a real increase in yield for the least amount of effort (Davies, 1964). In another approach the process has been automated and its validity demonstrated by the isolation of several new thermosensitive DNA replication mutants of *E. coli* (Sevastopoulos, Wehr & Glaser, 1977).

In cases where appropriate genetic linkage data is available, muta-tional methods based on *N*-methyl-*N'*-nitro-*N*-nitrosoguanidine (MNNG) (Adelberg, Mandel & Chen, 1965; Delic, Hopwood & Friend, 1970) can be useful as there have been several reports of its ability to give mutations in clusters around the replicating fork of the bacterial chromosome (Hirota *et al.*, 1968; Calender *et al.*, 1970; Guerola, Ingraham & Cerda-Olmedo, 1971; Randazzo *et al.*, 1973). If one select-able mutation is induced in bacteria, co-mutations in several closely linked genes are observed. Hence selection for mutations in a known gene can lead to the isolation of unselectable mutations in adjacent genes of interest (and hence to strain improvement).

With the advent of gene cloning and rapid DNA-sequencing methods (Maxam & Gilbert, 1977; Sanger, Nicklen & Coulson, 1977) it may be possible in the future, using the technique of site-directed mutagenesis (Domingo, Flavell & Weissmann, 1976), to produce a series of directed mutations within the gene, operon or locus of choice, and select for improvements in function in a more direct manner.

Mutation versus more rational approaches

The preference in industrial strain-improvement programmes for the essentially random mutation process over methods based on under-standing the genetic and biochemical make-up of the organism of choice can be traced to three factors.

(1) The process has a proven 'track record' from the early days of penicillin yield improvements to present success in amino acid and antibiotic production. Typically yields can be improved from a few mg per litre to 20–30 g per litre.

(2) Success can be achieved relatively rapidly. In situations where competition for a lucrative market is fierce this factor can dominate all others.

(3) In general, the organism will have been selected via a product screen, and in most cases little will be known of its genetics and bio-chemistry. Only in cases where the market price is sensitive to pro-duction cost and where the product or organism is of clear long-term importance to the industry in question, will it be sensible to follow up a mutation and selection programme with one based on a more long-term genetics study. A good example of where this is being done is the area of Streptomycetes genetics. A good many important antibiotics derive from this genus, and as a consequence it seems appropriate to analyse specific strains genetically and to develop modern genetic manipulation systems to improve their productivity. In the case of *Streptomyces coelicolor* Hopwood and co-workers (e.g. Bibb, Freeman & Hopwood, 1977; Hopwood *et al.*, 1977; Friend, Warren & Hopwood, 1978) have been developing recombination, conjugation and cell fusion systems. Doubtless many industrial producers of antibiotics are following suit.

The future

The ability to transfer genetic information with increasing facility between different micro-organisms means that the fermentation technologist is not necessarily faced with the daunting prospect of genetically analysing his preferred organism in great detail. Instead he should be able to perform a good many of his genetic manipulations in *E. coli*, where the technology is more fully developed, then transfer the manipulated genetic material to his organism of choice, provided the host organism has been conditioned to accept and express incoming genetic material.

The following sections review the state of the art in gene-transfer technology and the attempts being made to develop gene-transfer systems for organisms other than *E. coli*.

GENE-TRANSFER TECHNOLOGY

In-Vitro methods

The basic requirements for the in-vitro transfer and expression of foreign DNA in a host bacterium are outlined below:

(1) A 'vector' DNA molecule (plasmid or bacteriophage) capable of entering the host cell and replicating within it. Ideally the vector should be small, easily prepared and must contain at least one site where integration of foreign DNA will not destroy an essential function.

(2) A method of splicing foreign genetic information into the vector. The finding and characterization of restriction enzymes that generate cohesive ends on functional DNA fragments were crucial to the development of recombinant DNA technology.

(3) A method of introducing the vector/foreign DNA recombinants into the host cell and selecting for their presence. Commonly used simple characteristics include drug resistance, immunity, plaque formation, or an inserted gene recognizable by its ability to complement a known auxotroph.

(4) A method of assaying for the 'foreign' gene product of choice from the population of recombinants created.

A typical process is shown in Fig. 1.

Vector molecules

Two broad categories of vector molecules have been developed as vehicles for gene transfer; plasmids and lambdoid phages. Plasmids are being found in an increasingly wide range of organisms but have been most studied in Gram-negative bacteria. Of the phages only λ, which has a very limited *E. coli* host range, has been developed to any degree of sophistication. Recent reports also indicate that the double-stranded replicative form (RF) of the filamentous phage M13 can be used as a vector (Hofschneider, 1978). This system has a number of interesting features which may be exploited in the future. The phage can replicate to greater than 200 copies per cell while not lysing it. Also there would appear to be no limit to the size of DNA fragment which could be cloned.

Both plasmids and lambdoid phage vectors have replication systems that are independent of the host cell chromosome. They exist mostly as independent, circular, double-stranded DNA molecules. Plasmids vary considerably in size and copy number. Some of the relatively large

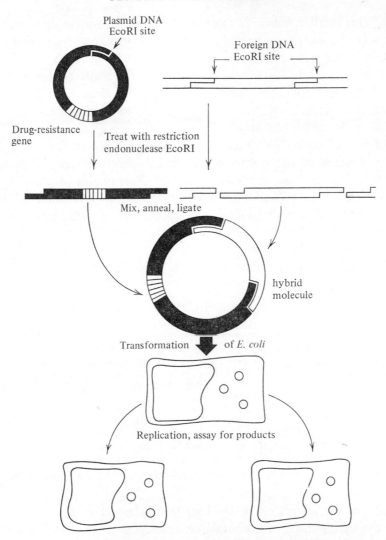

Fig. 1. A summary of the steps in in-vitro genetic recombination. Both plasmid vector and foreign DNA are cut by the restriction endonuclease, EcoRI, producing linear double-stranded DNA fragments with single-stranded cohesive projections. EcoRI recognizes the oligonucleotide sequence $^{-GAATTC-}_{-CTTAAG-}$ and will cut any double-stranded DNA molecule to yield fragments with the same cohesive ends $^{-GAATT}_{-C}$, $^{C-}_{TTAAG-}$. On mixing vector and foreign DNA, hybrids form into circular molecules which can be covalently joined using DNA ligase. Transformation of *E. coli* results in the low-frequency uptake of hybrid molecules whose presence can be detected by the ability of the plasmid to confer drug resistance on the host.

plasmids (molecular wt about 3×10^7) carry a set of genes to promote sexual DNA transfer between organisms by the process known as conjugation. Plasmids with this property are designated conjugative plasmids. The smaller plasmids tend to be non-conjugative, but usually their sexual transfer can be promoted by the presence of a conjugative plasmid in the same cell.

As well as coding for sexual gene transfer, naturally occurring plasmids have been found to carry information for drug or heavy-metal resistance, antibiotic production, or the ability to metabolize or degrade specific hydrocarbon substrates (for example, see Chakrabarty, 1976). In the case of the colicinogenic plasmid, Col E1, replication continues in the presence of chloramphenicol, an inhibitor of bacterial protein synthesis. Under these circumstances 1000–2000 copies of the circular DNA form of the plasmid will accumulate per cell. Because of their high copy-number, Col E1-derived plasmids show some promise for producing high yields of desirable gene products (Hershfield *et al.*, 1976). Some of the more commonly used non-conjugative plasmid cloning vehicles for *E. coli* are shown in Table 1.

Table 1. *Some commonly used plasmid cloning vehicles*

Plasmid	Molecular weight $(\times 10^{-6})$	Selectable marker[a]	Single restriction sites
pSC101	5.8	Tc^r	BamHI, EcoRI, HindIII, HpaI, SalI, SmaI
Col E1	4.2	Col^{imm}	EcoRI
pMB9	3.6	Tc^r, Col^{imm}	BamHI, EcoRI, HindIII, HpaI, SalI, SmaI
pBR313	5.8	Tc^r, Ap^r, Col^{imm}	BamHI, EcoRI, HindIII, HpaI, SalI, SmaI
pBR322	2.6	$Tc^r Ap^r$	BamHI, EcoRI, HindIII, PstI, SalI

[a] Tc^r: tetracycline resistance. Ap^r: ampicillin resistance. Col^{imm}: colicin immunity.

Much of the attraction of working with phage λ as a vector stems from the vast amount of accumulated knowledge of its genetics. Phage λ vectors suitable for cloning fragments of DNA have been described by a number of authors (Rambach & Tiollais, 1974; Thomas, Cameron & Davies, 1974; Murray & Murray, 1975; Enquist *et al.*, 1976; Murray, Brammar & Murray, 1977). The first λ vectors were developed with HindIII and EcoRI restriction sites. λ vectors have now been developed for cloning SalI, BamHI and Xma restriction fragments. The use of oligonucleotide adaptor fragments (Heynecker *et al.*, 1976) will certainly extend this range. One of the most exciting developments from the industrial standpoint has been manipulation of the phage genome to enhance the level of expression of inserted genes. Operations of this sort have been used to improve the production of T4 DNA ligase

(Murray, N. E., personal communication), *E. coli* DNA ligase (Pana-senko *et al.*, 1977), DNA polymerase and DNA restriction and modification enzymes (Murray, 1977). Yields of protein can be very high. A λ transducing phage containing the *trp* operon of *E. coli* (inserted by in-vivo methods) has produced 50 % of the cell's soluble protein as products of these genes (Moir & Brammar, 1976).

Whether a plasmid or bacteriophage vector is used is often a function of the experience individual workers possess, though each system does have its own advantages (Table 2). Indeed attempts are now being made to combine the best qualities of phage and plasmid systems (see, for example, Helinski *et al.*, 1977).

Table 2. *Relative merits of plasmid and λ phage vector systems*

Phenomenon	Plasmid	Phage λ
Host range	Can be narrow or wide depending on plasmid used	Limited to a few *E. coli* strains
Selection	Most commonly used selectable markers are drug-resistance genes	Range of alternative markers available including plaque morphology
Size of DNA insert	No theoretical limit on size	λ/'foreign' DNA hybrid can only be 105% of wild-type λ chromosome
Splicing methods (see text)	Plasmids available for a wide range of restriction enzymes. Hybrids can also be produced by the dA.dT connector method	Limited to a few restriction enzymes. Cannot yet be readily used with the dA.dT procedure
Enhanced regulation of gene expression	Amplification of gene products has been achieved but not to levels as high as in the phage λ system	Possible to amplify gene products derived from inserted DNA sequences to very high levels
Storage	Each hybrid plasmid must be stored in an individual, pure, bacterial colony	Hybrid phage can be stored as a pool of infectious phage particles

Splicing genes

The most significant advances which allowed the construction of hybrid DNA molecules *in vitro* came from the discovery that site-specific restriction endonucleases produce specific DNA fragments which can be joined to any other similarly treated DNA molecule using DNA ligase.

Restriction enzymes are not a new phenomenon. It was in the early 1950s that Luria and co-workers (Luria & Human, 1952) first presented data showing that bacteriophages grown on one strain of bacteria displayed a wide variation in their ability to plate on other cells. They further showed that this was a property of the bacterial cell and not the bacteriophage. Later, studies by Arber & Dussoix (Arber & Dussoix, 1962; Dussoix & Arber, 1962) led them to propose a model of restriction and modification to account for these observations. Many restriction and modification systems have now been identified (Arber, 1974). The first biochemical studies on the class I restriction enzymes began in 1968 (Meselson & Yuan, 1968), but the important discovery of the class II restriction enzymes* did not come until 1970 when a restriction enzyme from *Haemophilus influenzae* was described (Smith & Wilcox, 1970). The number of new restriction enzymes has now risen in excess of 80 (Roberts, 1976) although not all have been fully characterized with respect to recognition sequence, and only a fraction are in general use in recombinant DNA research because of the lack of suitable vector molecules.

Some of the most frequently used restriction enzymes (Table 3) have a hexanucleotide recognition site and produce DNA fragments with a four-base single-stranded cohesive projection. These are preferred over enzymes with a penta- or tetranucleotide recognition sequence because the average size of fragment they produce is more than long enough to contain an average-sized gene. The frequency of any given hexanucleotide sequence occurring in a polynucleotide containing equal proportions of the four standard bases is 1 in 4^6 (1 in 4096, i.e. sufficient genomic information to code for a protein of molecular weight 136 500).

Table 3. *Recognition sites and cleavage patterns of some commonly used restriction enzymes*

EcoRI	G↓AATT C	SalI	G↓TCGA C
	C TTAA G↑		C AGCT G↑
HindIII	A↓AGCT T	HaeIII	GG↓CC
	T TCGA A↑		CC GG↑
BamHI	G↓GATC C	SmaI	CCC↓GGG
	C CTAG G↑		GGG CCC↑

* Class I enzymes recognize a specific nucleotide sequence but cut at non-specific sites away from the recognition sequence, while class II enzymes both recognize and cut a specific nucleotide sequence.

Restriction endonucleases like HaeIII and SmaI have come into common use since the observation that blunt-ended DNA fragments can be joined by a combination of T4 DNA and RNA ligases (Sgaramella, Bursztyn-Pettegrew & Ehrlich, 1977; Sugino *et al.*, 1977). The reaction requires a high concentration of termini to proceed at a reasonable rate.

Apart from serving as a host defence system against unmodified incoming DNA, a recent observation suggests that restriction enzymes may be involved in genetic recombination processes *in vivo* (Chang & Cohen, 1977). Introduction of the plasmid pACYC 184 (Fig. 2), containing an EcoRI insertionally-inactivated, chloramphenicol-resistance gene, into an *E. coli* strain containing the EcoRI-producing plasmid resulted in a few colonies acquiring resistance to the antibiotic. This is the first indication that such enzymes may play an important role in the evolution of plasmids and chromosomal genes.

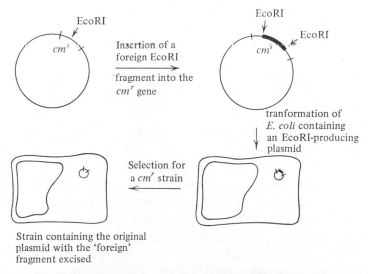

Fig. 2. Demonstration of in-vivo activity of EcoRI endonuclease.

Biologically functional DNA hybrids can also be constructed by the dA.dT 'terminal transferase' method (Jackson, Symons & Berg, 1972) (Fig. 3). Because the vector molecule cannot reform a circle without the presence of the 'foreign' DNA fragment with the complementary homopolymeric tail, this procedure gives a high yield of recombinant DNA circles *in vitro*.

In addition to its use in joining DNA molecules from a restriction

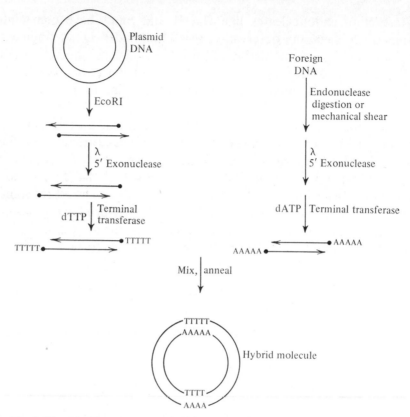

Fig. 3. The dA.dT connector method for construction of hybrid DNA molecules.

enzyme digest, the method can be used when the 'foreign' DNA fragments are produced by physical shearing. When the appropriate
quantities of DNA are used, this approach gives a high probability of
creating a population of hybrids containing at least one copy of every
gene from the donor organism. With restriction enzymes there is always
a finite chance that a gene of interest may contain a recognition site for
the particular enzyme being used. Using a combination of physical DNA
shearing and the dA.dT connector method, Clarke & Carbon (1976)
constructed a bank of hybrid DNA molecules containing sequences
representative of the entire *E. coli* genome.

One disadvantage of this technique is that once a gene is cloned, it is
not as readily excisable for transfer to other vectors. Possible solutions
include excision based on melting out the dA.dT regions followed by
single-strand nuclease cleavage (Hofstetter *et al.*, 1976), or the use of
blunt-ended ligation of the sheared fragments to synthetic linker molecules (Fig. 4) (Heyneker *et al.*, 1976).

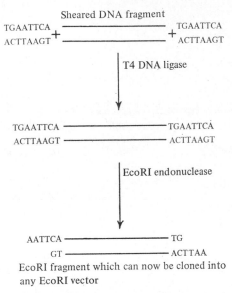

Fig. 4. The use of synthetic linker molecules in DNA cloning.

Transformation/transfection of the host bacterium and the identification of hybrid molecules

Introduction of the hybrid DNA into suitable host cells is accomplished by the process of transformation. Discovered as long ago as 1944 (Avery, Macleod & McCarty, 1944), transformation of cells by naked DNA is a very inefficient procedure and selectable markers are needed so that the small proportion of successfully transformed cells can be readily isolated. In some cases a plasmid can contain two drug-resistance markers with a specific restriction enzyme site in one of them. Cloning into this site facilitates the identification of cells which receive hybrid molecules. In Fig. 5, for example, cells which acquire either uncut or re-joined plasmid will be resistant to both kanamycin and tetracycline, while those which receive hybrid will be resistant to tetracycline but sensitive to kanamycin.

Transformation with phage λ DNA (transfection) is also an inefficient process but the in-vitro packaging method described by Hohn & Murray (1977) has increased the recovery of phage λ hybrids by orders of magnitude.

A variety of methods exist for the detection of λ hybrid DNA. In the replacement vector approach (Murray et al., 1977) a phage λ variant has two targets for HindIII in a non-essential part of its genome. Removal of the DNA fragment between these targets and rejoining the

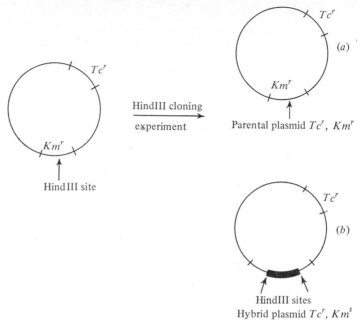

Fig. 5. Identification of hybrid molecules by insertional inactivation of drug resistance: (a) parental plasmid without insert, Tc^r, Km^r; (b) hybrid plasmid with insert, Tc^r, Km^s .

two outer fragments leaves a molecule that is too small to be packaged into a mature phage. Additional DNA must be inserted, therefore, into the space between the two outer fragments of the phage DNA for the phage to be viable. If the original fragment contains a sequence with a readily recognizable phenotype then the hybrid can be recognized by its loss. A second useful screening method with λ hybrids involves insertion into the immunity region of the vector such that hybrids produce a different plaque morphology to parental phage (Murray *et al.*, 1977). The presence of clear rather than turbid plaques signals the presence of recombinant phage.

Detecting the presence of cloned sequences

In some cases the molecular cloning of simple prokaryotic genes can be accomplished by direct transformation of *E. coli* auxotrophic strains with hybrid DNA, selecting directly for complementation of a host chromosomal mutation by the hybrid (Clarke & Carbon, 1975). A less wasteful process for the identification and cloning of a number of genes from the same organism is the colony bank approach of Clarke & Carbon (1976). Here a series of recombinant DNAs are made using a vector which can be mobilized by the presence of another plasmid. For example, *E. coli* strains carrying the conjugative plasmid F along with the non-conjugative plasmid Col E1 will transfer both plasmids with high efficiency to an F⁻ recipient. Therefore, if an F⁺ strain of *E. coli* is transformed with Col E1 hybrid plasmids, a number of F⁻ auxotrophic strains can be screened simply by mating individual members of the F⁺ transformants with the F⁻ auxotrophs.

In situations where the product of a cloning experiment cannot be identified by complementation analysis, alternative assay procedures have to be developed. One approach is to use a reaction that gives a colour change in the presence of a particular enzyme function. β-Galactosidase activity can be detected by the formation of blue colonies or plaques and is based upon the release from the medium of a non-diffusing blue pigment by hydrolysis of 4-chloro-3-indolyl-β-D-galacto-side (XGal). Similarly the presence of β-lactamase is signalled by the decoloration of iodine by penicilloic acid, the product of penicillin hydrolysis (Sherratt & Collins, 1973). When the desired protein can be isolated in reasonable purity and high-titre monospecific antibody raised, immunological screening methods can be developed. Such methods have been used to detect the presence of β-galactosidase from the *lac⁺* transducing derivative of phage λ (Sanzey *et al.*, 1976). In the absence of expression of the cloned DNA, methods are available which allow the detection of colonies or plaques containing a specific DNA sequence (Grunstein & Hogness, 1975; Benton & Davis, 1977). In these methods, replicates of hybrid colonies/plaques are produced and hybridized with radiolabelled mRNA or cDNA probes. In this way it should be possible to isolate unique genes by screening all the recombinants produced by shotgun cloning of the entire genome.

In-vivo methods

Methods of gene transfer and isolation based on in-vivo techniques pre-date by some years the gene-splicing technology so far described. Much early work involved the use of various transducing phages and *E. coli*, but more recently, systems have been developed which effect gene transfer within a range of Gram-negative bacteria.

Normally lysogeny of *E. coli* K12 by bacteriophage λ involves the insertion of the phage into a specific site adjacent to a cluster of genes determining the fermentation of galactose (*gal*), (Lederberg & Lederberg, 1953). If this *gal⁺* culture of *E. coli* K12, lysogenic for phage λ, is induced by u.v. light, a small proportion of the emerging phage particles carry the bacterial *gal⁺* genes. On subsequent infection of a *gal⁻* host these λ *dgal* transducing phages generated *in vivo* give rise to *gal*⁺ transductant clones (Hayes, 1968). In the absence of this normal site of integration, phage λ will insert, with decreased frequency, into many other chromosomal locations (Shimada, Weisberg & Gottesman, 1972), and hence any part of the *E. coli* chromosome can be brought into close proximity to the phage attachment site and subsequently incorporated into a transducing particle. More recent work which allows integration of the phage at random sites on the *E. coli* chromosome has involved derivatives of the phage, carrying segments of the bacteriophage Mu (Howe & Bade, 1975). Bacteriophage Mu has also been shown to transpose segments of the *E. coli* chromosome on to F′ episomes after infection of the cell with Mu or heat induction of a Mu *cts* prophage (Faelen & Toussaint, 1976). The F′ episome containing the transposed DNA can be identified by conjugation into a suitably marked recipient strain.

The limited host range of bacteriophage Mu (Howe & Bade, 1975) can be extended by combination with the wide host range plasmid RP4. These RP4::Mu derivatives have been shown to transpose chromosomal genes on to the plasmid and to promote intergeneric transfer by conjugation (Dénarié *et al.*, 1977). Introduction of RP4 into a Mu *cts* Klebsiella lysogen, partial induction of Mu at 37 °C and mating with a *his*⁻ strain of *K. pneumoniae* resulted in the transfer of the *his*D marker. In similar fashion, Klebsiella strains were able to transfer a range of markers, including the *nif* genes controlling nitrogen fixation, to *E. coli*.

Originally isolated from *Pseudomonas aeruginosa* (Datta *et al.*, 1971). RP4, like other plasmids of the P incompatibility group, can undergo transfer to a wide range of Gram-negative bacteria (Datta & Hedges, 1972). The development of RP4::Mu as a gene-transfer system (Dénarié *et al.*, 1977) opens the way for its use as a tool for the study of a range of Gram-negative bacteria whose genetics are poorly understood, and the construction of strains with improved performance in a more controlled way. No such systems exist for the Gram-positive bacteria, but the fact that transducing phages (e.g. Takahashi, 1961), plasmids (e.g. Ehrlich, 1977) and transposons (Shapiro, Adhya & Bukhari, 1977) have all been reported suggests that similar developments will follow for this family of organisms.

One technique that is better developed for Gram-positive bacteria is that of protoplast fusion. Successful fusions have been reported for the industrially important genus, *Streptomyces* (Hopwood *et al.*, 1977), for *Bacillus subtilis* (Schaeffer, Cami & Hotchkiss, 1976) and for *Bacillus megaterium* (Fodor & Alfoldi, 1976). In the short term, bacterial cell fusion offers those concerned with strain improvement the possibility of back-crosses between highly mutated, high-yielding organisms, containing perhaps several undesirable mutations, with strains further back in the mutation sequence where these genes are still intact. In the longer term, fusion may provide a method of re-assorting whole groups of genes between different strains of micro-organisms, providing some selection can be devised for desirable combinations in the hybrids. For the present the technique is limited to the Gram-positive bacteria (by virtue of their simpler cell wall architecture), a few strains of fungi (Anne & Peberdy, 1976; Kevei & Peberdy, 1977) and yeast (Van Solingen & Van der Plaat, 1977). In at least one case, the yeast hybrid showed increased production of a metabolite (Ferenczy & Maraz, 1977).

For cells derived from the higher eukaryotes, the use of cell fusion as a route to hybrid cells with a combination of desirable properties is a reality. The initial observation (Kohler & Milstein, 1975, 1976) that a non-dividing antibody-secreting mouse spleen cell could be fused with a dividing mouse myeloma cell to produce a dividing and antibody-secreting hybrid has spawned a whole series of similar reports. This technique has now been used to create a range of 'immortal' cell clones producing antibodies against haptens, soluble proteins, rat MHC antigens, *Streptococcus* carbohydrate (Robertson, 1977) and a repertoire of antiviral antibodies (Gerhard *et al.*, 1978). In addition, in a recent report, it was claimed that a line of cells capable of secreting insulin and growing in suspension culture had been generated by cell fusion (Wacker, Kaul & Eichler, 1978). The extension of this work to other systems will bring mammalian cell biology into a new exciting era.

EXPRESSION OF FOREIGN DNA IN *E. COLI*

One of the aims of this paper is to explore whether *E. coli* could be used both as a vehicle for the genetic analysis of more commercially important micro-organisms and as an intermediate for the selection of desirable genes prior to transfer to a suitable industrial host. Having briefly reviewed basic developments in gene-transfer technology, it is now worth exploring what level of success has been achieved in the expression of foreign DNA in *E. coli*. Examples are given in Table 4.

Confirmed cases of successful expression include genes from both Gram-positive and Gram-negative bacteria and from the lower eukaryotes, yeast and fungi. Reports of possible expression of *Drosophila* genes have yet to be confirmed. Care must be taken in generalizing from

Table 4. *Examples of 'foreign' DNA sequences cloned and expressed in* E. coli

DNA cloned	Activity detected	Reference
E. coli	Various	Clarke & Carbon (1976)
K. pneumoniae	*nif* genes	Dénarié *et al.* (1977)
P. aeruginosa	*arg⁻* complementation	Hedges & Jacob (1977)
Bacillus licheniformis	Penicillinase	Murray (1977)
B. subtilis	*leu⁻* complementation	Mahler & Halvorson (1977)
Streptomyces sp.	No complementation of *met⁻, thi⁻, pro⁻* E. coli auxotrophs	Johnson (1978)
Saccharomyces cerevisiae	Various enzymes, e.g. trp synthetase	Carbon *et al.* (1977)
S. cerevisiae	Imidazole-glycerophosphate dehydratase (*his* 3)	Struhl & Davies (1977)
Neurospora crassa	Dehydroquinase	Alton *et al.* (1978)
Drosophila melanogaster	New proteins detected	Rambach & Hogness (1977)
Bombyx mori (silk fibroin cDNA)	No translation reported	Ohshima & Suzuki (1977)
Sea urchin histone genes	No translation reported	Kedes *et al.* (1975)
Globin cDNA (various sources)	No evidence for polypeptide formation	e.g. Maniatis *et al.* (1976)
Chick ovalbumin cDNA		Breathnach, Mandel & Chambon (1977)
Mouse immunoglobulin cDNA		Mach *et al.* (1977)
Rat Insulin cDNA		Ullrich *et al.* (1977)
Human somatostatin sDNA (chemically synthesized and inserted into the lac operon cloned in pBR322)	Composite lac–somatostatin polypeptide produced. CNBr cleavage released somatostatin	Itakura *et al.* (1977)

the few examples quoted to date. For example, in their yeast experiments, Carbon *et al.* (1977) showed that out of fifteen complementations attempted, only four were successful. Such experiments have yet to be performed in the fungal case.

So far there is little evidence of any mammalian gene expression. A range of DNA reverse transcripts of the appropriate mRNA have been successfully cloned in plasmid vectors, but to date no expression has been reported. It will be interesting to see whether cloning into eukaryotic host-vector systems (see later) will bring more positive results.

The only example of expression of a higher organism 'gene' in *E. coli* has come from the chemically synthesized DNA fragment corresponding to the 14 amino acid sequence of human somatostatin. Here, the synthesis and molecular biology were elegantly designed so as to maximize the chance of effective processing by the host cell's transcriptional and translational machinery (Itakura *et al.*, 1977). In view of the recent discovery that the coding sequences in a number of eukaryotic genes are discontinuous (Jeffreys & Flavell, 1977), it is likely that those interested in the expression of mammalian genes in bacteria will approach the problem by inserting chemically or biochemically synthesized DNA into well defined bacterial genes or control sequences. From the industrial viewpoint, it is encouraging that the DNA of industrial fermentation organisms (bacteria, yeast and fungi) can be transcribed and translated in *E. coli*. However, it is equally important that, when required, suitably selected genetic information can be readily introduced into the industrial host of choice. The next section summarizes work in progress in developing gene transfer systems for alternative hosts.

NEW HOST-VECTOR SYSTEMS

Work on the development of cloning vehicles for organisms other than *E. coli* is increasing at a rapid rate, as Tables 5, 6 and 7 indicate. Some of the results summarized are based on conference abstracts and reports and are unpublished at the time of writing.

Most progress, in terms of host range, is being shown with the Gram-negative bacteria where, as discussed earlier, the conjugative P-group plasmids have been shown to be transferable to a wide range of recipients, e.g. *Salmonella, Shigella, Proteus, Pseudomonas, Rhizobium, Acinetobacter* etc. (Datta & Hedges, 1972; Olsen & Shipley, 1973).

No plasmid with such a wide host range is yet available for Gram-positive bacteria, although the *Streptococcus*-derived plasmids appear to enter and replicate with some facility in a range of *Streptococcus* strains. Also encouraging, is the report that plasmid pC194, derived from *Staphylococcus aureus*, can be maintained in *B. subtilis* and that a hybrid of this plasmid with the *E. coli* plasmid pBR322 is maintained in both hosts. Although the *B. subtilis* genes in this vector are expressed in an *E. coli* host, the reverse situation apparently does not hold. The generality of this observation has yet to be put to the test.

Eukaryotic host-vector systems are much further away from general use although systems based on the tumour inducing plasmid from *Agrobacterium tumefaciens* for plant cells, a 2 μm circle from yeast and SV40 virus in mammalian cells are being developed. In addition several polydisperse, plasmid-like circles have been discovered in Hela cells (Smith & Vinograd, 1972), *Drosophila* (Stanfield & Helinski, 1976) and tobacco cells (Wong & Wildman, 1972). It is possible that these could prove useful as gene transfer agents in specific circumstances. One real

Table 5. *Possible plasmid vector systems for other Gram-negative organisms*

Vector	Original host	Copy no.	Cloning sites	Molecular weight ($\times 10^{-6}$)	Selectable markers[a]	Comments	References
RP4	*P. aeruginosa*	3	EcoRI, HindIII BamHI, BglII	36	Ap^r, Tc^r, Km^r		Datta *et al.* (1971)
R1033	*P. aeruginosa*	3	Unknown	45	Ap^r, cm^r, Gm^r Km^r, Su^r	Conjugative transfer between most Gram-negative bacteria[b]	Matthew & Hedges (1976)
R702	*P. mirabilis*	3	Unknown	46	Km^r, Sm^r, Su^r Hg^r, Tc^r		Hedges (1975)
R751	*K. aerogenes*	3	Unknown	30	Tp^r		Jobanputra & Datta (1974)
R906	*Bordetella bronchiseptica*	3	Unknown	35	Ap^r, Sm^r, Su^r Hg^r		Hedges *et al.* (1974)

[a] Key: Ap^r, ampicillin resistance; Tc^r, tetracycline resistance; Km^r, kanamycin resistance; Su^r, sulphonamide resistance; cm^r, chloramphenicol resistance; Sm^r, streptomycin resistance; Gm^r, gentamycin resistance; Hg^r, mercury resistance; Tp^r, trimethoprim resistance.

[b] Besides *E. coli* a transformation system has been developed for *Pseudomonas putida* (Chakrabarty *et al.*, 1975).

Table 6. *Possible plasmid vector systems for Gram-positive bacterial[a] hosts*

Vector	Original host	Copy no.	Cloning sites	Molecular weight ($\times 10^{-6}$)	Selectable markers[b]	Comments	References
pUB110	*B. subtilis*	50	EcoRI, BamHI	2.8	*Nm*r	Inserted *B. subtilis*, *B. licheniformis* genes expressed in *B. subtilis*	Lovett *et al.* (1978) Keggins, Lovett & Duvall (1978)
SCP 2	*Streptomyces coelicolor*	low	–	18–20	*Mm*s	Sex factor activity	Bibb *et al.* (1977)
–	Group B streptococci	–	–	15–17	*Lm*r, *Sg*r	Sex factor activity with other streptococci Transformable into *S. sanguis*	Hershfield *et al.* (1978)
pR1405	*Streptococcus faecalis*	–	–	16.4	*Em*r	Sex factor activity	van Embden, Soedirman & Engel (1978)
pC194	*Staphylococcus aureus*	–	HindIII	1.8	*cm*r	Replicates in *B. subtilis*	Ehrlich & Goze (1978) Ehrlich (1978)
pHV14	Hybrid of pC194 and an *E. coli* plasmid	–	EcoRI, BamHI, SalI, PstI	–	*Ap*s, *cm*r	Replicates in *B. subtilis* and *E. coli.* Genes from *B. subtilis* expressed in *E. coli.* The reverse not the case	Ehrlich & Goze (1978) Ehrlich (1978)

[a] Gene transfer by transformation unless otherwise stated.

[b] Key: *Nm*r, neomycin resistance; *Mm*s, methylinomycin A synthesis; *Lm*r, lincomycin resistance; *Sg*r, streptogramin resistance; *Em*r, erythromycin resistance; *cm*r, chloramphenicol resistance.

Table 7. *Some eukaryotic host-vector systems under development*

Host	Vector	Cloning site	Molecular weight ($\times 10^{-6}$)	Selectable marker	Comments	Reference
Yeast	pYEleu and pYEhis Col E1 containing either the yeast *his* or *leu* 2 genes	–	10–15	Yeast *leu* or *his* complementation	Transformation and complementation of his⁻ or leu⁻ yeast	Fink (1978)
	2 μm circle from *S. cerevisiae*		3.9	Oligomycin^r	–	Beggs, Guerineau & Atkins (1976)
Plant cells	Ti plasmid from *Agrobacterium tumefaciens*	–	95–156	Octopine or nopaline metabolism	In-vivo transfer from *Agrobacterium* to any dicotyledonous plant. Possible successful transformation of Petunia protoplasts	Davey *et al.* (1978)
Mammalian cells	Defective SV40 virus	EcoRI, HpaII	2.2	Plaque formation with helper virus	Phage λ proteins in SV40 not detected. No attempts to express mammalian genes yet reported	Goff & Berg (1976, 1977)

surprise was the report that the *E. coli*-derived Col El plasmid containing the yeast *leu2* gene was able to complement a *leu⁻* derivative of *S. cerevisiae* by simple transformation (Fink, 1978).

CONCLUSION

Although there is still a considerable way to go before gene transfer between different micro-organisms becomes a matter of routine, progress has been sufficiently impressive to capture the attention of academia and industry alike. It is hoped that this review has at least identified some of the gaps in our present techniques, which, when bridged, will make the process of genetic analysis and manipulation much easier.

Acknowledgements

We wish to thank our many colleagues for stimulating discussion during the preparation of this manuscript and, in particular, David Pioli and Peter Barth for their helpful comments on the completed version. We would also like to thank Brian Symmington for help with the figures. Our thanks also go to Mrs Audrey Greaves for her patience and understanding during the typing of the manuscript.

REFERENCES

ADELBERG, E. A., MANDEL, M., CHEN, G. C. C. (1965). Optimal conditions for mutagenesis by *N*-Methyl-*N'*-nitro-*N*-nitrosoguanidine in *Escherichia coli* K12. *Biochemical and Biophysical Research Communications*, **18**, 788–95.

ALTON, N. K., KUSHNER, S. R., HAUTALA, J. A., JACOBSON, J. W., GILES, N. H. & VAPNEK, D. (1978). Expression of cloned eukaryotic DNA in *Escherichia coli*. In *Abstracts from the EMBO workshop: Plasmids and Other Extrachromosomal Elements*. Max Planck Institute for Molecular Genetics, W. Berlin, April 1–5, 1978.

ANNE, J. & PEBERDY, J. F. (1976). Induced fusion of fungal protoplasts following treatment with polyethylene glycol. *Journal of General Microbiology*, **92**, 413–17.

ARBER, W. (1974). DNA modification and restriction. *Progress in Nucleic Acid Research and Molecular Biology*, **14**, 1–37.

ARBER, W. & DUSSOIX, D. (1962). Host specificity of DNA produced by *Escherichia coli*. I. Host controlled modification of bacteriophage lambda. *Journal of Molecular Biology*, **5**, 18–36.

AVERY, O. T., MACLEOD, C. M. & MCCARTY, M. (1944). Studies on the chemical nature of the substance inducing transformation of Pneumococcal types. I. Induction of transformation by a deoxyribonucleic acid fraction isolated from pneumococcus type III. *Journal of Experimental Medicine*, **79**, 137–58.

BEGGS, J. D., GUERINEAU, M. & ATKINS, J. F. (1976). A map of the restriction targets in yeast two micron plasmid DNA cloned on bacteriophage lambda. *Molecular and General Genetics*, **148**, 287–94.

BENTON, W. D. & DAVIS, R. W. (1977). Screening λ gt recombinant clones by hybridization to single plaques *in situ*. *Science*, **196**, 180–2.

Bibb, M. J., Freeman, R. F. & Hopwood, D. A. (1977). Physical and genetical characterization of a second sex factor, SCP2, for *Streptomyces coelicolor* A3(2). *Molecular and General Genetics*, **154**, 155–66.

Breathnach, R., Mandel, J. L. & Chambon, P. (1977). Ovalbumin gene is split in chicken DNA. *Nature, London*, **270**, 314–19.

Calender, R., Lindqvist, B., Sironi, G. & Clark, A. J. (1970). Characterization of REP⁻ mutants and their interaction with P2 phage. *Virology*, **40**, 72–83.

Carbon, J., Ratzkin, B., Clarke, L. & Richardson, D. (1977). The expression of yeast DNA in *Escherichia coli*. In *Molecular Cloning of Recombinant DNA*, Miami Winter Symposia, vol. 13, eds W. A. Scott & R. Werner, pp. 59–72. London, New York: Academic Press.

Chakrabarty, A. M. (1976). Plasmids in *Pseudomonas*, *Annual Review of Genetics*, **10**, 7–30.

Chakrabarty, A. M., Mylroie, J. R., Friello, D. A. & Vacca, J. G. (1975). Transformation of *Pseudomonas putida* and *Escherichia coli* with plasmid-linked drug-resistance factor DNA. *Proceedings of the National Academy of Sciences, USA*, **72**, 3647–51.

Chang, S. & Cohen, S. N. (1977). In vivo site-specific genetic recombination promoted by the EcoRI restriction endonuclease. *Proceedings of the National Academy of Sciences, USA*, **74**, 4811–15.

Clarke, L. & Carbon, J. (1975). Biochemical construction and selection of hybrid plasmids containing specific segments of the *Escherichia coli* genome. *Proceedings of the National Academy of Sciences, USA*, **72**, 4361–5.

Clarke, L. & Carbon, J. (1976). A colony bank containing synthetic col E1 hybrid plasmids representative of the entire *Escherichia coli* genome. *Cell*, **9**, 91–9.

Cohen, S. N., Cabello, F., Chang, A. C. Y. & Timmis, K. (1977). DNA cloning as a tool for the study of plasmid biology. In *Recombinant Molecules: Impact on Science and Society*, eds R. F. Beers Jr & E. G. Bassett, pp. 91–105. New York: Raven Press.

Datta, N. & Hedges, R. W. (1972). Host range of R factors. *Journal of General Microbiology*, **70**, 453–60.

Datta, N., Hedges, R. W., Shaw, E. J., Sykes, R. B. & Richmond, M. H. (1971). Properties of an R factor from *Pseudomonas aeruginosa*. *Journal of Bacteriology*, **108**, 1244–9.

Davey, M. R., Cocking, E. C., Freeman, J., Pearce, N. & Tudor, I. (1978). Experiments using plant protoplasts to determine whether transformation is possible using *Agrobacterium* plasmids. In *Abstracts from the EMBO Workshop: Plasmids and Other Extrachromosomal Genetic Elements*. Max Planck Institute for Molecular Genetics, W. Berlin, April 1–5, 1978.

Davies, O. L. (1964). Screening for improved mutants in antibiotic research. *Biometrics*, **20**, 576–91.

Delic, V., Hopwood, D. A. & Friend, E. J. (1970). Mutagenesis by *N*-methyl-*N*′-nitro-*N*-nitrosoguanidine (NTG) in *Streptomyces coelicolor*. *Mutation Research*, **9**, 167–76.

Dénarié, J., Rosenberg, C., Bergeron, B., Boucher, C., Michel, M. & Barate De Bartalmo, M. (1977). Potential of RP4::Mu plasmids for *in vivo* genetic engineering. In *DNA Insertion Elements, Plasmids and Episomes*, eds A. I. Bukhari, J. A. Shapiro & S. L. Adhya, pp. 507–20. Cold Spring Harbor Laboratory.

Domingo, E., Flavell, R. A. & Weissmann, C. (1976). In vitro site-directed mutagenesis: generation and properties of an infectious extracistronic mutant of bacteriophage Qβ. *Gene*, **1**, 3–25.

Dussoix, D. & Arber, W. (1962). Host specificity of DNA produced by *Escherichia coli* II. Control over acceptance of DNA from infecting phage lambda. *Journal of Molecular Biology*, **5**, 37–49.

EHRLICH, S. D. (1977). Replication and expression of plasmids from *Staphylococcus aureus* in *Bacillus subtilis*. *Proceedings of the National Academy of Sciences, USA*, **74**, 1680–2.

EHRLICH, S. D. (1978). DNA cloning in *Bacillus subtilis*. *Proceedings of the National Academy of Sciences, USA*, **75**, 1433–6.

EHRLICH, S. D. & GOZE, A. (1978). Use of *Bacillus subtilis/Escherichia coli* hybrid plasmids to study heterospecific gene expression. In *Abstracts from the EMBO Workshop: Plasmids and other Extrachromosomal Genetic Elements*. Max Planck Institute for Molecular Genetics, W. Berlin, April 1–5, 1978.

ENQUIST, L., TIEMIEIR, D., LEDER, P., WEISBERG, R. & STERNBERG, N. (1976). Safer derivatives of bacteriophage λ gt λ C for use in cloning of recombinant DNA molecules. *Nature, London*, **259**, 596–8.

FAELEN, M. & TOUSSAINT, A. (1976). Bacteriophage Mu I: A tool to transpose and to localise bacterial genes. *Journal of Molecular Biology*, **104**, 525–39.

FERENCZY, L. & MARAZ, A. (1977). Transfer of mitochondria by protoplast fusion in *Saccharomyces cerevisiae*. *Nature, London*, **268**, 524–5.

FINK, G. R. (1978). Yeast: A host for hybrid DNA. Reported to the International symposium on Genetic Engineering: Scientific developments and practical applications. Milan, March 29–31, 1978.

FODOR, K. & ALFOLDI, L. (1976). Fusion of protoplasts of *Bacillus megaterium*. *Proceedings of the National Academy of Sciences, USA*, **73**, 2147–50.

FRIEND, E. J., WARREN, M. & HOPWOOD, D. A. (1978). Genetic evidence for a plasmid controlling fertility in an industrial strain of *Streptomyces rimosus*. *Journal of General Microbiology*, **106**, 201–6.

GERHARD, W., CROCE, C. M., LOPES, D. & KOPROWSKI, H. (1978). Repertoire of antiviral antibodies expressed by somatic cell hybrids. *Proceedings of the National Academy of Sciences, USA*, **75**, 1510–14.

GOFF, S. P. & BERG, P. (1976). Construction of hybrid viruses containing SV40 and lambda phage DNA segments and their propagation in cultured monkey cells. *Cell*, **9**, 695–705.

GOFF, S. P. & BERG, P. (1977). Construction of hybrid viruses containing SV40 and lambda phage DNA segments and their propagation in monkey cells. In *Recombinant Molecules: Impact on Science and Society*, eds R. F. Beers Jr & E. G. Bassett, pp. 285–98. New York: Raven Press.

GRUNSTEIN, M. & HOGNESS, D. S. (1975). Colony hybridization: A method for the isolation of cloned DNA's that contain a specific gene. *Proceedings of the National Academy of Sciences, USA*, **72**, 3961–5.

GUEROLA, N., INGRAHAM, J. L. & CERDA-OLMEDO, E. (1971). Induction of closely linked multiple mutations by nitrosoguanidine. *Nature New Biology, London*, **230**, 122–5.

HAYES, W. (1968). In *The Genetics of Bacteria and their Viruses*, ed. W. Hayes, pp. 627–49. Oxford: Blackwell Scientific.

HEDGES, R. W. (1975). R factors from *Proteus mirabilis* and *P. vulgaris*. *Journal of General Microbiology*, **87**, 301–11.

HEDGES, R. W. & JACOB, A. E. (1977). *In vivo* translocation of genes of *Pseudomonas aeruginosa* on to a promiscuously transmissable plasmid. *FEMS Microbiology Letters*, **2**, 15–19.

HEDGES, R. W., JACOB, A. E. & SMITH, J. T. (1974). Properties of an R factor from *Bordatella bronchiseptica*. *Journal of General Microbiology*, **84**, 199–204.

HELINSKI, D. R., HERSHFIELD, V., FIGURSKI, D. & MEYER, R. J. (1977). Construction and properties of plasmid cloning vehicles. In *Recombinant Molecules: Impact on Science and Society*, eds R. F. Beers Jr & E. G. Bassett, pp. 151–65. New York: Raven Press.

HERSHFIELD, V., BOYER, H. W., CHOW, L. & HELINSKI, D. R. (1976). Plasmid col E 1 as a molecular vehicle for cloning and amplification of DNA. *Proceedings of the National Academy of Sciences, USA*, 71, 3455–9.

HERSHFIELD, V., INAMINE, J. & TEUNISSEN, J. (1978). Conjugal plasmids in Group B Streptococcus. In *Abstracts from the EMBO Workshop: Plasmids and Other Extrachromosomal Genetic Elements*. Max Planck Institute for Molecular Genetics, W. Berlin, April 1–5, 1978.

HEYNECKER, H. L., SHINE, J., GOODMAN, H. M., BOYER, H. W., ROSENBERG, J., DICKERSON, R. E., NARANG, S. A., ITAKURA, K., LIN, S. & RIGGS, A. D. (1976). Synthetic lac operator DNA is functional in vivo. *Nature, London*, 263, 748–52.

HIROTA, Y., JACOB, F., RYTER, A., BUTTIN, G. & NAKAI, H. (1968). On the process of cellular division in *Escherichia coli*: I Asymmetrical cell division and production of deoxyribonucleic acid-less bacteria. *Journal of Molecular Biology*, 35, 175–92.

HOFSCHNEIDER, P. H. (1978). Reported at the Societe Francaise de Microbiologie: Genie Genetique structure expression des genes clones. Institut Pasteur, Paris, 8–9 June, 1978.

HOFSTETTER, H., SCHAMBOCK, A., VAN DEN BERG, J. & WEISSMANN, C. (1976). Specific excision of the inserted DNA segment from hybrid plasmids constructed by the dA dT method. *Biochimica et Biophysica Acta*, 454, 587–91.

HOHN, B. & MURRAY, K. (1977). Packaging recombinant DNA molecules into bacteriophage particles in vitro. *Proceedings of the National Academy of Sciences, USA*, 74, 3259–63.

HOPWOOD, D. A., WRIGHT, H. M., BIBB, M. J. & COHEN, S. N. (1977). Genetic recombination through protoplast fusion in *Streptomyces*. *Nature, London*, 268, 171–4.

HOWE, M. & BADE, E. (1975). Molecular Biology of bacteriophage Mu. *Science*, 190, 624–32.

ITAKURA, K., HIROSE, T., CREA, R., RIGGS, A. D., HEYNEKER, H. L., BOLIVAR, F. & BOYER, H. W. (1977). Expression in *Escherichia coli* of a chemically synthesized gene for the hormone somatostatin. *Science*, 198, 1056–63.

JACKSON, D. A., SYMONS, R. M. & BERG, P. (1972). A biochemical method for inserting new genetic information into SV4O DNA: circular SV4O DNA molecules containing lambda phage genes and the galactose operon of *Escherichia coli*. *Proceedings of the National Academy of Sciences, USA*, 69, 2904–9.

JEFFREYS, A. J. & FLAVELL, R. A. (1977). The rabbit β-globulin gene contains a large insert in the coding sequence. *Cell*, 12, 1097–1108.

JOBANPUTRA, R. S. & DATTA, N. (1974). Trimethoprim R factors in enterobacteria from clinical specimens. *Journal of Medical Microbiology*, 7, 169–77.

JOHNSON, I. S. (1978). Reported to the International Symposium on Genetic Engineering: Scientific Developments and practical applications. Milan, March 29–31, 1978.

JONES, K. W. & MURRAY, K. (1975). A procedure for detection of heterologous DNA sequences in lambdoid phage by *in situ* hybridization. *Journal of Molecular Biology*, 96, 455–60.

KEDES, L. H., CHANG, A. C. Y., HOUSEMAN, D. & COHEN, S. N. (1975). Isolation of histone genes from unfractionated sea urchin DNA subculture cloning in *Escherichia coli*. *Nature, London*, 255, 533–8.

KEGGINS, K. M., LOVETT, P. S. & DUVALL, E. J. (1978). Molecular cloning of genetically active fragments of Bacillus DNA in *Bacillus subtilis* and properties of the vector plasmid pUB110. *Proceedings of the National Academy of Sciences, USA*, **75**, 1423–7.

KEVEI, F. & PEBERDY. J F. (1977). Interspecific hybridization between *Aspergillus nidulans* and *Aspergillus rugulosus* by fusion of somatic protoplasts. *Journal of General Microbiology*, **102**, 255–62.

KOHLER, G. & MILSTEIN, C. (1975). Continuous cultures of fused cells secreting antibody of predefined specificity. *Nature, London*, **256**, 495–7.

KOHLER, G. & MILSTEIN, C. (1976). Derivation of specific antibody-producing tissue culture and tumour lines by cell fusion. *European Journal of Immunology*, **6**, 511–19.

LEDERBERG, E. M. & LEDERBERG, J. (1953). Genetic studies of lysogenicity in interstrain crosses in *Escherichia coli. Genetics*, **38**, 51–64.

LOVETT, P. S., KEGGINS, K. M., DUVALL, E. J. & TAYLOR, R. K. (1978). Molecular cloning of genetically active Bacillus DNA fragments in *Bacillus subtilis*. In *Abstracts from the EMBO Workshop: Plasmids and Other Extrachromosomal Genetic Elements*. Max Planck Institute for Molecular Genetics, W. Berlin, April 1–5, 1978.

LURIA, S. E. & HUMAN, M. L. (1952). A non-hereditary, host-induced variation of bacterial viruses. *Journal of Bacteriology*, **64**, 557–69.

MACH, B., ROUGEON, F., LONGACRE, S. & AELLEN, M. F. (1977). DNA cloning in bacteria as a tool for study of immunoglobulin genes. In *Molecular Cloning of Recombinant DNA*, Miami Winter Symposia, Vol. 13, eds W. A. Scott & R. Werner, pp. 219–35. New York, London: Academic Press.

MAHLER, I. & HALVORSON, H. O. (1977). Transformation of *Escherichia coli* and *Bacillus subtilis* with a hybrid plasmid molecule. *Journal of Bacteriology*, **131**, 374–7.

MANIATIS, T., KEE, S. G., EFSTRATIADIS, A. & KAFATOS, F. C. (1976). Amplification and characterization of a β-globin gene synthesized *in vitro. Cell*, **8**, 163–82.

MATTHEW, M. & HEDGES, R. W. (1976). Analytical isoelectric focusing of R factor determined β-lactamase: Correlation with plasmid compatibility. *Journal of Bacteriology*, **125**, 713–18.

MAXAM, A. M. & GILBERT, W. (1977). A new method for sequencing DNA. *Proceedings of the National Academy of Sciences, USA*, **74**, 560–4.

MESELSON, M. & YUAN, R. (1968). DNA restriction enzymes from *Escherichia coli. Nature, London*, **217**, 1110–14.

MOIR, A. & BRAMMAR, W. J. (1976). The use of specialized transducing phages in the amplification of enzyme production. *Molecular and General Genetics*, **149**, 87–99.

MURRAY, K. (1977). Applications of bacteriophage lambda in recombinant DNA research. In *Molecular Cloning of Recombinant DNA*, Miami Winter Symposia, vol. 13, eds W. A. Scott & R. Werner, pp. 133–54. London, New York: Academic Press.

MURRAY, K. & MURRAY, N. E. (1975). Phage lambda receptor chromosomes for DNA fragments made with restriction endonuclease III of *Haemophilus influenzae* and restriction endonuclease I of *Escherichia coli. Journal of Molecular Biology*, **98**, 551–64.

MURRAY, N. E., BRAMMAR, W. J. & MURRAY, K. (1977). Lambdoid phages that simplify the recovery of *in vitro* recombinants. *Molecular and General Genetics*, **150**, 53–61.

NAKAYAMA, K. (1972). Fermentation technology today. In *Proceedings of the IVth International Fermentation Symposium*, ed. E. Terui, pp. 443–68. Society of Fermentation Technology, Osaka, Japan.

OHSHIMA, Y. & SUZUKI, Y. (1977). Cloning of the silk fibroin gene and its flanking sequences. *Proceedings of the National Academy of Sciences, USA*, **74**, 5363–7.

OLSEN, R. H. & SHIPLEY, P. (1973). Host range and properties of the *Pseudomonas aeruginosa* R Factor R1822. *Journal of Bacteriology*, **113**, 772–80.

PANASENKO, S. M., CAMERON, J. R., DAVIS, R. W. & LEHMAN, I. R. (1977). Five-hundredfold overproduction of DNA ligase after induction of a hybrid lambda lysogen constructed *in vitro*. *Science*, **196**, 188–9.

PONTECORVO, G. (1976). Presidential Address. In *Second International Symposium on the Genetics of Industrial Micro-organisms*, ed. K. D. Macdonald, pp. 1–4. London, New York: Academic Press.

RAMBACH, A. & HOGNESS, D. S. (1977). Translation of *Drosophila melanogaster* sequences in *Escherichia coli*. *Proceedings of the National Academy of Sciences, USA*, **74**, 5041–5.

RAMBACH, A. & TIOLLAIS, P. (1974). Bacteriophage lambda having EcoRI endo-nuclease sites only in the non-essential region of the genome. *Proceedings of the National Academy of Sciences, USA*, **71**, 3927–30.

RANDAZZO, R., SERMONTI, G., CARERE, A. & BIGNAMI, M. (1973). Comutation in Streptomyces. *Journal of Bacteriology*, **113**, 500–1.

ROBERTS, R. J. (1976). Restriction endonucleases. *CRC Critical Reviews in Bio-chemistry*, November 1976, 123–64.

ROBERTSON, M. (1977). Immunoglobulin genes and the immune response. *Nature, London*, **269**, 648–50.

SANGER, F., NICKLEN, S. & COULSON, A. R. (1977). DNA sequencing with chain-terminating inhibitors. *Proceedings of the National Academy of Sciences, USA*, **74**, 5463–7.

SANZEY, B., MERCEREAU, O., TERNYNCK, T. & KOURILSKY, P. (1976). Methods for identification of recombinants of phage lambda. *Proceedings of the National Academy of Sciences, USA*, **73**, 3394–7.

SCHAEFFER, P., CAMI, B. & HOTCHKISS, R. D. (1976). Fusion of bacterial proto-plasts. *Proceedings of the National Academy of Sciences, USA*, **73**, 2151–5.

SEVASTOPOULOS, C. G., WEHR, C. T. & GLASER, D. A. (1977). Large-scale auto-mated isolation of *Escherichia coli* mutants with thermosensitive DNA repli-cation. *Proceedings of the National Academy of Sciences, USA*, **74**, 3485–9.

SGARAMELLA, V., BURSZTYN-PETTEGREW, A. & EHRLICH, S. D. (1977). Use of the T4 ligase to join flush-ended DNA segments in recombinant molecules. In *Recombinant Molecules: Impact on Science and Society*, eds R. F. Beers Jr & E. G. Bassett, pp. 57–68. New York: Raven Press.

SHAPIRO, J. A., ADHYA, S. L. & BUKHARI, A. I. (1977). Introduction: new pathways in the evolution of chromosome structure. In *DNA Insertion Elements, Plas-mids and Episomes*, eds A. J. Bukhari, J. A. Shapiro & S. L. Adhya, pp. 3–11. Cold Spring Harbor Laboratory.

SHERRATT, D. J. & COLLINS, J. F. (1973). Analysis by transformation of the peni-cillinase system in *Bacillus licheniformis*. *Journal of General Microbiology*, **76**, 217–30.

SHIMADA, K., WEISBERG, R. A. & GOTTESMAN, M. E. (1972). Prophage λ at unusual chromosomal locations. I. Location of the secondary attachment site and the properties of the lysogens. *Journal of Molecular Biology*, **63**, 483–503.

SMITH, C. A. & VINOGRAD, J. (1972). Small polydisperse circular DNA of Hela cells. *Journal of Molecular Biology*, **69**, 163–78.

SMITH, H. O. & WILCOX, K. W. (1970). A restriction enzyme from *Haemophilus influenza* I. Purification and general properties. *Journal of Molecular Biology*, **51**, 379–91.

STANFIELD, S. & HELINSKI, D. R. (1976). Small circular DNA in *Drosophila melanogaster*. *Cell*, **9**, 333–45.

STRUHL, K. & DAVIS, R. W. (1977). Production of a functional eukaryotic enzyme in *Eschericia coli*: Cloning and expression of the yeast structural gene for imidazole-glycerophosphate dehydratase (his 3). *Proceedings of the National Academy of Sciences, USA*, **74**, 5255–9.

SUGINO, A., GOODMAN, H. M., HEYNEKER, H. L., SHINE, J., BOYER, H. W. & COZZARELLI, N. R. (1977). Interaction of Bacteriophage T4 RNA and DNA ligases in joining of duplex DNA at base-paired ends. *Journal of Biological Chemistry*, **252**, 3987–94.

TAKAHASHI. (1961). Genetic transduction in *Bacillus subtilis*. *Biochemical and Biophysical Research Communications*, **5**, 171–5.

THOMAS, M., CAMERON, J. R. & DAVIS, R. W. (1974). Viable molecular hybrids of bacteriophage lambda and eukaryotic DNA. *Proceedings of the National Academy of Sciences, USA*, **71**, 4579–83.

ULLRICH, A., SHINE, J., CHIRGWIN, J., PICTET, R., TISCHER, E., RUTTER, W. J. & GOODMAN, H. M. (1977). Rat insulin genes: Construction of plasmids containing the coding sequence. *Science*, **196**, 1313–19.

VAN EMBDEN, J. D. A., SOEDIRMAN, N. & ENGEL, H. W. B. (1978). Transfer of drug resistance between streptococci of the groups A, B and D. In *Abstracts from the EMBO Workshop: Plasmids and Other Extrachromosomal Genetic Elements*. Max Planck Institute for Molecular Genetics, W. Berlin, April 1–5, 1978.

VAN SOLINGEN, P. & VAN DER PLAAT, J. (1977). Fusion of yeast spheroplasts. *Journal of Bacteriology*, **130**, 946–7.

WACKER, A., KAUL, S. & EICHLER, A. (1978). Insulin-synthese fusioniester tierischer Zellen in Suspensionskultur. *Naturwissenschaften*, **65**, 69–70.

WONG, F. Y. & WILDMAN, S. G. (1972). Simple procedure for isolation of satellite DNA's from tobacco leaves in high yield and demonstration of minicircles. *Biochemica et Biophysica Acta*, **259**, 5–12.

animal feeds: biomass from treatment of waste for, 12, 21, 226, 227; stillage from industrial alcohol distillation for, 321

anions, microbial uptake of, 288–9

antibiotics: catabolite regulation of production of, 165–81, and of interconversion of, 168; commercial production of, 6; in cow's milk, from treatment of mastitis, 210–11; limit of usefulness of? 151; produced by lactic acid bacteria, 190

antigens: information about, and measurement of, in vaccine production, 153; of lactic streptococci, 191; new, produced by vaccinia and measles viruses on release from cells, 159–60

antimony: bacterial recovery of, as sulphide, 294; released from sulphide by bacteria, 269, 271, 275

arsenic, released from sulphide by bacteria, 271, 277

Arthrobacter globiformis, substrate supply and yield of, 40

Arthrobacter spp., co-operative degradation of insecticide Diazinon by *Streptomyces* and, 242

Arthrobacter viscosus, polysaccharide produced by, 129

asparaginase: carbon dioxide supply, and production of, by *E. coli*, 87

Aspergillus niger: citrate production by, 20, 67; shear effect in cultures of, 89

ATP: in ammonia assimilation, 62; in ammonia-limited cells, 63; in calculating expected yield of SCP, 34–5; in methanogenesis, 365; phosphate for, 64

Aureobasidium pullalans, polysaccharide produced by, 128–9

Azotobacter indicum, polysaccharide produced by, 124

Azotobacter sp., lead at cell surface of, 291

Azotobacter vinelandii: alginases of, 138; alginate produced by, 114, 122, 123, 131; growth rate, and nucleic acid content of, 46; RNA/protein ratio in, 46

Bacillus anthracis, vaccines for, 153, 154

Bacillus cereus: growth temperature, and products of, 75; oxygen supply, and toxin production by, 87; surface adsorption of metals by, 290

Bacillus circulans, glycerol as preferred carbon source for butirosin production by, 167

Bacillus globigii, surface adsorption of metals by, 290

Bacillus licheniformis: genes of, transferred to *E. coli*, 394; glucose repression of bacitracin formation by, 165–6; peptide antibiotics of, as iron scavengers? 67

Bacillus macerans, surface adsorption of metals by, 290

Bacillus megaterium: growth rate, and nucleic acid content of, 46; protoplast fusion in, 392; RNA/protein ratio in, 46; surface adsorption of metals by, 289, 290; uptake of divalent cations by, 286

Bacillus polymixa, surface adsorption of metals by, 290

Bacillus spp.: cAMP not found in, 164; in treatment of waste, 240

Bacillus subtilis: accumulation of metals by, 284, 286, 290, 291; carbon dioxide concentration, and production of α-amylase and inosine by, 39; genes of, transferred to *E. coli*, 394, 395; protoplast fusion in, 392; RNA/protein ratio in, 47; substrate uptake, and products, on glucose with different limiting nutrients, 66

Bacillus thuringiensis, molybdenum-complexing peptide produced by, 296

bacitracin, 67; inhibitory effect of glucose catabolites on production of, 165–6

bacteria: Gram-negative, glutamic acid in, 62; Gram-negative, plasmid vector systems for, 392, 395, 396; Gram-positive, plasmid vector systems for, 397; Gram-positive, protoplast fusion in, 392–3; maintenance coefficients for, 37; production of vaccines for, 153–4; sulphur-containing amino acids in, 42; *see also individual species*

bacteriophages, of lactic acid bacteria, 187–8, 215; differentiation of, 208–9; in milk-product manufacture, 209–10, 211, 212; in raw milk, may survive pasteurization, 209; require calcium for penetration into host cells, 212; resistance to, 192, 208

bacteriophages, lambdoid, as vectors for transfer of foreign DNA to host bacteria, 382–5, 389–90, 391–2; relative merits of plasmids and, for transfer, 385

bacteriophages, Mu: transfer DNA in combination with plasmid, 392

bacteriorhodopsin, could theoretically serve as basis for photo-cell, 366

balance equations, in microbiology, micro and macro, 337–8, 344–5, 354; for conservative properties, 345–6; for growth of *Hyphomicrobium*, 349–52; for non-conservative properties, 346–9

beet sugar, melibiase-producing yeast required, to utilize raffinose in, for production of alcohol, 314

Beijerinckia sp. (nitrogen-fixer), in acid leaching of sulphides, 268

biodegradation, microbial, 232–5

biological oxygen demand, of waste effluents, 226, 230, 232, 321

DATE DUE